The Handbook of Environmental Chemistry

Volume 4 Air Pollution
Part G

O. Hutzinger
Editor-in-Chief

Springer-Verlag Berlin Heidelberg GmbH

Intercontinental Transport of Air Pollution

Volume Editor: A. Stohl

With contributions by
H. Akimoto · I. Bey · J. W. Bottenheim · O. R. Cooper
A. Dastoor · S. Eckhardt · A. E. Farrell · C. Forster
S.-L. Gong · K. Higuchi · H. Huntrieser · R. B. Husar
T. J. Keating · M. G. Lawrence · Y.-F. Li · D. D. Parrish
S. J. Piketh · P. Pochanart · H. Schlager · M. G. Schultz
N. Spichtinger-Rakowsky · A.Stohl · A. M. Thompson
N. M. Walton · J. J. West · O. Wild

Springer

Volume Editor

Dr. A. Stohl

Cooperative Institute for Research in the Environmental Sciences (CIRES)
University of Colorado
NOAA Aeronomy Laboratory
325 Broadway
Boulder, CO 80305
USA
astohl@al.noaa.gov

ISSN 1433-6855

ISBN 978-3-662-14529-6 ISBN 978-3-540-40037-0 (eBook)
DOI 10.1007/978-3-540-40037-0

Library of Congress Control Number: 2004104065

Production Editor: Christiane Messerschmidt, Rheinau
Cover Design: E. Kirchner, Springer-Verlag
Typesetting: Fotosatz-Service Köhler GmbH, Würzburg
Printed on acid-free paper 52/3020 – 5 4 3 2 1 0

The Handbook of Environmental Chemistry is also Available Electronically

For all customers who have a subscription to The Handbook of Environmental Chemistry, we offer the electronic version via SpringerLink free of charge. Please contact your librarian who can receive a password for free access to the full articles by registering at:

(http://www)springerlink.com

If you do not have a subscription, you can still view the tables of contents of the volumes and the abstract of each article by going to the SpringerLink Homepage, clicking on "Browse by Online Libraries", then "Chemical Sciences", and finally choose The Handbook of Environmental Chemistry.

You will find information about the

- Editorial Bord
- Aims and Scope
- Instructions for Authors
- Sample Contribution

at (http://www)springeronline.com using the search function.

Preface

Environmental Chemistry is a relatively young science. Interest in this subject, however, is growing very rapidly and, although no agreement has been reached as yet about the exact content and limits of this interdisciplinary discipline, there appears to be increasing interest in seeing environmental topics which are based on chemistry embodied in this subject. One of the first objectives of Environmental Chemistry must be the study of the environment and of natural chemical processes which occur in the environment. A major purpose of this series on Environmental Chemistry, therefore, is to present a reasonably uniform view of various aspects of the chemistry of the environment and chemical reactions occurring in the environment.

The industrial activities of man have given a new dimension to Environmental Chemistry. We have now synthesized and described over five million chemical compounds and chemical industry produces about hundred and fifty million tons of synthetic chemicals annually. We ship billions of tons of oil per year and through mining operations and other geophysical modifications, large quantities of inorganic and organic materials are released from their natural deposits. Cities and metropolitan areas of up to 15 million inhabitants produce large quantities of waste in relatively small and confined areas. Much of the chemical products and waste products of modern society are released into the environment either during production, storage, transport, use or ultimate disposal. These released materials participate in natural cycles and reactions and frequently lead to interference and disturbance of natural systems.

Environmental Chemistry is concerned with reactions in the environment. It is about distribution and equilibria between environmental compartments. It is about reactions, pathways, thermodynamics and kinetics. An important purpose of this Handbook, is to aid understanding of the basic distribution and chemical reaction processes which occur in the environment.

Laws regulating toxic substances in various countries are designed to assess and control risk of chemicals to man and his environment. Science can contribute in two areas to this assessment; firstly in the area of toxicology and secondly in the area of chemical exposure. The available concentration ("environmental exposure concentration") depends on the fate of chemical compounds in the environment and thus their distribution and reaction behaviour in the environment. One very important contribution of Environmental Chemistry to the above mentioned toxic substances laws is to develop laboratory test methods, or mathematical correlations and models that predict the environmental fate of new chemical compounds. The third purpose of this Handbook is to help in the basic understanding and development of such test methods and models.

The last explicit purpose of the Handbook is to present, in concise form, the most important properties relating to environmental chemistry and hazard assessment for the most important series of chemical compounds.

At the moment three volumes of the Handbook are planned. Volume 1 deals with the natural environment and the biogeochemical cycles therein, including some background information such as energetics and ecology. Volume 2 is concerned with reactions and processes in the environment and deals with physical factors such as transport and adsorption, and chemical, photochemical and biochemical reactions in the environment, as well as some aspects of pharmacokinetics and metabolism within organisms. Volume 3 deals with anthropogenic compounds, their chemical backgrounds, production methods and information about their use, their environmental behaviour, analytical methodology and some important aspects of their toxic effects. The material for volume 1, 2 and 3 was each more than could easily be fitted into a single volume, and for this reason, as well as for the purpose of rapid publication of available manuscripts, all three volumes were divided in the parts A and B. Part A of all three volumes is now being published and the second part of each of these volumes should appear about six months thereafter. Publisher and editor hope to keep materials of the volumes one to three up to date and to extend coverage in the subject areas by publishing further parts in the future. Plans also exist for volumes dealing with different subject matter such as analysis, chemical technology and toxicology, and readers are encouraged to offer suggestions and advice as to future editions of "The Handbook of Environmental Chemistry".

Most chapters in the Handbook are written to a fairly advanced level and should be of interest to the graduate student and practising scientist. I also hope that the subject matter treated will be of interest to people outside chemistry and to scientists in industry as well as government and regulatory bodies. It would be very satisfying for me to see the books used as a basis for developing graduate courses in Environmental Chemistry.

Due to the breadth of the subject matter, it was not easy to edit this Handbook. Specialists had to be found in quite different areas of science who were willing to contribute a chapter within the prescribed schedule. It is with great satisfaction that I thank all 52 authors from 8 countries for their understanding and for devoting their time to this effort. Special thanks are due to Dr. F. Boschke of Springer for his advice and discussions throughout all stages of preparation of the Handbook. Mrs. A. Heinrich of Springer has significantly contributed to the technical development of the book through her conscientious and efficient work. Finally I like to thank my family, students and colleagues for being so patient with me during several critical phases of preparation for the Handbook, and to some colleagues and the secretaries for technical help.

I consider it a privilege to see my chosen subject grow. My interest in Environmental Chemistry dates back to my early college days in Vienna. I received significant impulses during my postdoctoral period at the University of California and my interest slowly developed during my time with the National Research

Council of Canada, before I could devote my full time of Environmental Chemistry, here in Amsterdam. I hope this Handbook may help deepen the interest of other scientists in this subject.

Amsterdam, May 1980 *O. Hutzinger*

Twentyone years have now passed since the appearance of the first volumes of the Handbook. Although the basic concept has remained the same changes and adjustments were necessary.

Some years ago publishers and editors agreed to expand the Handbook by two new open-end volume series: Air Pollution and Water Pollution. These broad topics could not be fitted easily into the headings of the first three volumes. All five volume series are integrated through the choice of topics and by a system of cross referencing.

The outline of the Handbook is thus as follows:

1. The Natural Environment and the Biochemical Cycles,
2. Reaction and Processes,
3. Anthropogenic Compounds,
4. Air Pollution,
5. Water Pollution.

Rapid developments in Environmental Chemistry and the increasing breadth of the subject matter covered made it necessary to establish volume-editors. Each subject is now supervised by specialists in their respective fields.

A recent development is the accessibility of all new volumes of the Handbook from 1990 onwards, available via the Springer Homepage http://www.springer. de or http://Link.springer.de/series/hec/ or http://Link.springerny.com/ series/hec/.

During the last 5 to 10 years there was a growing tendency to include subject matters of societal relevance into a broad view of Environmental Chemistry. Topics include LCA (Life Cycle Analysis), Environmental Management, Sustainable Development and others. Whilst these topics are of great importance for the development and acceptance of Environmental Chemistry Publishers and Editors have decided to keep the Handbook essentially a source of information on "hard sciences".

With books in press and in preparation we have now well over 40 volumes available. Authors, volume-editors and editor-in-chief are rewarded by the broad acceptance of the "Handbook" in the scientific community.

Bayreuth, July 2001 *Otto Hutzinger*

Contents

Foreword

While writing this, I am sitting in an airliner taking me from Denver to Frankfurt within less than 10 hours. I am wondering whether in Frankfurt, or at my final destination in Munich, I will be breathing air that I have already breathed in the town of Boulder near Denver, where I live – or that I have polluted there with my car. Not long ago, air pollution was considered a local problem, but now we are beginning to realize that not only climate change or the ozone hole are problems of a global dimension, but air pollution is a global problem, too! In fact, the pollution I have produced in Boulder, under special meteorological conditions, may reach Europe in little more time than it takes me to get from my office in Boulder to my other home near Munich. Today is not that type of weather, but on other days, combined with one of the annoying delays in air travel, I would have a chance of breathing, when I leave the aircraft in Munich, the same air that I polluted in Boulder on my way to the airport. Chances are that the air pollution has taken a similar route as I have taken – up to the higher realms of the atmosphere, taking advantage of the jet stream just as airliners do, onto a great-circle track to Europe, and final descending somewhere over Central Europe.

It is easy to understand that, when such rapid transport occurs (and it does, though more often it is not quite that rapid), chemical transformations may have brought about relatively little change to the composition of the air mass, and dispersion may also have had little chance to dilute whatever was contained in the air in the first place. Even when air mass transport is less rapid, significant concentrations of air pollutants may still be contained in air masses arriving from an upwind continent. This has led to concerns that pollution control is not possible on a local scale alone. Even the largest local efforts to keep our air clean will not succeed if the air arrives already polluted.

This reminds us that we all are sharing one world with limited resources, whether we live in Africa, Asia, Europe, North America, or on the still cleaner southern hemisphere. We all share the responsibility of keeping (or making again) the air clean. Sitting in an airliner, I am wondering whether business travel can justify that I am producing much more air pollution than the average citizen of the world, even though I try to avoid activities causing excessive pollutant emissions as much as possible. People in the heavily industrialized world must realize that people in poorer nations have the same right to use Earth's natural resources as they have themselves. However, it is almost unthinkable that every person on Earth can cause the same amount of pollution as people in the richest nations are producing now, without causing severe harm to our common environment. Thus, we all share the responsibility of keeping our worldwide emissions at a sustainable level.

This book describes the rapid advancements made in a very young field of atmospheric research – intercontinental pollution transport. Reports of suspected intercontinental transport of air pollution were made decades ago, but only now are measurements and models good enough to quantify how much pollution goes from one continent to the other – at a time when global emissions of many chemical compounds are still increasing. I therefore believe that it is timely to summarize what we know about this topic and to build a basis for future studies.

In order to know how anthropogenic emissions affect the atmosphere, it is also necessary to know what nature does itself. Thus, three of the chapters describe the transport of natural emissions of harmful substances caused by desert dust storms and forest fires. Even these emissions are not entirely natural anymore, but are affected by humans like everything else on this planet. Seven chapters describe what we know about intercontinental transport of – mostly anthropogenic – emissions. Finally, the last chapter describes the policy implications of intercontinental pollution transport. It is not at all clear at this time how the nations of the world should and will manage intercontinental pollution transport in the future. But sooner or later management will be necessary, as an increasing world population is struggling to maintain or achieve a comfortable lifestyle.

It is my hope that this book will help to build the scientific basis for wise policy decisions in the future. I hope, that by selecting authors from different continents, I have maintained a broad consensual basis for the book – although I have the feeling I should have invited authors from Australia and South America as well. However, space is limited and the southern hemisphere, fortunately, is still less polluted, so people from these continents may forgive me for not giving these continents the otherwise well-deserved space in this book. It is my impression that the authors – and the scientific community in general – have already reached a relatively broad consensus about which continent pollutes which one and to what extent. Let us hope that the political leaders of this world will also find common ground on this issue and others of importance for humankind.

Boulder, USA, January 2004 Andreas Stohl

The Handbook of Environmental Chemistry Vol. 4, Part G (2004): 1–11
DOI 10.1007/978-3-540-40037-0

Intercontinental Transport of Air Pollution: An Introduction

Andreas Stohl[1] · Sabine Eckhardt[2]

[1] Cooperative Institute for Research in the Environmental Sciences, University of Colorado/NOAA Aeronomy Laboratory, 325 Broadway, Boulder, CO 80305, USA
E-mail: astohl@al.noaa.gov

[2] Department of Ecology, Technical University of Munich, Am Hochanger 13, 85354 Freising, Germany

Abstract This chapter introduces the reader to the topic of intercontinental air pollution transport. It starts by describing how air pollution problems started on the local scale many centuries ago, became regional-scale problems as emissions of pollutants continued to rise, and are now considered to be global-scale. Research into source-receptor-relationships for pollutants like ozone or aerosols can no longer be restricted to the continental scale, but must also consider emissions from upwind continents. The typical global transport pathways of air pollutants are briefly described in this chapter, distinguishing between man-made and natural emissions (dust storms, biomass burning), and implications for environmental policies are mentioned. This chapter also serves the purpose of directing the reader to the detailed topical chapters of this book.

Keywords Air pollution · Long-range transport · Air quality

1 From Local-Scale to Intercontinental-Scale Air Pollution Problems

Emissions of gaseous and particulate pollutants began when humans first burned wood for cooking and residential heating. Indoor pollution problems followed suit, but the impacts of man-made emissions on the outdoor atmosphere were negligible for a long time. Slash-and-burn agriculture was perhaps the most significant anthropogenic source of outdoor air pollutants before medieval times, but the effects of the pollutants on the environment or people were limited to a local scale. First reports of urban air pollution stem from the Roman times, and urbanization in medieval times also brought about high concentrations of "smoke" in northern cities in winter. In fact, as early as 1273 A.D. urban smoke became a subject of concern in England when King Edward I banned the burning of certain highly polluting coals in London.

The first industrial activities affected only relatively small areas, but this changed rapidly as industrialization gained momentum. Smoke no longer affected the health of factory workers alone, but put the whole urban population at risk. The unhealthy smoke frequently encountered in British cities was caused mainly by emissions of soot particles and sulfur dioxide. During wintertime inversion situations, fog frequently accompanied the smoke, leading Harold Antoine Des Voeux to coin the word "smog" at the beginning of the 20th century to characterize this soup. Sometimes these conditions prevailed for days or weeks. The most infamous air pollution episode in London occurred in December 1952, causing more than 4,000 human fatalities [1].

Another type of smog was discovered in the Los Angeles basin in the late 1940s and 1950s, when pollution layers formed almost daily and vegetable crops began to show damage. It was found that under conditions of strong solar insolation in summer, ozone could build up to toxic levels. The ozone was formed from the precursor substances nitrogen oxides and hydrocarbons [2, 3], both typically contained in car exhaust. While initially this was thought of as another urban-scale problem, it was soon realized that ozone maxima occurred more commonly in rural areas downwind of large cities rather than in the cities themselves [4]. This significantly extended the scale of air pollution problems from urban to regional. Concerns were also raised that anthropogenic emissions could increase ozone concentrations on a global scale.

The effects of wintertime air pollution were also found to extend far beyond urban areas. Regulatory efforts were successful in reducing particulate levels in cities, but other compounds such as sulfur dioxide and nitrogen oxides at first remained largely uncontrolled. Growing emissions and the practice of building high stacks to reduce local pollutant levels – one of the control strategies after the severe pollution episode in London in 1952 – both contributed to the shift of air pollution problems from the urban to the regional scale. Acid deposition of sulfur and nitrogen compounds was found to be a major environmental problem in large areas of Europe and North America in the 1970s [5]. Even in remote regions of Scandinavia and Canada eutrophication of lakes and soil acidification caused severe ecosystem damage. For the first time, it was also fully realized that air pollution could cross political boundaries and affect foreign countries. Environmental policy reacted by establishing the Convention on Long-Range Transboundary Air Pollution, which was signed by the European nations, the United States and Canada in 1979.

In the 1950s, large-scale haze layers were observed in the Arctic in winter and spring [6], a region almost void of local emissions and which was previously thought to be pristine. Over time the haze became more severe and in the 1970s it could be explained as the result of the long-range transport of air pollutants, mostly sulfur [7], from far-away industrial regions at lower latitudes. This was perhaps the first air pollution problem discovered that was truly intercontinental in scale and caused by anthropogenic emissions. It was recognized, for example, that a wide range of pollutants, including sulfur, nitrogen compounds, persistent organic pollutants, lead, and mercury, can be transported from Eurasia all the way to the North American Arctic. However, probably because of the small human population in the Arctic, environmental policies did not respond strongly to this problem. But the emission reductions brought about by the agreements to fight soil acidification in Europe and North America were also successful in reducing sulfur concentrations in the Arctic.

The accident in the nuclear power plant in Chernobyl in 1986, when large areas of Europe were blanketed by a radioactive cloud, was a particularly sad, but high-profile example of how far-ranging air pollution problems can be. One consequence of this accident was the development or improvement of models that could forecast the long-range transport and deposition of hazardous substances. While fortunately these models have not yet been used in such a severe emergency, they were perhaps the first generation of models that were in principle capable of accurately simulating – and forecasting – pollution transport over intercontinental distances. But in the 1980s intercontinental pollution transport was not yet identified as a major problem, even though global-scale effects of humankind on the atmosphere dramatically became clear in 1985 when the ozone hole was discovered [8] and traced back to the destructive effects of man-made chlorofluorocarbons. Concern also grew about the increasing concentrations of greenhouse gases in the atmosphere, another example of a global modification of the atmosphere by humans.

It was suspected in the 1980s that intercontinental transport of pollutants – apart from the Arctic Haze problem – might be possible and would have some effect on pollutant concentrations over downwind continents. However, clear evidence was lacking and the effects on acid deposition – the pollution problem considered most important at that time – were perhaps too small to be detectable. Observations of enhanced pollutant concentrations were occasionally related to intercontinental transport [9], but they were too rare to be of much significance and could not be well confirmed with models.

Interest in intercontinental transport of air pollutants, particularly of ozone, grew rapidly in the second half of the 1990s for several reasons. First, it was found that ozone concentrations over the North Atlantic could be strongly enhanced in the continental outflow from North America [10]. Second, ozone formation is much more efficient (per molecule of nitrogen oxide emitted) in the free troposphere than in the boundary layer. As it was known that significant amounts of ozone precursors could be exported from the boundary layer, it was suspected that ozone formation may continue en-route from one continent to the other [11]. Third, the strongly increasing emissions of ozone precursors in Asia gave rise to speculations that surface ozone concentrations over North America may increase as a result, thus rendering American ozone abatement efforts inefficient. Model studies indeed suggested a substantial influence of Asian emissions on North American ozone [12, 13]. Experimental evidence was also found that intercontinental transport really affects pollutant levels at the west coast of North America [14], and at the same time a "textbook example" of intercontinental transport of ozone from North America to Europe was found [15]. The transport mechanism identified in this study turned out to be the most important one for rapid intercontinental transport both from North America to Europe and from Asia to North America.

In the meantime, evidence for intercontinental transport of pollutants is mounting. Several aircraft measurement campaigns have been dedicated to the study of intercontinental transport events. Models have shown considerable skill in simulating these events and are now used operationally to forecast transport events for the flight planning during research campaigns.

Scientific research into the intercontinental transport of pollutants has progressed to the point that a review of this topic is now worthwhile. So which substances are covered in this book? The very-long-lived greenhouse gases (e.g., carbon dioxide, nitrous oxide, etc.) are not well covered, because they attain an almost uniform concentration throughout the troposphere. The details of the transport and chemical reactions occurring on time scales of a few days to a few weeks – the typical range of transport times from one continent to the other – are largely irrelevant for these substances. This does not mean that nothing can be learnt about pollution sources by considering small enhancements over background concentrations in continental-scale plumes. Some long-lived trace gases, e.g., carbon dioxide, are therefore occasionally used as tracers. Very-short-lived compounds with lifetimes less than a few hours (e.g., radicals) are also not directly addressed in this book, because they have no chance of being transported from one continent to the other. However, via chemical reactions, they may be important for other longer-lived substances and, thus, bear some relevance. Many substances have lifetimes that are comparable to the timescales of intercontinental transport: reactive nitrogen compounds, carbon monoxide, volatile organic compounds, ozone and aerosols are just a few of them, and are mentioned most often in this book.

Chapters 2 to 12 cover a wide range of topics related to intercontinental pollutant transport. The remainder of this chapter identifies cross-cutting links between them and refers the reader to the appropriate chapters for more detailed information.

2
Intercontinental Transport of Anthropogenic Pollution

Figure 1 shows the global distribution of anthropogenic emissions of carbon monoxide (CO) according to the EDGAR inventory for the year 1995 [16]. Carbon monoxide is produced by incomplete combustion processes (e.g., by cars, biofuel burning). The CO emission distribution largely reflects the energy consumption of day-to-day human activities and, thus, the emission distribution of other substances (at least for those associated with combustion, e.g., nitrogen oxides) is quite similar. The bulk of the CO emissions are concentrated in the Northern Hemisphere. In the Southern Hemisphere, an emission hot spot is found in South Africa, which is even more obvious for some other substances such as nitrogen oxides or sulfur dioxide, because of a high number of large power plants with relatively low CO emissions. In the Northern Hemisphere there are four regions with particularly high emissions: the North American east coast, Western and Central Europe, East Asia, and Southern Asia. The first three are in the middle latitudes, where westerly winds prevail during most of the year (at least at higher altitudes), while the fourth is located in tropical Asia, where the winds are do-

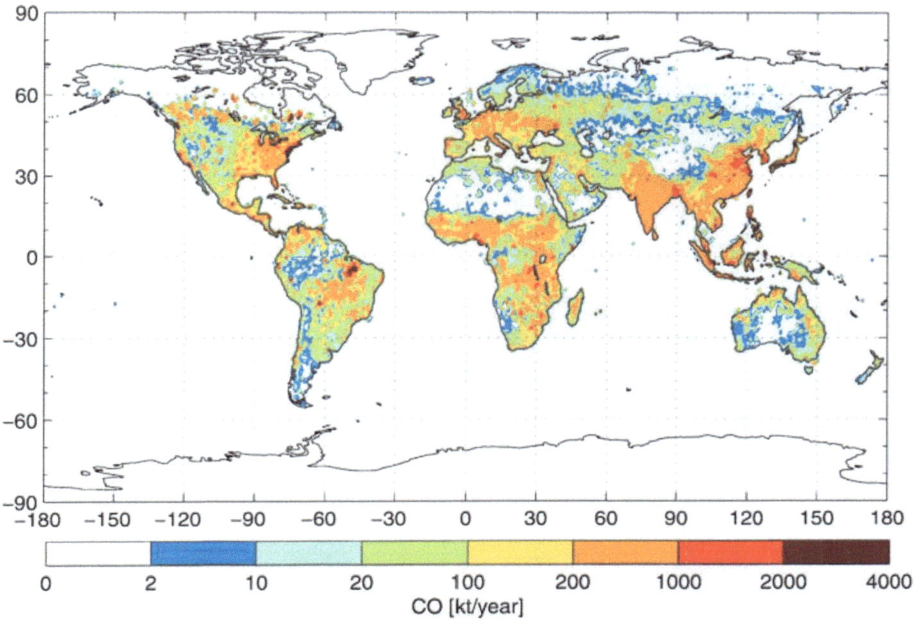

Fig. 1 Distribution of global anthropogenic emissions of carbon monoxide according to the EDGAR 1995 inventory

minated by the seasonally varying monsoons. It is therefore clear that the transport patterns of emissions from South Asia will be vastly different from those of the emissions occurring at more northerly latitudes. Convection plays a very important role during the monsoon season, when emissions from South Asia are carried high into the upper troposphere.

In the middle latitudes, two high-emission regions (eastern North America and East Asia) are located at the eastern seaboards of the continents, while Europe is located at the west side of the Eurasian continent and at higher latitudes than the other two regions. The first two regions are located close to the entrance to the North Atlantic and North Pacific stormtrack, while Europe is located at the exit of the North Atlantic stormtrack. Because air on average ascends at the beginning of a stormtrack and descends at its end [17], this has important consequences for the transport of the emissions from the three regions: Emissions from North America and Asia tend to be lifted into the upper troposphere, while emissions from Europe tend to remain in the lower troposphere [18]. The airstreams associated with extratropical cyclones are very important for the detailed dynamics of pollutant transport in the middle latitudes. These processes are explained in detail in Chapter 3.

Figure 2 summarizes the average transport of pollutants from the three Northern Hemisphere continents during the first 10 days after emission. This picture is a synthesis of a 15-year climatology investigation [19] for which a transport model was run with 6-hourly analyzed wind fields. Emissions were taken from

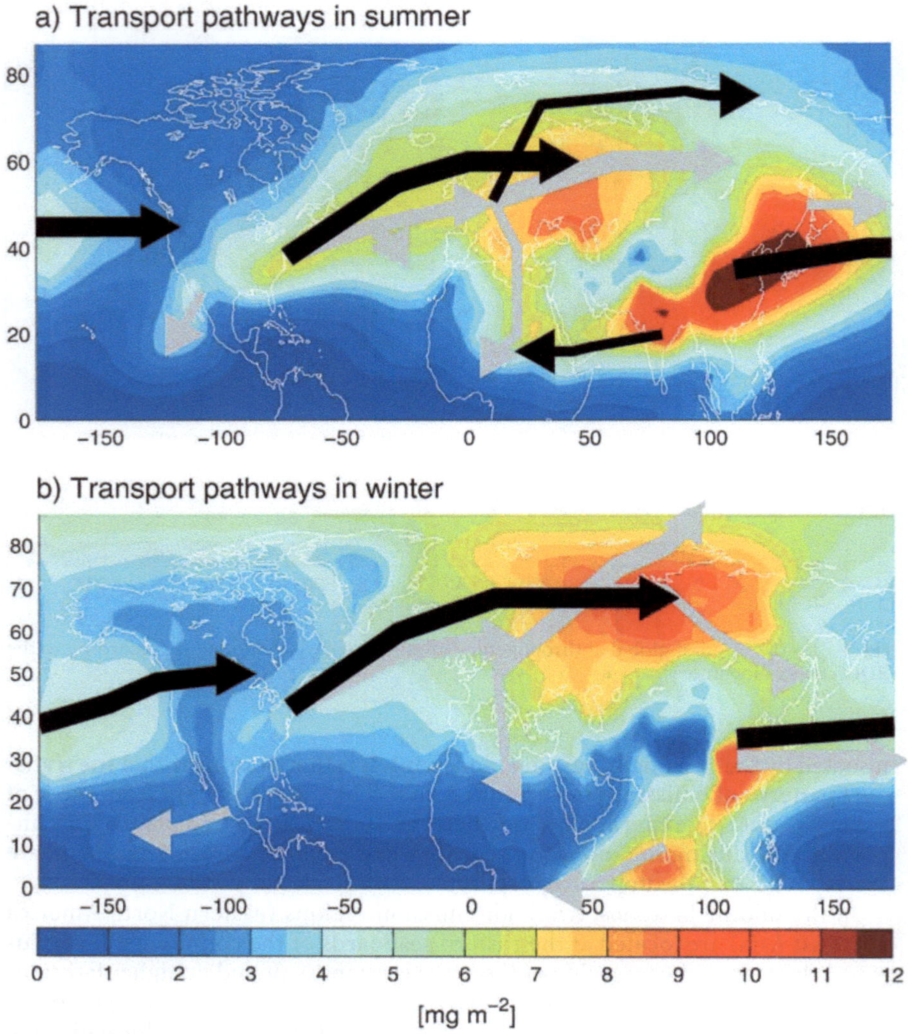

Fig. 2 Sketch of the global pathways of intercontinental pollution transport from the various continents. The contours show the 15-year average of the total columns of an anthropogenic CO emission tracer released over the Northern Hemisphere continents for an age class of 8–10 days. This means tracer is only shown 8–10 days after its emission. Four other age classes from 0–2, 2–4, 4–6, and 6–8 days after the emission were also used to qualitatively identify the most significant pathways that lead from the various emission maxima over the continents to the tracer's distribution after 8–10 days (arrows). *Gray arrows* show transport in the lower troposphere (below 3 km), whereas *black arrows* indicate transport in the middle and upper troposphere (above 3 km). The *upper panel* shows transport pathways in summer (June, July, August), the *lower panel* shows transport pathways in winter (December, January, February)

the inventory shown in Fig. 1 (but valid for 1990) and used to simulate a CO tracer up to 50 days after emission. Individual "tracer particles" were tagged with their release time and later binned into "age classes", which allowed the creation of a movie that shows the 15-year average transport. The color contours in Fig. 2 show the 8–10 day time frame of this movie, which was used together with the frames for 0–2, 2–4, 4–6 and 6–8 days to qualitatively depict the tracer dispersion over the first 10 days with arrows. Grey and black arrows distinguish low-level and upper-level transport, respectively, and the arrows' widths qualitatively indicates how much tracer mass is transported along the respective pathways.

CO emitted in the tropical region (especially in South Asia, but also small amounts of tracer emitted in the southernmost parts of North America), is transported eastwards, whereas CO emitted in the middle latitudes is transported westwards. Over South Asia the effect of the monsoon can be seen clearly: In summer, westward transport occurs in the upper troposphere (following rapid upward transport in the monsoonal convection), whereas in winter, low-level transport into the Indian Ocean towards the Intertropical Convergence Zone via the northeasterly trade winds predominates. During this season, large haze layers have been observed over the northern Indian Ocean as a result of anthropogenic emissions in India, with a possibly important influence on the regional climate [20]. More details about transport from South Asia can be found in Chapter 6.

In the middle latitudes, most of the CO is transported with the westerlies. However, there are distinct differences between the three continents. Emissions from North America and Asia are mostly transported in the upper troposphere, following upward transport with so-called warm conveyor belts, airstreams associated with extratropical cyclones. Transport there is rapid, especially in winter when there is a strong jet stream in the upper troposphere, and pollutants on average can cross the Atlantic and Pacific Ocean within the 10 days shown in Fig. 2 (and episodically within about three or six days). This has been confirmed by aircraft measurements over both North America and Europe, where distinct plumes from Asia and North America, respectively, were found. These plumes are characterized by strongly elevated concentrations of carbon monoxide and sometimes contain enhanced levels of aerosols and ozone. Low-level transport is much slower and, particularly in summer, is too slow to cross the Pacific Ocean within 10 days.

Emissions from Europe behave differently. In wintertime, practically no European pollutants are transported to the upper troposphere and even in summer low-level transport predominates. Furthermore, there is much more meridional transport than over the other two continents. In winter, most of the European pollutant outflow is directed towards the Arctic. A significant pathway, especially in summer, also leads into the Mediterranean and towards Africa, where it is overrun by pollutants from South Asia. The North Atlantic Oscillation, the dominant mode of climate variability in the North Atlantic region, has a strong influence on the pathways of pollutants from Europe: During its high phase, there is more meridional transport than during its low phase. Especially the transport towards the Arctic is strongly enhanced during the high phase of the North Atlantic Oscillation [19].

Four chapters describe intercontinental pollution transport in the northern middle and high latitudes, each with a different regional focus. Chapter 2 reviews Arctic air pollution problems, starting from the discovery of Arctic Haze. It describes transport patterns into and out of the Arctic, presents evidence from measurements in the Arctic, and discusses some actual Arctic pollution problems in more detail. For instance, persistent organic pollutants have complicated pathways as they can be transported both in the ocean and in the atmosphere and can switch phases. Furthermore, they condense in cold environments, leading to their accumulation in the Arctic.

Chapters 3, 4 and 5 describe the situations in North America, Europe and East Asia. Each chapter begins with a description of the regional emission situation, continues with a description of the typical flow patterns important for intercontinental transport in its region, and then describes both the export of pollution from the continent and the import of pollution from upwind continents. All three chapters concentrate on the available measurement data, but model results are also mentioned to the extent that is necessary to understand the measurement data.

Especially large anthropogenic emissions in the Southern Hemisphere are found over South Africa, with most of the emissions concentrated in a relatively small area in the industrial Highveld region. On a global scale, this region can almost be considered a pollutant point source (it is therefore also not well seen in Fig. 1; furthermore, carbon monoxide emissions are relatively low compared to other substances). The meteorological situation over South Africa is very complex. Most of the year, the area is under the influence of an anticyclone, which makes both export and removal of pollutants inefficient. Instead, high pollutant loads are found in several distinct haze layers, which can be transported out into the Atlantic or Indian Ocean, but are then often recirculated back to South Africa [21]. Efficient export of pollutants occurs mainly when low-pressure systems affect the area. Then, pollutants are swept out into the South Indian Ocean and may travel towards Australia. Chapter 7 describes the complex meteorological and chemical processes associated with pollutant transport over South Africa.

While several aircraft measurement campaigns have recently targeted intercontinental transport events, large gaps remain in our understanding of intercontinental transport. The few campaigns were all conducted during a particular season and results cannot be simply extrapolated to other times of the year. Furthermore, the most important impact of intercontinental transport is the enhancement of the background concentrations of air pollutants at the Earth's surface, because this is where people live and where air quality criteria may be violated. However, most of the evidence presented in the literature concerns pollutant plumes in the upper troposphere. In contrast, there are relatively few conclusive measurement data from surface stations. Therefore, most of what is known about the impact of intercontinental transport on the surface air quality stems from chemistry transport models. For instance, one model study suggested that 20% of the violations of the European Council air quality standard for ozone would not have occurred in the absence of North American emissions [22]. Chapter 8 describes the models that are available to simulate intercontinental pollution transport and presents their results.

3 Intercontinental Transport of Pollutants from Natural Sources

The very first observations of intercontinental atmospheric transport concerned desert dust. Episodic deposition of red sand over large parts of Europe occurs almost every year, and people were certainly aware of this phenomenon already a long time ago. At the beginning of the 20th century, the source-receptor-relationships were clearly established, and the techniques applied at this time to trace the dust back to its source region in the Sahara are reminiscent of modern studies of intercontinental transport. Chapter 11 gives credit to these early studies of intercontinental dust transport by comparing them to modern studies of similar events, for which detailed satellite observations are available. The most spectacular and well documented cases of intercontinental dust transport are probably those from Asian desert regions across the Pacific to the west coast of North America and beyond [23]. Occasionally trans-Pacific dust transport may contribute to violations of air quality standards for particulate matter in North America. These events are also well documented in Greenland ice for prehistoric times.

Dust transport in the atmosphere is important for a number of reasons. It leads to soil erosion in the source region, but can replenish soils at other locations or can fertilize the ocean. Dust is potentially a hazard to human health, it affects the radiative balance of the atmosphere, and it can modify cloud occurrence and provides surfaces for chemical reactions in the atmosphere. While dust storms are a natural phenomenon, humans influence dust source regions via land use changes, and climate change may affect the soil moisture availability and vegetation.

Emissions from biomass burning have also been known as an important pollution source for a long time. About 15 years ago, using satellite data it was discovered that biomass burning in the tropics causes a widespread distribution of high levels of ozone and carbon monoxide in the tropical southern hemisphere [24], confirming earlier studies which suggested a global influence from biomass burning [25]. The pollution was shown to be caused by large biomass burning emissions in Africa during the dry season from August to October. Tropical biomass burning also occurs on other continents and can cause strong seasonal enhancements of the concentrations of aerosols, ozone and carbon monoxide. The large-scale burning of biomass in Indonesia during 1997/1998, when El Niño conditions prevailed, caused severe air pollution problems over very large regions [26]. The fires were mostly started by humans, but were intensified by the drought conditions due to El Niño. With large on-going land use changes in the tropics, accompanied possibly by changes in the regional climate, it can be expected that biomass burning emissions will also change in the future. Chapter 9 describes the formation and transport of pollutants, particularly of ozone, from biomass burning in the tropics.

Large forest fires occur also in the boreal region, but it was not until recently that their hemispheric-scale impact on the atmospheric composition was fully recognized. In 2000, it was discovered that Canadian forest fires can strongly affect ozone concentrations in the southern United States [27], and soon after large enhancements in aerosols, carbon monoxide, nitrogen oxides and ozone were found in a Canadian forest fire plume over Europe [28]. Even larger areas burn in Siberia, leading to a strong impact of boreal forest fire emissions on the northern hemispheric trace gas budget [29]. However, in contrast to tropical biomass burning, there is a large interannual variability in fire occurrence. While many of these fires are ignited by lightning, a significant fraction are started by humans. Humans also influence boreal forest fire occurrence through changes in land use, fire management and fire suppression. Chapter 10 provides a review of boreal forest fire occurrence, burning conditions and related trace gas emissions, fire-induced deep convection, and the intercontinental transport of fire emissions.

4
Final Remarks

The interest in intercontinental transport of air pollution arose because of concerns that regional control strategies for air pollution may be inefficient in a situation where background concentrations of air pollutants increase due to rising emissions in other parts of the world. The efforts that are currently dedicated to studying intercontinental pollution transport may partly be driven by scientific curiosity. Another driving agent, however, is the hope that a better causal understanding may in the end help ameliorate the environmental problems that are caused by air pollution, of which climate change and air quality are the two most important.

Chapter 12 describes the "tightening vise" that air pollution managers in industrialized nations are currently facing: background concentrations of many pollutants (e.g., ozone) are rising, air quality standards are becoming more stringent, local controls are becoming more difficult to achieve after the cheapest emission reduction measures have already been taken, and emissions in other parts of the world are rapidly increasing. In this difficult situation, the best (i.e., most cost-efficient) way to improve the air quality may not be to reduce the local emissions, but instead to reduce the emissions somewhere else. Achieving compliance with air quality standards most cost-effectively may include concepts like emission trading, which is one of the instruments used for reducing emissions of carbon dioxide. However, compared to the long-lived carbon dioxide, the situation for most air pollutants is much more complicated. Emissions of nitrogen oxides in different regions of the globe, for instance, are not equally efficient in forming ozone, and the ozone formed is not equally distributed around the globe. Therefore, detailed understanding of source-receptor-relationships is needed in order to address such issues as where emission reductions are needed (or where they are most efficient) in order to improve the air quality at a particular location. The research described in this book is now beginning to provide this information, and Chapter 12 discusses how it can be used in the future for the international management of air pollution that knows no political or continental boundaries.

Acknowledgments We thank Owen Cooper for valuable comments on this chapter. S. Eckhardt was partly supported by the German Federal Ministry for Education and Research within the framework of the Atmospheric Research Program 2000 (project CARLOTTA).

5
References

1. Wilkins ET (1954) J R Sanitary Inst 74:1
2. Haagen-Smit AJ (1952) Industrial and Engineering Chemistry 44:1342
3. Leighton PA (1961) Photochemistry of Air Pollution. Academic Press, New York
4. National Research Council – Committee on Tropospheric Ozone Formation and Measurement (1991) Rethinking the ozone problem in urban and regional air pollution. National Academy Press, Washington, DC
5. Sehmel GA (1980) Atmos Environ 14:983
6. Mitchell Jr JM (1956) J Atmos Terrest Phys Special supplement 195
7. Shaw GE (1975) Bull Am Met Soc 76:2403
8. Farman JC, Gardiner BG, Shanklin JD (1985) Nature 315:207
9. Dutot AL, Colin P, Etienne A, Perros PE, Toupance G (1994) In: Borrell P, Borrell PM, Seiler W (eds) Proceedings of EUROTRAC Symposium '94: Transport and Transformation of Pollutants in the Troposphere. SPB Academic Publishing, The Hague, Netherlands, p 239
10. Parrish DD et al. (1993) Science 259:1436
11. Wild O, Law KS, McKenna DS, Bandy BJ, Penkett SA, Pyle JA (1996) J Geophys Res 101:29269
12. Berntsen TK, Karlsdottir S, Jaffe DA (1999) Geophys Res Lett 26:2171
13. Jacob DJ, Logan JA, Murti PP (1999) Geophys Res Lett 26:2175
14. Jaffe D et al. (1999) Geophys Res Lett 26:711
15. Stohl A, Trickl T (1999) J Geophys Res 104:30445
16. Olivier JGJ, Berdowski JJM (2001) Global emissions sources and sinks. In: Berdowski J, Guicherit R, Heij BJ (eds) The Climate System. AA Balkema Publishers/Swets & Zeitlinger Publishers, Lisse, The Netherlands. ISBN 90 5809 255 0, p 33
17. Stohl A (2001) J Geophys Res 106:7263–7279
18. Stohl A, Eckhardt S, Forster C, James P, Spichtinger N (2002) J Geophys Res 107:4684
19. Eckhardt S, Stohl A, Beirle S, Spichtinger N, James P, Forster C, Junker C, Wagner T, Platt U, Jennings SG (2003) Atmos Chem Phys 3:1791
20. Lelieveld J et al. (2001) Science 291:1031
21. Held G, Scheifinger H, Snyman G (1994) South African J Science 90:91
22. Li Q et al. (2002) J Geophys Res 107: doi: 10.1029/2001JD001422
23. Husar RB et al (2001) J Geophys Res 106:18317
24. Fishman J, Fakhruzzaman K, Cros B, Nganga D (1991) Science 252:1693
25. Crutzen PJ, Heidt LE, Krasnec JP, Pollack WH, Seiler W (1979) Nature 282:253
26. Thompson AM, Witte JC, Hudson RD, Guo H, Herman JR, Fujiwara M (2001) Science 291:2128
27. Wotawa G, Trainer M (2000) Science 288:324
28. Forster C et al. (2001) J Geophys Res 106:22887
29. Wotawa G, Novelli PC, Trainer M, Granier C (2001) Geophys Res Lett 28:4575

The Handbook of Environmental Chemistry Vol. 4, Part G (2004): 13–39
DOI 10.1007/978-3-540-40037-0

Long Range Transport of Air Pollution to the Arctic

Jan W. Bottenheim[1] · Ashu Dastoor[2] · Sun-Ling Gong[1] · Kaz Higuchi[1] · Yi-Fan Li[1]

[1] Meteorological Service of Canada, 4905 Dufferin Street, Toronto, Ontario, Canada M3H 5T4
[2] Meteorological Service of Canada, 2121 Trans Canada Highway, Dorval, Quebec, Canada H9P 1J3

Abstract Transport of air pollution to the Arctic is reviewed with emphasis on what has been measured. General atmospheric circulation patterns especially favor transport of pollution from European and Central Asian ("Eurasian") sources in early winter and spring. Many anthropogenic pollutants such as aerosols, smog components, gaseous mercury and POPs are observed at that time in surprisingly high concentrations and current understanding of their origin is reviewed. Evidence for outflow of these pollutants from the Arctic to mid-latitude regions is also reviewed.

Keywords Arctic · Transport · Air pollution

1 Introduction

The Arctic is generally considered to be the part of the world to the far North. One way to define its boundaries is the Arctic circle at 66°32′N, the southern boundary of the midnight sun. Other, climate-determined, definitions are possible, and for the purpose of this chapter it is useful to consider the location of

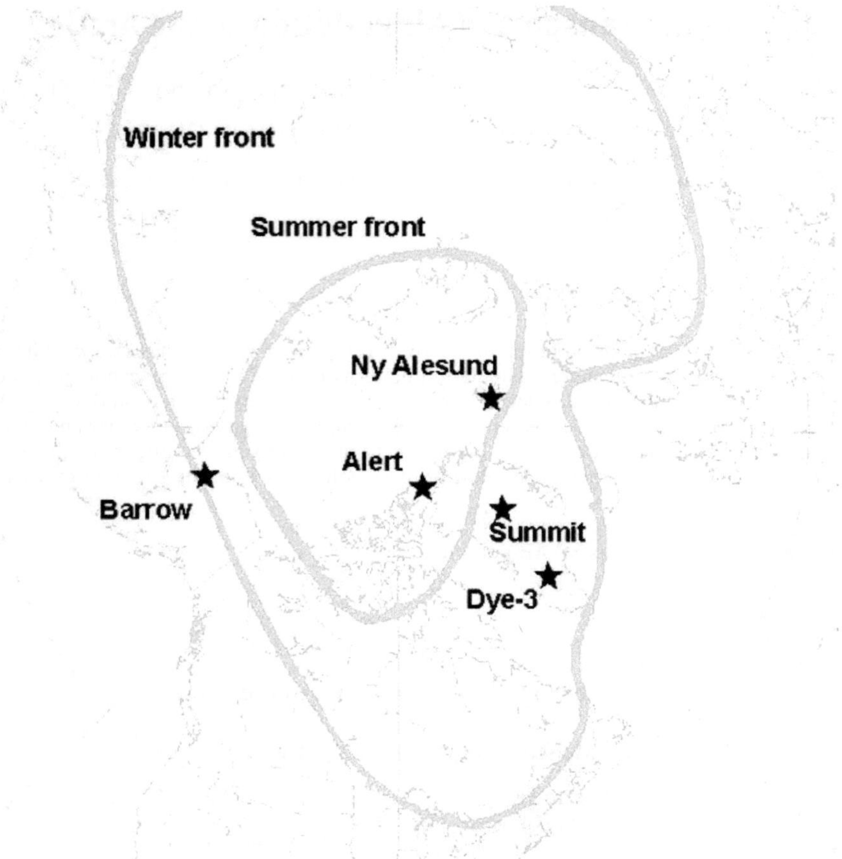

Fig. 1 Location of the Arctic front in winter and summer. Also indicated are the locations of some of the major air observatories

the Arctic front. Figure 1 shows its average location in January and July. It shows spatial fluctuations from about 40°N polewards in the winter, enveloping the snow covered North American and Eurasian continents. In the summer it is confined to an area roughly north of the 70°N latitude. This is a region with a harsh climate: winters are cold and dark, often with clear skies and very little precipitation. Summer temperatures usually reach above the freezing level and there are a high percentage of days with drizzling marine stratus. It is sparsely inhabited except for a few regions on the fringes such as upper Siberia and Alaska. There are few sources of local air pollution, primarily mining related point sources such as the Prudhoe Bay area in Alaska, the Kola peninsula in European Russia and the Taymir peninsula in Siberia (Norillsk). Truly a pristine, remote area of the world – a haven of clean environments?

In fact this picture is wrong. There are occasional references to the presence of a haze layer in the official reports of the Arctic explorers in the 19th century

("The northern horizon had a slight haze over it, which made the distant land indistinct", noted Nares in 1875 [1], and at the beginning of the 20th century, MacMillan commented on the observation of a dark haze by the Inuit who called it "poo-jok" (MacMillan, as quoted by Schnell [2]). Other early references speak of the occurrence of an "ice crystal fog" [3], or an "Arctic haze" [4] in the spring but it was not until the1970s that the studies of Shaw (e.g. [5]) and Rahn (e.g. [6]) revealed the true nature of this haze as due to long range transport of mostly sulfur pollution. Comprehensive studies followed with the establishment of long term measurement programs such as in Alaska (Point Barrow), the Canadian high Arctic (Alert) and Svalbard (Ny Ålesund), see Fig. 1. Other manifestations of pollution that could only have been brought to the Arctic via atmospheric transport have since been discovered such as man-made pesticides in the snowpack [7] and unusually high oxidant concentrations (PAN levels of up to 0.5 ppbv in the early spring [8]).

Questions therefore arise as to what the origin of these contaminants is, and what are the physical processes that bring them to the Arctic. In this chapter, we will first address the atmospheric circulation patterns that are responsible for the observed pollution episodes and then turn our attention towards observations at some of the more prominent long-term measurements stations. We will focus on the transport of air pollution to and from the Arctic, and pay limited attention to the atmospheric chemistry in the Arctic itself such as ozone and mercury depletion, or the chemistry in the snow pack [9–11]. Also, we will present the topic of Arctic air pollution from the perspective of what has been observed at the surface and in the lower boundary layer, and not discuss transport to mid to upper tropospheric air. This limitation also implies that we will pay scant attention to measurements on the Greenland ice sheet since those measurements, made at >2500 m altitude, have little relevance to lower tropospheric air pollution in the Arctic. Several excellent reviews exist on the topic of Arctic air pollution and its origin, such as the early reviews by Rahn [12] and Barrie [13], the monograph on Pollution of the Arctic Atmosphere, edited by Sturges [14] and most notably the recent exhaustive assessment report of the Arctic Monitoring and Assessment Program [15].

In these times of limited resources and decreasing budgets, the question is frequently raised as to what is the relevance of studying Arctic air pollution and its origin. There are several compelling reasons for this. Firstly, the Arctic troposphere is a perfect, large scale, chemical reactor, i.e., a natural laboratory for studying atmospheric processes at low temperatures and for long periods of time without the presence of solar radiation. The observation of *tropospheric* ozone depletion in the spring led to a vast reassessment of the importance of BrO_x chemistry. Similarly, the more recent discovery of air–snow chemical interactions and their impact on air quality was made in this natural laboratory with potential implications for snow-covered regions outside the Arctic. Secondly, the Arctic plays a crucial role in understanding past and future climates. Ice cores yield insights into past climate variations. On the other side of the time scale is the often-quoted phrase that "the Arctic is the best early warning system for future global change". But there is also a more immediate societal urgency to try and understand the impact of air pollution on the Arctic ecology. The recent discovery of mercury depletion and deposition of its oxidation products to the surface [16] where they may become available for humans and other organisms as illustrated by increasing concentrations in marine mammals, is an example of our limited understanding of the relationships between anthropogenic inputs and the natural system in the Arctic and its potential consequences.

2 Circulation Patterns Driving Transport to the Arctic

Lower tropospheric transport in and around the Arctic region has a very recognizable seasonal signature. Figure 2a and b show, respectively, climatological circulation patterns for winter and summer, as reflected in the 1000 hPa geopoten-

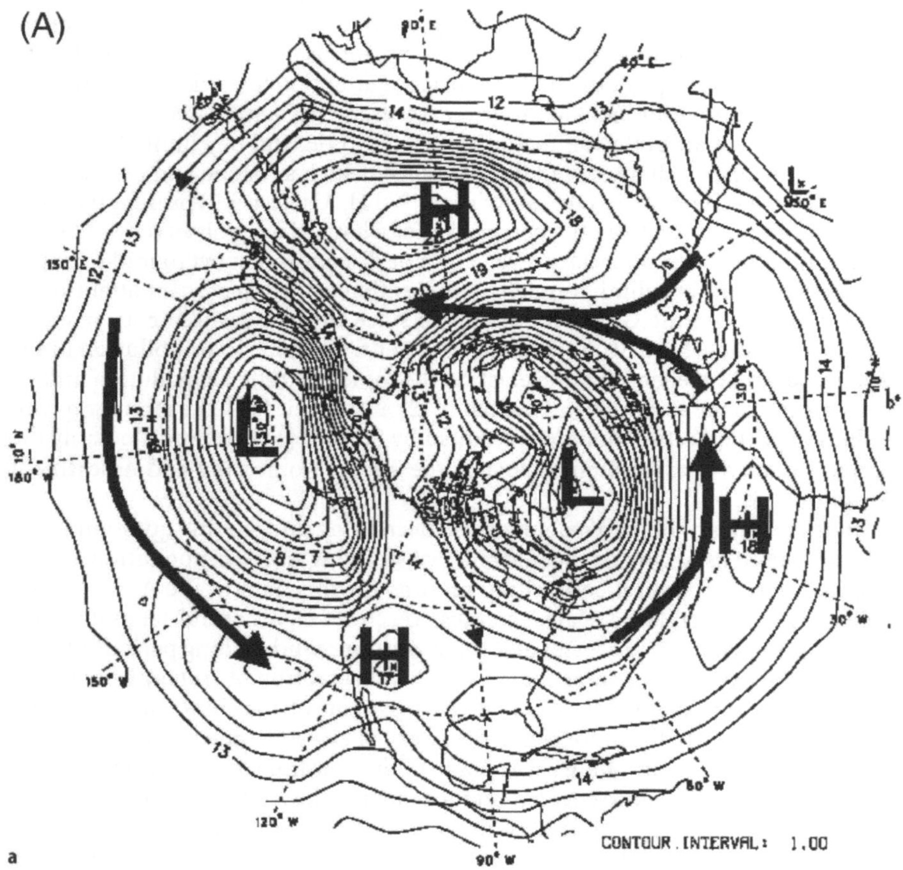

Fig. 2 1000 hPa climatological height field calculated from the CMC archived meteorological data for **a** winter (1947–1995) and **b** summer (1946–1995). Figures courtesy of Amir Shabbar, MSC

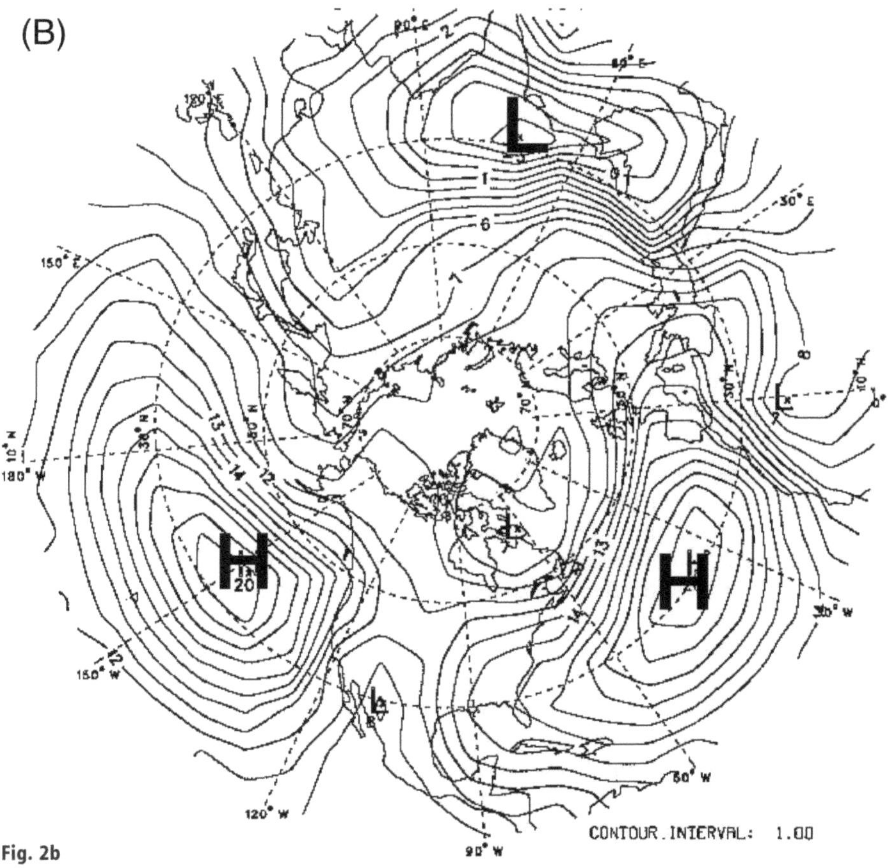

Fig. 2b

tial height fields derived from the climate data archive at the Canadian Meteorological Centre (CMC). The winter season (defined here as December, January and February) is marked very distinctly by three major height centers (corresponding to pressure centers on a sea-level pressure chart). These centers consist of two low centers located in sub-polar North Pacific Ocean, just south of the Bering Sea area, and southeast of Greenland in the North Atlantic Ocean (the Icelandic Low), and a high situated over the Mongolia region. Less intense highs are located over western North America and the mid North Atlantic ocean (the Subtropical High). The summer (defined as June, July and August) is characterized by a much weaker circulation regime than the winter, while spring and fall are slow transition periods that are less clearly defined.

The impact of these pressure systems on transport in the winter can be summarized as follows (see Fig. 2a). The cyclonic circulation around the Icelandic Low in combination with the Mongolian/Siberian High will transport material from source regions in northern Europe and Siberia into the Arctic. The North

Pacific Low induces eastward transport of material from East Asia to North America along its southern portion over the mid to subtropical latitudes, and then along western Canada and Alaska into the western Arctic. The same circulation may also transport chemical species produced in western North America into the Arctic. Central and eastern North American pollutants will flow along the southern portion of the Icelandic Low to northern Europe.

Transport along marine pathways will usually experience extensive scavenging by clouds and precipitation [17]. Hence both East Asian and North American pollution will be largely removed from the air before reaching the Arctic. This is largely a consequence of the fact that pollutants emitted by East Asian and North American sources are subject to relatively strong lifting (see chapters 3 and 5). In addition, lifting transports these pollutants to higher levels of the troposphere, further limiting their impact on boundary layer air in the Arctic. European emissions on the other hand remain mostly in the lowest few kilometers of the troposphere, particularly in winter.

The strong cross-Arctic transport from the northern European and Siberian source regions impacts on regions in the Canadian Arctic and Alaska, and in combination with the high over the North American continent, may extend to the eastern parts of southern Canada and the U.S. This cross-Arctic transport has for instance been demonstrated to be quite effective in moving CO_2 from northern Europe and Siberia to the Canadian Arctic Archipelago [18]. Outflow from the Arctic to eastern Siberia and Japan is also a distinct possibility.

Compared to the winter circulation pattern, the climatological summer pattern shows an overall weaker height gradient field in the Northern Hemisphere, resulting in considerably less intense airflow movements (Fig. 2b). The Icelandic Low has weakened and shifted westward to the southern part of Baffin Bay, between northeastern Canada and southern Greenland. The Subtropical High has moved northward, and is located in mid-latitude North Atlantic. A high center is now located over the northeastern North Pacific Ocean, just to the south of the Gulf of Alaska. Although these major systems have a significant influence on the pattern of atmospheric transport, the reduced intensity (compared to winter) and the relative spatial positioning of the systems around the Arctic region reduce the chances for significant advection of materials from mid-latitude regions into the Arctic. In addition, the higher summer temperatures lead to the formation of extensive drizzling marine stratus removing most contaminants from the air before reaching the Arctic [15]. Convection over the source regions in eastern Asia and North America is particularly strong in the summer, again transporting pollution from those regions effectively to higher levels in the troposphere.

The circulation patterns shown in Fig. 2 are useful as a climatological reference. However, atmospheric circulation does vary on many time scales, from as short as daily [19, 20] to as long as millennia [21], and this will impact on transport of chemical species into the Arctic. With the exception of data from ice cores, the observational record of air pollution in the Arctic is a few decennia old, and hence variability associated with multi-decadal periods cannot yet be detected. However there are variations on shorter time scales that merit special attention.

At a scale of 5–15 days, mid-latitude atmospheric blocking can have a significant impact on the transport of air pollution to the Arctic. A relationship between blocking and elevated levels of air pollution was first noted by Raatz [22]. Blocking is a quasi-stationary feature characterized by a high-pressure cell centered around 60°N with a low-pressure cell to the south of it lasting up to 15 days [23]. As a result of this dipole structure, there is a split in the westerly jet stream with a branch transporting a significant air mass meridionally to the North. Blocking is predominantly observed at two locations, the NE Atlantic and the NE Pacific. Data show approximately 2.5 times more blocking events in the Atlantic compared to the Pacific and a winter/spring maximum in their frequency. Numerical modeling studies of the long-range transport of sulfur to the Arctic by Iversen [24] and Dastoor and Pudykiewicz [25] indicated a clear connection between blocking events and high Arctic air-pollution episodes.

On a somewhat longer timescale, two major low-frequency variability modes have an impact: the Pacific-North America (PNA) and the North Atlantic Oscillation (NAO) teleconnection patterns [26, 27]. Although there is some relationship between the NAO and the PNA modes of oscillation [28, 29], positive and negative phases of the PNA and the NAO can occur in different combinations with various intensities. Both of these variability modes are dominant during the winter season. Comprehensive reviews of the NAO have been given recently by Wanner et al. [30] and Hurrell et al. [31]. It fluctuates on multiple timescales, from monthly to decadal, and with different frequencies contributing more at certain times [32]. In contrast to the NAO, the PNA teleconnection pattern is influenced by boundary forcing [33], and is usually associated with the occurrence of a moderate to strong ENSO (El Niño-Southern Oscillation) event [26, 27, 34]. The PNA oscillates with a quasi-periodicity, ranging from 2 to 4 years.

Recent studies indicate that the NAO and PNA do in fact influence transport of atmospheric pollutants to the Arctic. The impact of atmospheric transport can be seen quite clearly, for example, in atmospheric CO_2, a greenhouse gas that has an atmospheric lifetime measured in decades [35]. The positive phase of the NAO, associated with a strengthened Icelandic Low, provides a very strong cyclonic circulation over the eastern Arctic, with a resulting tendency to enhance transport of atmospheric CO_2 from Europe and Siberia into the Arctic. The negative phase of the PNA has a similar effect. On the other hand, during the negative phase of the NAO (or positive phase of the PNA), the weaker Icelandic Low results in a decrease of the biospheric CO_2 influence from Europe and Siberia in favor of an increased influence from the North American sources. The inter-annual variability in the atmospheric CO_2 concentration (usually referred to as the growth rate change) is related to the NAO and PNA with statistically significant correlations of –0.35 and 0.40 at Alert, and –0.25 and 0.42 at Pt. Barrow, respectively [35]. The relationship is graphically shown in Fig. 3.

The impact of the NAO and the PNA has also been observed in other atmospheric chemical species. Their source functions are likely to be different from CO_2 in terms of spatial and temporal distributions, and hence the influence of low-frequency circulation variability modes is expected to be somewhat different from that on CO_2. However, the fundamental way in which the advection of various atmospheric chemicals into the Arctic is influenced by the NAO and PNA

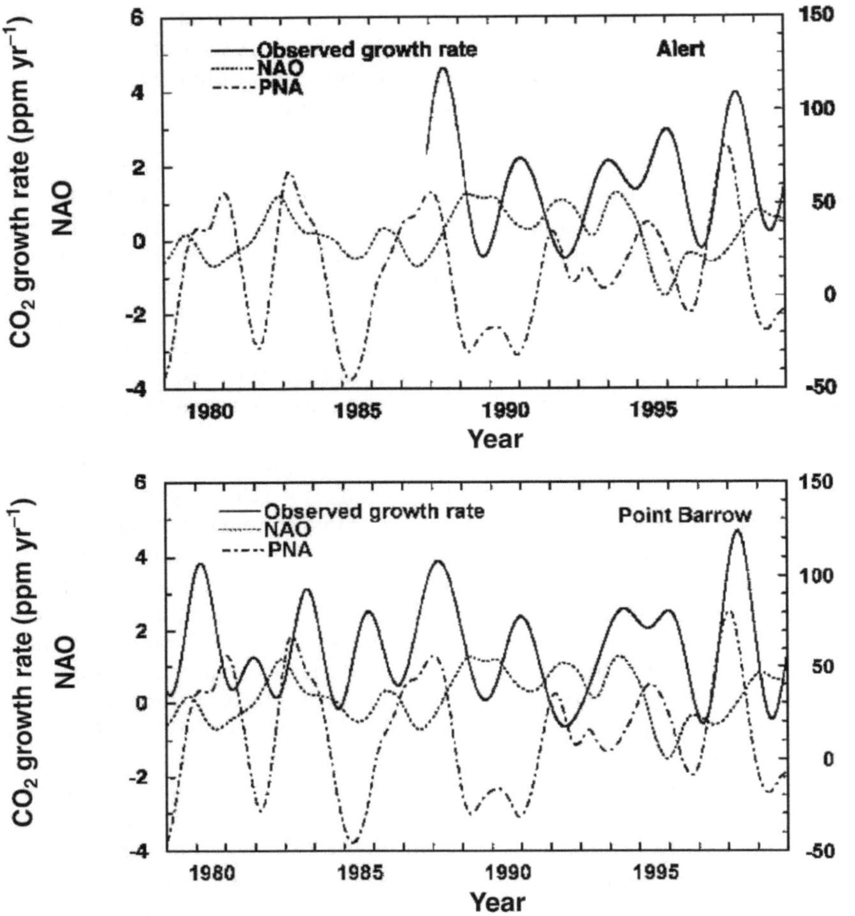

Fig. 3 Normalized PNA and NAO indices plotted against the observed atmospheric CO_2 growth rates at Alert and Pt. Barrow. From [35]

is the same. Eckhardt et al. [36] found that an enhanced transport of atmospheric pollutants (NO_2, CO, BC) into the Arctic from Europe occurs during the positive phase of the NAO in winter and spring. A similar finding was documented by Christensen et al. [37]. This is in part due to the location of the European air pollution sources on the eastern sector of the predominantly cyclonic flow (Fig. 2a) that is enhanced during the positive phase of the NAO [38]. In addition, Eckhardt et al. showed that at this time, due to the enhanced Icelandic Low, a stronger westerly flow leads to more effective transport of North American pollutants across the North Atlantic to Europe where they then merge with the pollutants produced in Europe and get transported into the Arctic.

The effect of the NAO and PNA variability modes on long-range atmospheric transport has also been shown to be important for deposition of materials on the

Summit of the Greenland ice sheet. Kahl et al. [39] performed an extensive 44-year 10-day back trajectory analysis, to establish seasonally dependent relationships between trajectory residence times over specific regions (Europe, Asia, Pacific, North America, Atlantic, Arctic) and Summit, mediated by the PNA and the NAO. The influence of the NAO has also been detected in the GISP2 isotope data [40]. These and other studies provide increasing and consistent evidence for a significant impact of large-scale atmospheric oscillation modes on the variability of the atmospheric chemical measurements in the Arctic.

3 Evidence from Measurements in the Arctic

The previous section made it clear that transport of air pollution to the Arctic should be expected to occur mostly in the winter and spring. This is indeed what has been observed and in this section we will look at this experimental evidence. We do not intend this to be an exhaustive review but rather will present examples from several classes of air pollutants that have been observed in the Arctic atmosphere, and will focus on recent advances in these fields.

While atmospheric circulation patterns drive the transport of these pollutants to the Arctic and hence have a major impact on short and long term trends, there are other important factors to consider when discussing Arctic air pollution. Low temperatures and the absence of direct sunlight for a large part of the year create a stable troposphere in the winter. It leads to the creation of highly stable stratification with very distinct layers of polluted air. Cloud formation is minimal and precipitation is largely limited to ice crystal showers; furthermore, photochemistry is turned off. As a result, these polluted layers do have a tendency to remain cohesive and chemically unmodified for long periods of time. It follows that these layers retain a strong memory of their origin and this has been used to great advantage in tracing the sources of Arctic air pollution.

3.1 Aerosols and Arctic Haze

Impaired visibility was reported as early as the 19th century. While the most visible expression of Arctic air pollution, we cannot be sure that poo-jok was actually sulfate aerosol of anthropogenic origin. However, evidence from ice core studies suggests that aerosol acidity was relatively constant during the first half of the 20th century and then increased rapidly after the 2nd World War, roughly in agreement with estimates of SO_2 emission data during the same time [41]. In any case, by 1980 the nature of Arctic Haze was summarized by Rahn and Heidam [42] as follows: (i) It is probably Arctic wide and of distant origin, (ii) Most originates from Eurasian sources, and (iii) Its composition is predominantly SO_4^{2-}. The studies of the 1970s spawned the establishment of a host of surface measurement sites (see e.g. the compilation by Rahn in 1981 [43]). Several of these sites are still in operation as of today, now permitting long-term trend analysis. Furthermore, the Arctic wide extent of the occurrence of the haze has been established beyond a doubt by many intensive aircraft field campaigns such as the four AGASP (Arctic Gas and Aerosol Sampling Program) campaigns between 1983 and 1992, and the Russian/German Arctic Haze studies in 1993–1995.

Detailed physical and chemical analysis of the composition of the haze particles, in particular metals and soot has been used to investigate the origin of the aerosols. For instance Lowenthal and Rahn [44] developed a seven-element tracer system to deduce the source regions for aerosols observed at Barrow in 1979–1980. Their analysis suggested that ca. 70% of most tracer elements came from the (then) USSR, but that SO_4^{2-} was about equally due to the USSR and Europe which they claimed to be in agreement with expectations from SO_2 emissions at that time. Maenhaut et al. [45] applied an absolute principal component analysis (APCA) on an extensive set of elemental concentration data and determined the origin of polluted air masses at the Norwegian stations Ny Ålesund and Vardø. While this method clearly showed the impact of the industrial activities at the Kola Peninsula on the observations at Vardø, it was not possible to extend this observation to distinguish between other source regions. Use of the 7-element tracer system of Lowenthal and Rahn in a CMB (chemical mass balance) analysis was more successful in that respect and showed that at Ny Ålesund the tracer elements could be attributed about 60/40 between the USSR and Europe.

The measurement record since 1980 at Alert, Canada, has recently been used in an extensive principal component analysis (PCA) to determine the seasonal variation and long term trend of Arctic aerosols measured at that location [46, 47] and it is instructive to look at the results in some detail. Among the ten factors that explained 93% of the variance, five had a clearly identifiable signature, three of anthropogenic (ANTHRO, SMELTER and PHOTO-S) and two of natural origin (SOIL and SEASALT). With the exception of the SOIL factor, all showed a strong seasonal cycle with a maximum in winter/spring and a minimum in summer/fall, reflecting the change in circulation patterns discussed in Sect. 2. Furthermore, removal processes are important: aerosol residence times are much longer in winter (~3 to 7 weeks) than in summer (~3 to 7 days) due to stable thermal stratification in the atmospheric surface boundary layer and the less than 10-mm H_2O/month of precipitation in the cold half of the year.

Metals such as Pb, V, Mn and Zn were mostly linked with the ANTHRO factor representing anthropogenic pollution in general, which peaked from January to March. This factor also included non-acidic salts of sulfate with ammonium as the dominant cation. Since no detailed emission signatures exist for all of these compounds, the specific contributions from fuel combustion or other sources could not be distinguished, but the PCA identified two interesting additional factors of anthropogenic origin. A second factor, dominated by Zn and Cu, was associated with smelter emissions (SMELTER), with a broad peak ranging from November to March. While this factor clearly represented primary emissions from smelters in the Arctic air shed (e.g. the Norillsk region in Siberia), it did not include a significant contribution from sulfate. Sulfate in the Arctic atmosphere is dominated by formation from conversion of sulfur dioxide and this resulted in a separate factor (PHOTO-S) [46] with a maximum later than the other anthropogenic factors (March/April). A large difference between winter/spring and summer/fall, displayed by SEASALT factor (consisting of Na, K and Mg) reflected the long-range transport from surrounding oceans and hence the changing

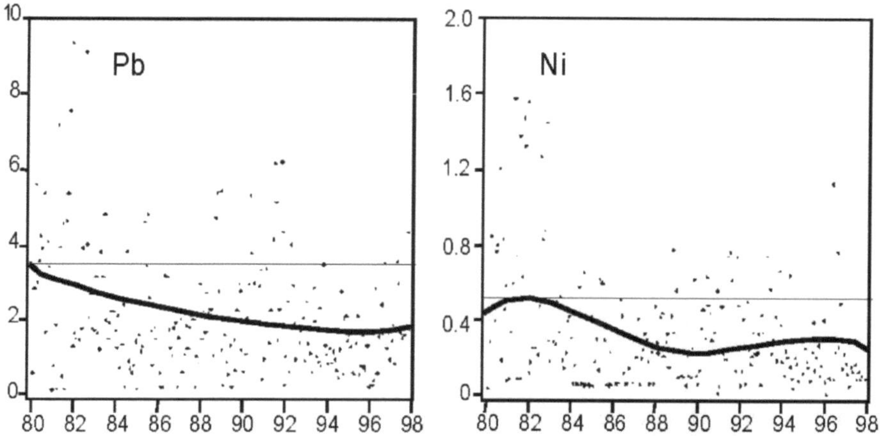

Fig. 4 Long-term trends for lead and nickel from winter only data. The concentration is in ng m^{-3}. From [47]

circulation patterns. Only metals associated with the SOIL factor (Mg, Al, Ca, Ba, Ti and Fe) had a weak seasonal variation. This observation is believed to be due to the fact that there are two contributions to the soil components: long-range transport in the spring, and local natural emissions in the fall.

The long-term trends were analyzed for 12 metals. As expected, no long-term trends were observed for Na (reflecting SEASALT) and Ca (reflecting SOIL erosion). Zn and Cu (reflecting SMELTER emissions) showed a slight increase from 1980 to 1985 and then, beginning around 1991, a decrease to the end of the current measurement period. Because of the strong seasonality (see above) the power to detect changes on Arctic aerosol contaminants is enhanced if data from winter/spring only are used since it eliminates most BDLs (below detection limits). Clear trends were obtained this way for the ANTHRO components Pb and Ni, see Fig. 4.

Sulfate levels exhibited a decadal increase in the 1980s and peaked around 1990 (Fig. 5a), followed by a steady decrease since then. The decline of sulfates in aerosols since 1990 was attributed to the collapse of the former Soviet Union and the trend is similar to the smelter emissions which are largely attributed to Northern Siberian sources. It is also mirrored by the trend in sulfur emissions from the former Soviet Union up to 1989 and all of its separated nations since then (Fig. 5b) [48]. Lowenthal and Rahn [44] showed that about 50% of the sulfate observed at Barrow came from European sources outside the then USSR, and a similar impact of non-USSR emission sources might have been anticipated for Alert. Since European emissions have decreased since the early 1980s [49] a different trend than indicated by Fig. 5a would therefore have been expected. It is unclear at this time whether the apparent correlation suggested by Fig. 5a and b is an indication for a much less important contribution from non-USSR emission sources to sulfate levels at Alert. We note that a similar correspondence between NO_x emissions from sources in the region of the former USSR and observed concentrations of PAN at Alert is found, as discussed in Sect. 3.2.

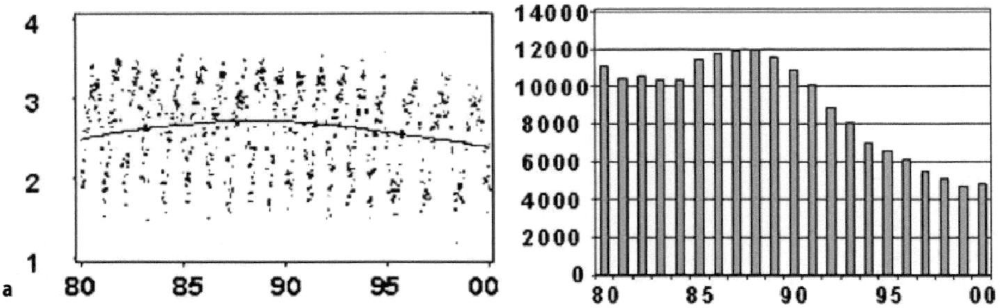

Fig. 5 **a** Long-term trend of sulfate aerosol in the last 20 years in log (ng m^{-3}) and **b** emission data from the former USSR (see text) during the same period in kTonnes sulfur. From [48]

Only a few 3-D Eulerian chemical transport model studies have been reported on the transport of haze components, in particular sulfate, to the Arctic [24, 25, 50]. Instead, most studies have relied on Lagrangian back trajectory models to verify the conclusions from measurement programs. Most noted in this respect are the isobaric [51] and isentropic models [52, 53] developed by Harris and coworkers at NOAA, and the 3-dimensional AES-trajectory model of Voldner et al. [54]. One interesting application of the AES model is the study of Barrie et al. [55] to assess the anthropogenic sources of sulfur to the Arctic for the year 1979–1980. The model was essentially a linear sulfur model that was extensively used for acid rain scenario calculations in North America. Using observed winds and precipitation data as well as best estimates of S emission in the Northern hemisphere, they found that about 4–8% of anthropogenic sulfur released in North America and Eurasia entered the Arctic, but 94% of this amount was from Eurasia and only 4% from North America.

The meteorological observation network in the Arctic is quite sparse. Hence results from trajectory calculations, spatially diverging at the best of times, will have a very large uncertainty when applied for the Arctic, and the results should be treated with caution especially when they are used to try and explain short term (episodic) data [56]. A comparison between the early NOAA and AES models suggested that a 5-day back trajectory might have an uncertainty in the 800 to 1000 km range [57]. While a 10-day back trajectory may be desirable since 5-day trajectories are seldom long enough to diagnose for potential source regions, their divergence will be even larger. However, Harris and Kahl [53] have argued that it is reasonable to use such trajectories in a climatological sense because of cancellation of random wind interpolation errors.

3.2
Greenhouse and Oxidant Related Gases

Long-term data records exist for such greenhouse gases as CO_2, CO, CH_4, and the CFCs at the main Arctic chemistry observatories. While these measurements are principally made to establish their long-term trends in the context of climate change programs, their short-term variations do reflect long-range transport of pollution from mid-latitude regions. As with the aerosol data back trajectory in-

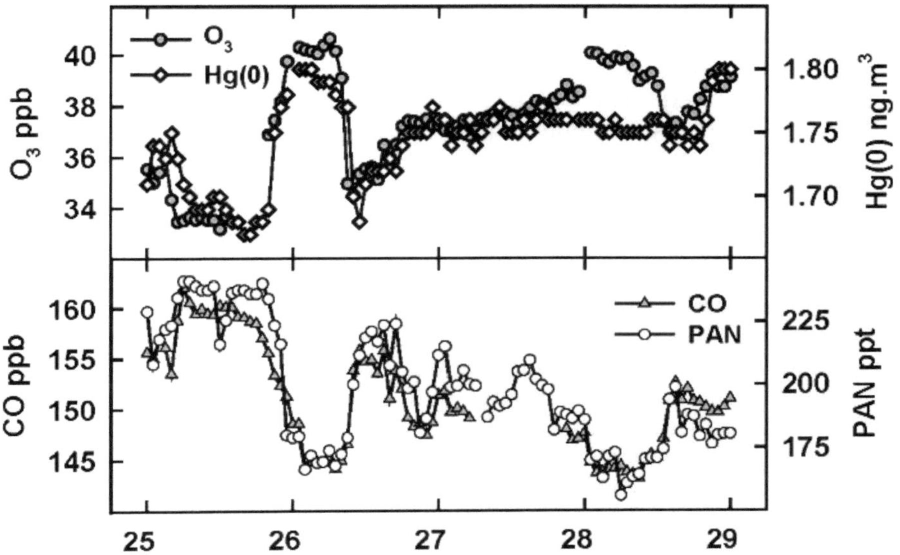

Fig. 6 Time series plot of CO, PAN, Hg^0 and O_3 in late January 2001 at Alert. From [20]

formation has been combined with concentration data to establish the origin of these gases. For example, Worthy et al. [58] reported a close correlation between in-situ measurements of CO_2, CH_4, PAN, as well as BC and CN, during January–February 1992. Using a 3-dimensional back trajectory model [54] they found that elevated levels of these compounds, typically indicative of combustion processes, were observed in air parcels that originated directly from Central Siberia. The correlation appeared to become weaker in March/April, coinciding with the time that biogenic influences would be expected to have a stronger bearing on the C-cycle compounds. Furthermore, the prevailing winter circulation patterns bringing mid-latitude pollution to the Arctic would be expected to weaken at that time.

During these winter time pollution episodes O_3 is generally found to be reduced and using the Barrow data, Harris et al. [59] have ascribed this to titration by NO in industrial plumes. The authors assumed that O_3 titration would occur largely in the dark, and hence about 1.5–2 O_3 molecules would be lost per NO molecule. This then permitted an estimation of the emission ratios for $\Delta NO_x/\Delta CO_2$ and $\Delta NO_x/\Delta CO$, which were found to agree reasonably well with published emission ratios. A ten-day isentropic back trajectory analysis suggested that most pollution plumes originated again from Siberia, or regions upwind thereof.

There are complicating factors with this picture that become clear when the Alert data are considered [20]. Firstly, while identical correlations are seen as at Point Barrow, in addition PAN is measured at Alert, and its concentration is found to increase during reduced O_3 episodes in winter; it is hard to see how PAN would be formed if the O_3 titration took place essentially in the dark. Secondly, as early as late January the decrease in O_3 is often accompanied by a decrease in gaseous elementary mercury (Hg^0), see Fig. 6. The covariance between O_3 and Hg^0

in late winter – early spring in polar regions has been the subject of many studies since its discovery in the mid 1990s at Alert [60]; the current consensus is that this is a result from reactive halogen chemistry, notably involving BrO_x. Again, BrO_x formation is not expected to occur to any extent in the dark. Clearly, as noted by Harris et al., these observations require further analysis and interpretation by a chemical transport model [59], which has not been performed to date.

PAN measurements have been made at the Alert station since 1987. We alluded above to the fact that in the winter PAN and O_3 show a good (negative) correlation on a day-to-day basis. The seasonal cycle of PAN is however much more distinct than for O_3, resembling the seasonal cycle in SO_4^{2-}, with very low concentrations in the summer, and a maximum in the spring (Fig. 7; note that in this figure the time period shown only runs from November to June, as the PAN mixing ratios drop to below detection levels during the summer and early fall). Clearly its concentration is strongly influenced by the seasonally changing circulation patterns, as well as by the fact that its atmospheric lifetime is much shorter in the summer due to higher temperatures. PAN constitutes about 70–80% of total oxidized nitrogen (NO_y) in the Arctic, likely due to more efficient deposition (wet and dry) of inorganic compounds such as HNO_3 and nitrate aerosol [61]. Therefore, if PAN is mostly due to transport of polluted air from mid-latitude regions, and if the atmospheric circulation patterns are determining its seasonal cycle, it may be expected that the long term trend in PAN reflects changes in the Eurasian emission patterns of nitrogen oxides. This is born out by the data as shown in Fig. 8 [8], where the long-term trend in PAN at Alert is compared with recent estimates of NO_x emissions from Russian sources [49].

Measurements of hydrocarbons have also been reported for most of the major Arctic chemistry observatories, showing a strong seasonal cycle (see, e.g., [62–65] and references therein). As with PAN, winter concentrations appear often to be comparable to observations in polluted mid-latitude regions, decreas-

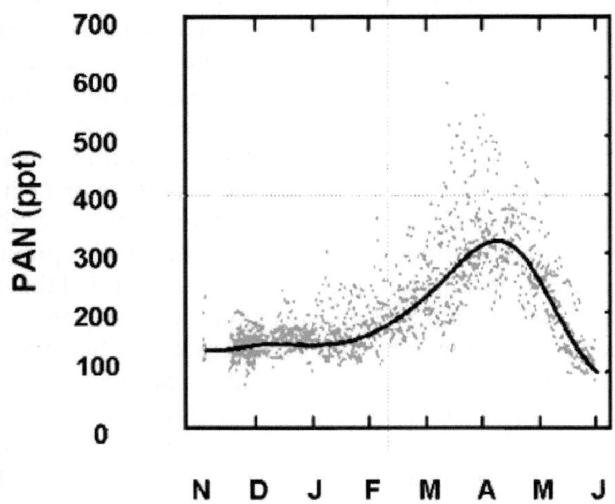

Fig. 7 Seasonal cycle of PAN at Alert from 1987–2002. From [8]

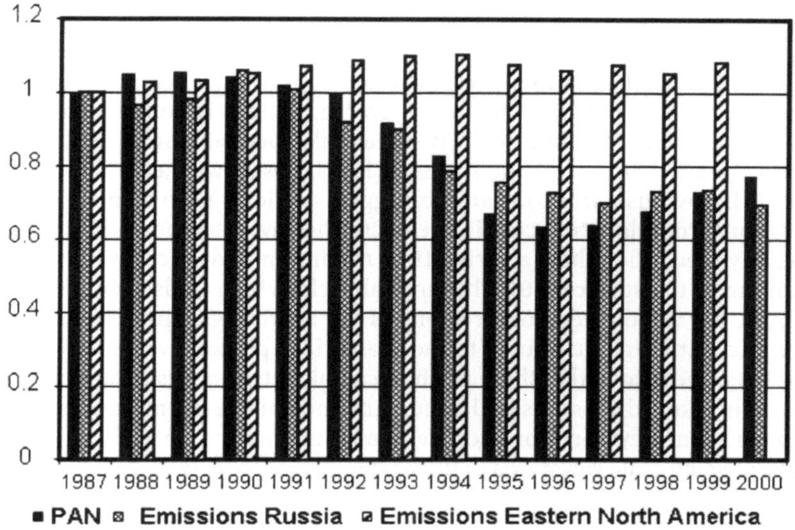

Fig. 8 Trend in the annual average concentration of PAN at Alert, and the average emission data from Russia and North America (data normalized to the values for 1987). From [8]

ing to very low levels in the summer. The seasonal cycle has been interpreted as largely reflecting OH chemistry [65, 66]. Using stable carbon isotope ratio data, Rudolph [67] has derived a distinct impact of the OH chemistry on the origin of these compounds. Ethane has a sufficiently long lifetime that its concentration reflects emissions at both mid and high northern latitudes year-round, while benzene with about an order of magnitude lower lifetime, is almost exclusively due to high latitude emissions in the summer. Recent analysis of a seven-year database hints at a large reduction in emission sources in the early 1990, most likely again reflecting events in the former USSR. We note that much attention has been directed to episodic variations in the spring during ozone depletion episodes, but it has been shown that these variations do not reflect differences in pollutant origin but rather the impact of Cl and Br atom chemistry [64].

3.3 Mercury

Mercury (Hg) is emitted from natural as well as anthropogenic sources but there are virtually no sources within the Arctic ecosystem. Pacyna and Pacyna [68] estimated the global total anthropogenic emissions in 1995 at about 2000 tons, representing about 40% of the total (this number is highly uncertain due to the poor understanding of the natural emissions). It appears that two to three times as much mercury is now cycling through the atmosphere and the upper oceans compared to pre-industrial times [69] and long term lake sediment records show an increase of mercury fluxes to the Arctic ranging from 3 to 7 fold during this period [70, 71]. Mercury levels in humans and animals in the Arctic are also elevated over historical levels [72]. The dominant part of global mercury emissions is in the form of gaseous elemental mercury, Hg^0, and its lifetime in the atmosphere is estimated at up to two years before deposition. Hence it is well mixed in the troposphere [73] with an annual average concentration of ca. 1.6 ng m^{-3} and little seasonal variation. Such levels have also been measured at the surface at several stations in the Arctic since the late 1990s as well as in the boundary layer [73]. Obviously it is too early to suggest any long-term trends in air concentrations from data collected in the Arctic, however, seasonal differences can be readily seen. They show a spring time minimum and an early summer time maximum at the surface [60]. This annual cycle, which is also observed in the Antarctic but not outside polar regions, is not reflecting transport processes to the Arctic, but rather attributed to the dramatic depletion events in the spring. It has been speculated that there might be a link between this depletion and the increase in lake sediments as well as humans and animals mentioned before [74] but experimental evidence or a coherent mechanism for such a link has yet to be reported. Small scale variations in the winter appear to be related to direct transport events (e.g., see Fig. 6) but this topic has also not been explored in any detail until now.

The implications of these findings have recently led to the development of Hg specific models for the Arctic. Using trajectory calculations in combination with a potential source contribution function (PSCF) model, Lin et al. [75] concluded that the *maximum* of Hg^0 in the Arctic summer is not from direct anthropogenic input but rather has its origin in evasion of Hg^0 from surface sources. Eulerian models have also been used such as the Danish Eulerian Hemispheric Model (DEHM) [50]. Figure 9 shows an example of this model indicating the importance of the halogen driven Hg^0 depletion process in depositing mercury to the surface in the spring [76]. Using the Eulerian global multi-scale, high-resolution atmospheric mercury model (GRAHM) Dastoor and Laroque [77] observed frequent episodes of mercury transport from Europe to North America in the winter by the circulations created by Icelandic Lows and Siberian High (Sect. 2); the descending air mass from the Arctic into North America was found to be significantly responsible for the high concentrations of mercury over Canada in winter. Travnikov and Ryaboshapko [78] modelled the influence of different regions and continents on the Arctic mercury contamination and estimated that approximately 55% of deposition comes from global surface emissions (of natural and re-emission origin). Direct anthropogenic mercury input to the Arctic was estimated to be a significant part of the total deposition (40%) with highest contributions from Asia followed by Europe. Most recently, Ariya et al. [79] investigated the role of halogen chemistry in accumulating the globally transported gaseous elemental mercury into the Arctic. By using an extended version of the GRAHM model including the physical and chemical processes leading to Arctic mercury depletion events, they estimated a 44% increase of mercury deposition into the Arctic biosphere (north of 60°N) implying a net accumulation of ca. 325 tons Hg per year.

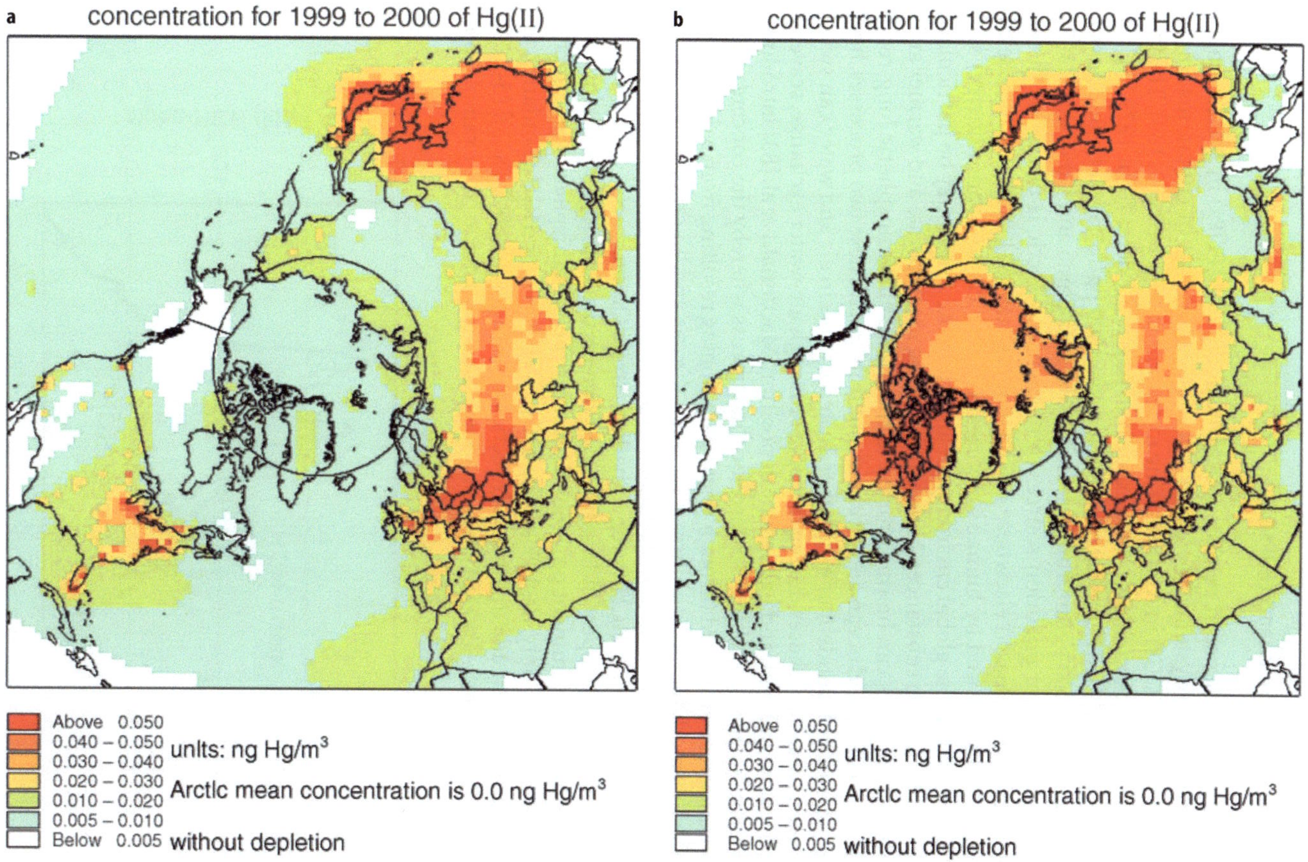

Fig. 9 Modeling result of the distribution of mercury deposition in the Arctic using the Danish Eulerian Hemispheric Model [50]. **a** without, and **b** with including halogen driven depletion chemistry in the Arctic. From Christensen, with permission [76]

3.4 Persistent Organic Pollutants (POPs)

Interest in the presence of persistent organic pollutants (POPs) in Arctic aquatic and terrestrial ecosystems has arisen since the first report of contamination by organochlorines (OCs) in Arctic marine mammals in the late 1960s [80]. The first report of HCHs (hexachlorocyclohexanes) in Arctic air dates from 1979 [81] indicating the total HCH concentration in air was 1000 pg/m^3 over the Bering Sea, while averaging 3600 pg/m^3 from research cruises off China and Japan in 1975–77. Without significant pollution sources of its own, the Arctic is obviously the receptor of POPs released elsewhere on the globe. Ottar [82] suggested that semi-volatile POPs released by human activities (e.g., agriculture and industry) seek cold locations in the global environment and, thus, the Arctic may be particularly vulnerable to these substances. Mainly originating in temperate and warmer areas of the world, POPs can be transported to the Arctic via atmospheric long-range transport (LRT) and ocean and river current transport. These two processes, however, are not independent of each other. Instead, POPs in air and water can switch between these two phases during transport through air–water exchange, scavenging by snow and rain, and particulate deposition. Macdonald et al. [83] introduced the term "solvent switching" to describe this process whereby the concentrations of POPs can be enhanced through phase partitioning resulting from a disturbance of the thermodynamic equilibrium between the gas and liquid phases in the environment. We will illustrate the impact of solvent switching with the tale of the transport of α- and β-HCH to the Canadian Arctic Ocean.

Technical HCH, a pesticide introduced during the 1940s, is a prominent organochlorine contaminant now found in all northern hemisphere water. Technical HCH contains only 10–12% of the active insecticide, γ-HCH (Lindane), and is predominantly made up of non-insecticidal isomers α (60–70%), β (5–12%), δ (6–10%), and ε (3–4%) [84], which appear also to be toxic (for example, both

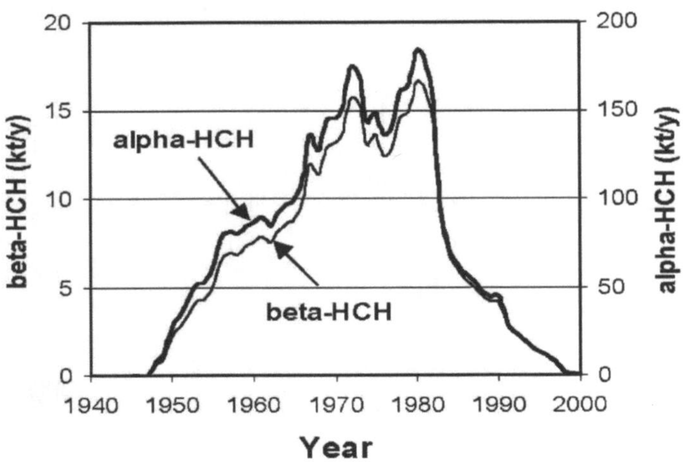

Fig. 10 Annual global emissions of α-HCH and β-HCH between 1945 and 2000. From [90]

α- and β-HCH have recently been shown to disrupt endocrine processes [85,86]). The long-term temporal trends of global emissions of α- and β-HCH from 1945 to 2000, shown in Fig. 10 are almost identical with the exception of their scales. Emissions increased steadily from the 1940s until they reached a peak in the early 1970s. The ensuing decrease in emissions is most likely due to the ban of use of technical HCH in USA (1976), Canada (1971), Japan (1972), and many European countries [87]. Another peak happened in the early 1980s followed by a sharp decrease due to the ban of technical HCH in China in 1983.

A coherent understanding of α-HCH based on a reasonably dense set of spatial and temporal measurements has been achieved [88]. A speculative interpretation of β-HCH has also been provided [89] even though the data for this compound are sparse.

3.4.1
Pathway of α-HCH Transport to the Arctic

The atmosphere plays a significant role in the global distribution of α-HCH as demonstrated by two key pieces of evidence: the close link between α-HCH concentration in Arctic air and the global α-HCH emissions from 1979 to the early 1990s [90, 91], and the significant increase of α-HCH with latitude in the Pacific and Arctic Oceans. When air concentration data for α-HCH, collected since 1979 by various groups at differing locations in the western Arctic, are collated into a time series (Fig. 11), two significant decreases are seen, clearly corresponding to

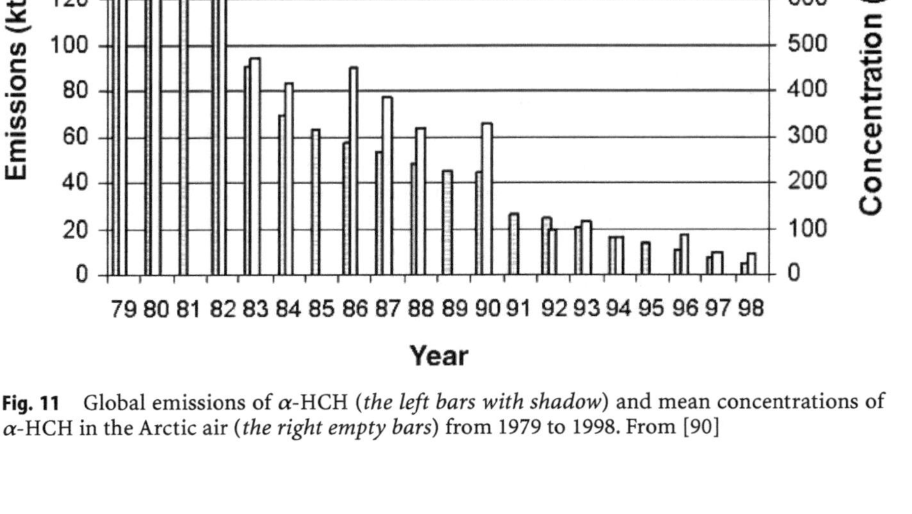

Fig. 11 Global emissions of α-HCH (*the left bars with shadow*) and mean concentrations of α-HCH in the Arctic air (*the right empty bars*) from 1979 to 1998. From [90]

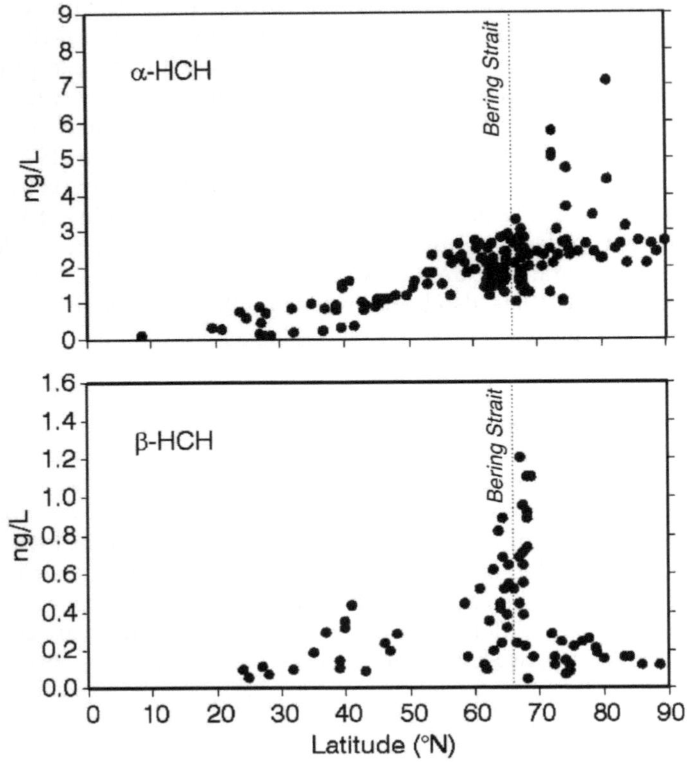

Fig. 12 Concentrations of α-HCH and β-HCH in surface water of the North Pacfic and western Arctic Ocean measured between 1988–99 (from [89]). Although these data span a 10 year period and are collected from a wide range in longitudes, they illustrate well the general increase in α-HCH in cool, northern waters with exceptionally high values under the pack ice of the Canada Basin in contrast to β-HCH, which shows the highest concentrations centered on Bering Strait

documented prominent emission reductions in1982–83 (China abandoning technical HCH use) and 1990–92 (reductions in technical HCH use in India and the former USSR). The rapid response of α-HCH concentrations in Arctic air to the change of global (mainly Asian) emissions shows that atmospheric long-range transport provides rapid dispersion of α-HCH from its release point into the Arctic. In fact, with $r^2 = 0.90$, a reasonable estimate of historical air concentrations of α-HCH before 1990 in the western Arctic can be inferred from the global emission data [90].

The second piece of evidence, the relatively high concentrations of α-HCH in northern surface waters of the Pacific and in the Arctic Ocean (Fig. 12 top panel), also illustrates the importance of atmospheric transport especially in the early "ocean loading" stage. Concentrations of α-HCH are lower in the eastern Arctic Ocean but, nevertheless, increase significantly with latitude from 0.4 ng/L in the Barents Sea at 73–79°N to 1 ng/L over the Lomonosov Ridge at 85–88°N [92, 93].

The increase of α-HCH northward into the Arctic Ocean cannot be explained solely by a water transportation mechanism since that would produce dilution away from the prominent sources off India and Asia but instead is due to solvent switching. Partitioning favors water – perhaps 95% or more of the α-HCH partitions from the air into the water beneath it. The ocean surface concentrations and the atmospheric measurements imply that α-HCH was distributed quickly in the atmosphere from which it partitioned preferentially into cold water due to a Henry's law constant that decreases appreciably with water temperature [88]. Hence, α-HCH moved northward by air and slowly built a large reservoir in the frigid surface waters of the Arctic Ocean. The dramatic downturn in atmospheric concentration of α-HCH in the 1980s (Fig. 11) then produced a situation where previously undersaturated surface water achieved near-equilibrium in some areas and supersaturation in others during the early 1990s. The net exchange then reversed, and the western Arctic Ocean became a source of α-HCH to the atmosphere [94]. By the mid-1990s, α-HCH in Arctic air was determined by the ocean as demonstrated by the atmospheric enantiomeric composition over ice-free regions, which displayed depletion in the same enantiomers as surface water.

3.4.2
Pathway of β-HCH Transport to the Arctic

The concentrations of both α- and β-HCH in the Arctic air at the Alert station for six years are given in Fig. 13 which shows that in contrast to α-HCH the trend of β-HCH air concentration at Alert does not reflect the global emission of this isomer. The largest decrease happened in 1994, three years after the decrease of α-HCH. Furthermore, while β-HCH has been found in the surface ocean of the

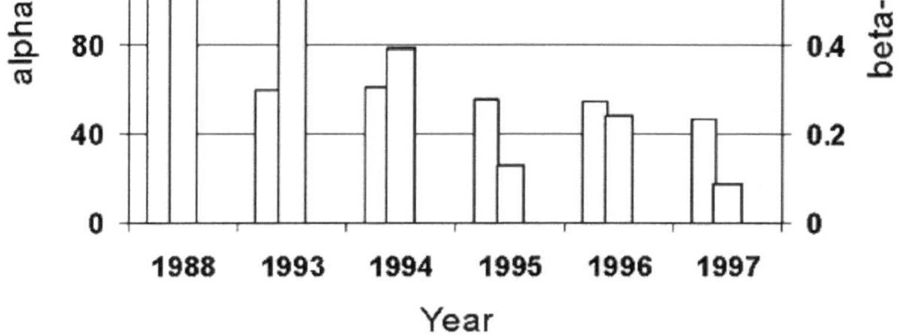

Fig. 13 Arithmetic mean concentrations of α- (*shaded bars*) and β-HCH (*blank bars*) in the Arctic air (Alert station), in pg m^{-3}

North Pacific and Arctic (Fig. 12, bottom panel), its spatial distribution differs from that of α-HCH. Whereas α-HCH increases steadily with latitude along a transect from southeast Asian seas and the North Pacific to the polar cap, β-HCH peaks in the Bering-Chukchi region and then decreases northward into the Arctic Ocean interior.

The observations indicate that β-HCH has not behaved at all in the same way as α-HCH. Apparently β-HCH was not transported very efficiently via the atmosphere to the Arctic Ocean despite its molecular similarity and similar release pattern (source and temporal trend). Instead, it arrived there with some delay via ocean currents through the Bering Strait. The crucial physical difference is the much lower Henry's Law constant of β-HCH, which caused it to partition about twenty times more strongly into water than α-HCH. In fact, Li et al. [89] concluded that β-HCH was entirely deposited in the North Pacific Ocean through precipitation and solvent switching *before* reaching the Arctic. As a result, almost all of the β-HCH entered into the Arctic via ocean currents through the Bering Strait and from there the water transportation mechanism was dominant: β-HCH dilutes away from the Bering Strait. It is a body of water separating Asia from North America, connecting the Bering Sea with the Arctic Ocean. The narrowest part of the strait is only 82 km wide, with a net ocean current of about 5–10 cm s^{-1} flowing Northwards from the Bering Sea to the Chukchi Sea. The observed β-HCH concentration from the Bering Strait to the North Pole (Fig. 12, bottom panel) strongly hints at the existence of a bottleneck, which is proposed as being the main reason that the concentration of β-HCH in the Arctic Ocean decreases from the Chukchi Seas towards the North Pole.

4 Outflow from the Arctic

One aspect that, in comparison, has received little attention is the question of the fate of pollutants, observed in the Arctic. This question is of some relevance in view of their high stability in the air (which is one reason for their often surprisingly high concentration in the air in the first place). There are a few studies that appear relevant to this topic.

Rahn et al. [95] collected aerosol samples for 6 months at Okushiri Island, west of Hokkaido, Japan in 1984–85. Their goal was to obtain source signature information of metals that were suspected to originate from Siberian sources for which direct emission data could not be obtained at that time (pre-1990). They concluded that in the winter several episodes of transport from Siberian sources occurred. Trajectory calculations suggested a route either direct or via the Arctic, as graphically indicated in Fig. 2a. Significantly, there seemed to be little evidence for a contribution from sources in China or Korea.

Penkett et al. [96] reported extensive measurements of nonmethane hydrocarbons from aircraft studies over the North Atlantic Ocean. Their analysis showed the seasonal variation discussed in Sect. 3.2. Interestingly, they noted a big difference in hydrocarbon concentrations in the winter between air originating directly from the North, compared to air sampled a few days earlier originating from the North American continent. Jobson et al. [97] similarly concluded from their measurements at a remote site in northern Ontario, Canada, that most hydrocarbons at that location reflected a polluted Arctic air mass. They observed winter hydrocarbon concentrations similar in magnitude to those they observed in Toronto, a city with over 2 million inhabitants.

Elevated sulfur pollution episodes measured at rural mid-continental North American sites in the winter and spring are also known to arrive from the North, indicating the influence of the polluted Arctic air mass [98]. Barrie and Bottenheim [98] therefore raised the question whether similar transport of PAN with its late winter/early spring maximum in the Arctic might at least partially contribute to the spring maximum of PAN and possibly O_3 at mid-latitude regions.

Honrath et al. [99] have addressed this question in detail for the possible impact on the North Atlantic troposphere. Using an 11-year climatology of isentropic forward trajectories and the available information on NO_y and hydrocarbon levels in the Arctic troposphere they estimated that in the spring the flux from the Arctic might amount to about 7.3–10 GgN/month, which would amount to approximately 4–33% of the direct flux from North America. While this could predominantly originate from Eurasian sources, they noted that Levy and Moxim suggested that in reality a significant fraction probably came from sources further upwind [100]. Honrath et al. also suggested that this outflow from the Arctic could affect the Pacific region and remote regions of North America and Asia. In a follow-up paper, Hamlin and Honrath [101] combined a detailed photochemical model with a few representative trajectories from their climatology. They concluded that while the Arctic outflow would lead to elevated levels of pollutants, on a seasonal basis the impact on the budgets of NO_x and O_3 is probably small.

Based on the TOPSE data, two well-known chemical transport models, HANK (driven by MM5) and MOZART (driven by ECMWF) have recently been used to try and answer the question whether the O_3 maximum in the spring is due to transport from the stratosphere or chemical production [102]. While it was concluded that net chemical production drives the spring maximum, this study did not answer the question to what extent transport from the Arctic plays a role.

5 Conclusions

There is no doubt that the Arctic lower troposphere is polluted in the winter and spring. Since there are virtually no local sources the source of this pollution has to be found outside the Arctic. Atmospheric circulation patterns are such that the dominant pathway from the Northern hemispheric mid-latitude regions is along the Siberian high. Hence it is to be expected that Eurasian pollution sources are the primary contributors. Pollution from North American or East Asian origin is either removed by precipitation before reaching this major highway into the Arctic lower troposphere, or transported to higher levels in the troposphere and thus resides mostly above the Eurasian pollution.

Extensive measurements since the late 1970s have conclusively shown that this expectation is correct. Aerosol particles that make up Arctic haze have signatures of heavy metals that can be traced predominantly to Eurasian sources. A large fraction of the aerosol mass consists of SO_4^{2-} and its trend, while not exactly following sulfur emission data, suggests, nevertheless, a preponderance of an Eurasian origin. It is also clear that in the winter this pollution may be transported in plumes that appear to have been hardly processed in a chemical sense. Often, such plumes display decreased concentrations of O_3 ostensibly due to titration by NO, although there are additional data that appear to be at odds with this explanation.

The Arctic is also a collector of potentially ecologically damaging chemicals such as POPs and Hg. Here a direct transport pathway is not the dominant factor, and in principle all potential sources in the northern hemisphere contribute. POPs are particularly persistent in the Arctic due to the low ambient temperatures that cause these compounds to condense into the Arctic oceans. After polar the sunrise, Hg is deposited on to the snow in the Arctic presumably due to chemical reactions with reactive halogens.

The extensive observational evidence for the transport mechanisms of these pollutants to the Arctic has been supported by an assortment of modeling studies. Much of this work has been based on Lagrangian trajectory analysis introducing an inherently large spatial uncertainty due to the need to consider as much as 10 days travel before potential source regions can be discerned. Hence it appears prudent to interpret such model results with caution. More sophisticated 3-D chemical transport models have seen limited application to date but developments in the last few years suggest that this situation is going to change. Processes after polar sunrise such as the boundary layer O_3 and Hg depletion, the chemistry in surface snow and ice, and the connection with the acidic aerosols that make up Arctic haze, require study with comprehensive modeling platforms and this is just emerging. Such studies will also enable an assessment of the impact of these chemical processes on the radiative balance of the Arctic atmosphere, or the reverse question what the impact of climate change on the Arctic boundary layer chemistry will be.

6 References

1. Arctic Blue Books (1877) Sessional Papers, House of Commons, Accounts and Papers, 1877, No./C.-1636./, vol LVI, p 348
2. Schnell RC (1984) Geophys Res Lett 11:359
3. Greenaway KR (1950) Publications of the Royal Meteorological Society, Canadian Branch. Toronto, Canada
4. Mitchell Jr JM (1956) J Atmos Terrest Phys Special Supplement: 195
5. Shaw GE (1975) Bull Am Met Soc 76:2403
6. Rahn KA, McCaffrey RJ (1980) Ann NY Acad Sci 338:486
7. McNeely R, Gummer WD (1984) Arctic 37:210
8. Bottenheim JW, Gallant AJ, Sirois A (2002) Long term measurements of PAN in the Arctic: trends, covariances and implications. Presented at the 10th CACGP/7th IGAC joint conference, Hersonissos, Crete, Greece
9. Barrie LA, Bottenheim JW, Hart WR (1994) J Geophys Res 99:25313
10. Barrie LA, Platt U (1997) Tellus 49B:450
11. Bottenheim JW, Shepson PB, Fuentes JD, Barrie LA (2002) Arctic atmospheric chemistry: the Polar Sunrise Experiments (PSE). Presented at the 10th CACGP/7th IGAC joint conference. Available at http://atlas.chemistry.uch.gr/IGAC2002, Hersonissos, Crete, Greece

12. Rahn KA (1985) Atmos Environ 19:1987
13. Barrie LA (1986) Atmos Environ 20:643
14. Sturges WT (1991) Elsevier Applied Science, London
15. AMAP (1998) Arctic Monitoring and Assessment Programme, Assessment Report: Arctic Pollution Issues. Oslo, Norway
16. Lu JY, Schroeder WH, Barrie LA, Steffen A, Welch HE, Martin K, Lockhart L, Hunt RV, Boila G, Richter A (2001) Geophys Res Lett 28:3219
17. Raatz WE (1991) The climatology and meteorology of Arctic air pollution. In: Sturges WT (ed) Pollution of the Arctic Atmosphere. Elsevier Science, London, p 13
18. Yuen CW, Higuchi K, Trivett NBA, Cho H-R (1996) J Meteor Soc Japan 74:781
19. Raatz WE, Shaw GE (1984) J Clim Appl Meteor 23:1052
20. Bottenheim JW, Worthy DEJ, Anlauf KG, Schroeder WH, Sharma S, Sirois A (2003) Episodes of very long-range transport of pollution in the Arctic boundary layer in winter as observed at the Canadian high Arctic station Alert. Presented at the EGS/AGU Joint meeting, Nice, France
21. Mayewski PA, Meeker LD, Twickler MS, Whitlow S, Yang Q, Lyons WB, Serreze MC, Prentice M (1997) J Geophys Res 102:26345
22. Raatz WE (1984) Geophys Res Lett 11:449
23. Austin JF (1980) Q J Royal Meteor Soc 106:327
24. Iversen T (1989) Atmos Environ 23:2571
25. Dastoor AP, Pudykiewicz J (1996) Atmos Environ 30:1501
26. Wallace JM, Gutzler DS (1981) Mon Weather Rev 109:784
27. Barnston AG, Livezey RE (1987) Mon Weather Rev 115:1083
28. Rogers JC (1984) Mon Weather Rev 112:1999
29. Huang J-P, Higuchi K, Shabbar A (1998) Geophys Res Lett 25:2707
30. Wanner H, Brönnimann S, Casty C, Gyalistras D, Luterbacher J, Schmutz C, Stephenson DB, Xoplaki E (2001) Surveys Geophys 22:321
31. Hurrell JW, Kushnir Y, Ottersen G, Visbeck M (2003) The North Atlantic Oscillation: Climate Significance and Environmental Impact. Geophysical Monograph Series, vol 134, p 279
32. Higuchi K, Huang J-P, Shabbar A (1999) Int J Climatol 19:1119
33. Barnett TP (1984) J Atmos Sci 42:478
34. Horel JD, Wallace JM (1981) Mon Weather Rev 784
35. Murayama S, Taguchi S, Higuchi K (2004) J Geophys Res (109:doi:10.1029/2003JD003729)
36. Eckhardt S, Stohl A, Beirle S, Spichtinger N, James P, Forster C, Junker C, Wagner T, Platt U, Jennings SG (2003) Atmos Chem Phys 3:1769
37. Christensen JH, Brandt J, Frohn LM, Geels C, Hansen KM (2003) Model studies of the influence from natural climate variations on transport of air pollution to the Arctic. Poster presentation at the EGS/AGU Joint meeting, Nice, France
38. Stohl A, Eckhardt S, Forster C, James P, Spichtinger N (2002) J Geophys Res 107:4684
39. Kahl JDW, Galbraith J, Martinez DA (1999) Geophys Res Lett 26:481
40. Barlow LK, Rogers JC, Serreze MC, Barry RG (1997) J Geophys Res 102:26333
41. Barrie LA, Fisher D, Koerner RM (1985) Atmos Environ 12:2055
42. Rahn KA, Heidam NZ (1981) Atmos Environ 15:1345
43. Rahn KA (1981) Atmos Environ 15:1349
44. Lowenthal DH, Rahn KA (1985) Atmos Environ 19:2011
45. Maenhaut WG, Cornille JM, Pacyna JM, Vitols V (1989) Atmos Environ 23:2551
46. Sirois A, Barrie LA (1999) J Geophys Res 104:11599
47. Gong SL, Barrie LA (2003) Other heavy metals in the Arctic atmosphere. In: Bidleman T, Macdonald R, Stow J (eds) Sources, Occurrences, Trends and Pathways in the Physical Environment, Canadian Arctic Contaminants Assessment Report (CACAR) II. Indian and Northern Affairs, Ottawa, Canada, pp 148–152
48. Stern DI (2003) Chemosph 47 (in press). See also: http://www.rpi.edu/~sternd/datasite.htm
49. Vestreng V, Klein H (2002) Emission data reported to UNECE/EMEP: Quality assurance and trend analysis & Presentation of WebDab. Oslo, Norway
50. Christensen JH (1997) Atmos Environ 31:4169

51. Harris JM (1984) Geophys Res Lett 11:453
52. Herbert GA, Harris JM, Bodhaine BA (1989) Atmos Environ 23:2521
53. Harris JM, Kahl JD (1994) J Geophys Res 99:25845
54. Voldner EC, Olson MP, Oikawa KK, Loiselle M (1981) J Geophys Res 86:5334
55. Barrie LA, Olson MP, Oikawa KK (1989) Atmos Environ 23:2505
56. Kahl JD (1993) Atmos Environ 27a:3037
57. Kahl JD, Harris JM, Herbert GA, Olson MP (1989) Tellus 41B:524
58. Worthy DEJ, Trivett NBA, Hopper JF, Bottenheim JW, Levin I (1994) J Geophys Res 99:25329
59. Harris JM, Dlugokencky EJ, Oltmans S, Tans PP, Conway TJ, Novelli PC, Thoning KW, Kahl JDW (2000) J Geophys Res 105:17267
60. Schroeder WH, Anlauf KG, Barrie LA, Lu YJ, Steffen A, Scheeberger DR, Berg T (1998) Nature 394:331
61. Bottenheim JW, Barrie LA, Atlas E (1993) J Atmos Chem 17:15
62. Doskey PV, Gaffney JS (1992) Geophys Res Lett 19:381
63. Solberg S, Schmidbauer N, Semb A, Stordal F, Hov Ø (1996) J Atmos Chem 23:310
64. Jobson BT, Niki H, Yokouchi Y, Bottenheim J, Hopper F, Leaitch R (1994) J Geophys Res 99:25355
65. Gautrois M, Brauers T, Koppmann R, Rohrer F, Stein O, Rudolph J (2003) J Geophys Res 108:doi:10.1029/2002JD002765
66. Rasmussen RA, Khalil MAK (1983) Geophys Res Lett 10:144
67. Rudolph J (2003) The seasonal cycles of stable carbon isotope ratios of non-methane hydrocarbons in the polar atmosphere. In: Worthy DEJ (ed) Canadian baseline program summary of progress to 2002, Toronto, Canada, pp 80–84
68. Pacyna EG, Pacyna JM (2002) Water Air Soil Poll 137:149
69. Mason RP, Fitzgerald WF, Morel FM (1994) Geochim Cosm Act 58:3191
70. Landers DH, Gubala C, Verta M, Lucotte M, Johansson K, Vlasova T, Lockhart WL (1998) Atmos Environ 32:919
71. Lockhart WL, Wilkinson P, Billeck BN, Hunt RV, Wagemann R, Brunskill GJ (1995) Water Air Soil Poll 80:603
72. AMAP (2003) Human Health in the ArcticAMAP Assessment 2002. Arctic Monitoring and Assessment Programme(AMAP). Oslo, Norway
73. Banic CM, Beauchamp ST, Tordon RJ, Schroeder WH, Steffen A, Anlauf KA, Wong HKT (2003) J Geophys Res 108:doi:1029/2002JD002116
74. Lindberg SE, Brooks SB, Lin C-J, Scott KJ, Landis MS, Stevens RK, Goodsite M, Richter A (2002) Environ Sci Technol 36:1245
75. Lin C-J, Cheng M-D, Schroeder WH (2001) Atmos Environ 35:1141
76. Christensen JH, Goodsite M, Skov H, Brooks S, Lindberg S (2002) http://www.chem.purdue.edu/arctic/ArcticWorkshop.htm, West Lafayette, Indiana
77. Dastoor AP, Laroque Y (2003) Atmos Environ 37:147
78. Travnikov O, Ryaboshapko A (2002). Meteorological Synthesizing Centre – East, Moscow, Russia
79. Ariya PA, Dastoor AP, Amyot M, Schroeder WH, Barrie L, Anlauf K, Raofie F, Ryzhkov A, Davignon D, Lalonde J, Steffen A (2003) Tellus (submitted)
80. Holden AY (1970) In: Ruivo M (ed) Marine pollution and sea life. Fishing News Books Led, pp 266–272
81. Tanabe S, Tatsukawa R (1980) J Ocean Soc Japan 36:217
82. Ottar B (1981) Atmos Environ 15:1439
83. Macdonald RW, Mackay D, Hickie B (2002) Environ Sci Technol 36:A457
84. Iwata H, Tanabe S, Sakai N, Tatsukawa R (1993) Environ Sci Technol 27:1080
85. Kendal R, Dickerson R, Giesy J, Suk W (1998) Principles and Processes for Evaluating Endocrine Disruption in Wildlife. SETAC Press, Pensacola
86. Willett K, Ulrich E, Hites RA (1998) Environ Sci Technol 32:2197
87. Li YF (1999) Sci Tot Environ 232:123
88. Macdonald RW, Barrie LA, Bidleman TF, Diamond ML, Gregor DJ, Semkin RG, Strachan WMJ, Li YF, Wania F, Alaee M, Alexeeva LB, Backus SM, Bailey R, Bewers JM, Gobeil C,

Halsall CJ, Harner T, Hoff JT, Jantunen LMM, Lockhart WL, Mackay D, Muir DCG, Pudykiewicz J, Reimer KJ, Smith JN, Stern GA, Schroeder WH, Wagemann R, Yunker MB (2000) Sci Tot Environ 254:93
89. Li YF, Macdonald RW, Jantunen LMM, Harner T, Bidleman TF, Strachan WMJ (2002) Sci Tot Environ 291:229
90. Li YF, Bidleman TF (2003) J Environ Inf 1:52
91. Li YF, Bidleman TF, Barrie LA, McConnell LL (1998) Geophys Res Lett 25:39
92. Carmack EC, Aagaard K, Swift JH, Macdonald RW, McLaughlin FA, Jones EP, Perkin RD, Smith J, Ellis K, Kilius L (1997) Deep-Sea Res 44:1487
93. Harner T, Kylin H, Bidleman TF, Strachan WMJ (1999) Environ Sci Technol 33:1157
94. Jantunen LM, Bidleman TF (1995) Environ Sci Technol 29:1081
95. Rahn KA, Lowenthal DH, Harris JM (1989) Atmos Environ 23:2597
96. Penkett SA, Blake NJ, Lightman P, Marsh ARW, Anwyl P, Butcher G (1993) J Geophys Res 98:2865
97. Jobson BT, Wu Z, Niki H, Barrie LA (1994) J Geophys Res 99:1589
98. Barrie LA, Bottenheim JW (1991) Sulfur and nitrogen pollution in the Arctic Atmosphere, chap 6. In: Sturges WT (ed) Pollution of the Arctic Atmosphere. Elsevier Applied Science, London, p 155
99. Honrath RE, Hamlin AJ, Merrill JT (1996) J Geophys Res 101:29335
100. Levy_II H, Moxim WJ (1989) Tellus 41B:256
101. Hamlin AJ, Honrath RE (2002) J Geophys Res 107: doi:10.1029/2001JD000453
102. Emmons LK, Hess P, A.Klonecki, Tie X, Horowitz L, Lamarque J-L, Kinnison D, Brasseur G, Atlas E, Browell E, Cantrell C, Eisele F, Mauldin RL, Merrill J, Ridley B, Shetter R (2003) J Geophys Res 108:doi:10.1029/2002JD002665

The Handbook of Environmental Chemistry Vol. 4, Part G (2004): 41–67
DOI 10.1007/978-3-540-40037-0

Air Pollution Export from and Import to North America: Experimental Evidence

Owen R. Cooper[1] · David D. Parrish[2]

[1] Cooperative Institute for Research in Environmental Sciences, University of Colorado/ NOAA Aeronomy Laboratory, R/AL4, 325 Broadway, Boulder, CO 80305, USA
E-mail: ocooper@al.noaa.gov
[2] NOAA Aeronomy Laboratory, R/AL7, 325 Broadway, Boulder, CO 80305, USA
E-mail: dparrish@al.noaa.gov

Abstract North America (Canada and the USA) contains only 5% of the world's population but accounts for 15–23% of the world's emissions in terms of trace gases important for the radiative processes of the atmosphere, ozone formation and acid rain. As a result, North America has a strong impact on global atmospheric chemistry. The bulk of North American trace gas and particulate matter emissions originate in the heavily populated eastern region, especially in the USA. The emission regions, coupled with the formation of deep convective systems and mid-latitude cyclones determines the export pathways of North American pollution plumes. Deep convection can rapidly loft surface pollutants to the upper troposphere over relatively small spatial areas. In contrast mid-latitude cyclones loft pollutants via slantwise ascent over several days, but can potentially vent the boundary layer of a region as large as the eastern seaboard in just one or two days. The warm conveyor belt airstream of mid-latitude cyclones is the mechanism most strongly associated with rapid intercontinental transport of pollution; this and other transport pathways are discussed in terms of transport from Asia to North America and from North America to Europe. The impacts of North American emissions on tropospheric chemistry, and the implications of emissions from other continents on North American air quality are also discussed.

Keywords Surface emissions · Deep convection · Warm conveyor belts · Intercontinental transport · Photochemistry

Abbreviations and Symbols

ABL	Atmospheric boundary layer
CCB	Cold conveyor belt
CO	Carbon monoxide
CO_2	Carbon dioxide
CTM	Chemical transport model
DA	Dry airstream
Gmol	Giga-mole, 1×10^9 mol
GOME	Global Ozone Monitoring Experiment
hPa	hectopascal
MOPITT	Measurements of Pollution in the Troposphere
NA	North America
NARE	North Atlantic Regional Experiment
N_2O	Nitrous oxide
NOAA	National Oceanic and Atmospheric Administration
NO_3^-	Nitrate ion
NO_x	Sum of nitrogen monoxide and nitrogen dioxide
NO_y	Total reactive nitrogen
NMVOCs	Non-methane volatile organic compounds
O_3	Ozone
PAN	Peroxyacetyl nitrate
PCFA	Post cold front airstream
ppbv	Parts per billion by volume
pptv	Parts per trillion by volume
$PM_{2.5}$	Particulate matter with diameter less than 2.5 micrometers
PM_{10}	Particulate matter with diameter less than 10 micrometers
SO_2	Sulfur dioxide
SO_4^{2-}	Sulfate ion
Tg	Teragram, 1×10^{12} grams
TOMS	Total Ozone Mapping Spectrometer

TTM	Tracer transport model
USA	United States of America
US EPA	United States Environmental Protection Agency
WCB	Warm Conveyor Belt
WNAO	Western North Atlantic Ocean
WP-3D	Type of research aircraft
$\mu g\ m^{-3}$	micrograms per cubic meter
µm	micrometer

1 Introduction

1.1 The North American Contribution to Global Emissions

The continent of North America (NA) physically extends from the northern tip of Greenland to the southern border of Panama. In this chapter's discussion of North America's relationship to intercontinental transport of air pollution we focus on just the regions within the political boundaries of Canada and the United States of America (USA). We make this choice for several reasons. These two countries account for the bulk of NA emissions, for example 89% of CO emissions from the burning of fossil fuel [1]. These countries have relatively similar economies and environmental regulations, so per-capita emissions are fairly similar. The major emission regions of both countries are in the mid-latitudes and are subject to mid-latitude transport pathways, while Mexico, the Caribbean Sea and Central America are subject to tropical weather patterns. Finally, relatively few studies have measured pollutants in the free troposphere of tropical NA making it difficult to draw conclusions on the importance of this region to intercontinental pollutant transport.

Recent emission inventories summarized in Table 1 place NA in the global context. While NA contains only 5% of the world's population it accounts for 15–23% of the world's emissions in terms of trace gases important for the radiative processes of the atmosphere, ozone (O_3) formation and acid rain. The bulk of NA emissions are from the USA, which has 9 times the population of Canada. Between 1990 and 2000 the USA population grew by 11%, while control strategies are credited with emission decreases of 5% and 24% for NMVOCs and SO_2, respectively. Meanwhile other emissions increased by 10%–17%, either keeping pace with or outpacing the increase in population. Therefore, despite efforts to reduce emissions NA will continue to exert a major influence on the chemistry of the atmosphere and remains a major focus of research on intercontinental pollutant transport.

The time scale for the transport of air masses between continents is on the order of a few days to weeks. From an air quality point of view, intercontinental transport is most significant for pollutants with lifetimes similar to this transport time. The first three species listed in Table 1 (CO_2, CH_4, N_2O) are long-lived (>1 year) so that they are circulated globally, and obtain a reasonably uniform

Table 1 A comparison of NA and global emission inventories (all sources)

	NA emissions[a] (Tg yr^{-1})	NA percentage of global emissions[a] (1995)	Percent change in USA emissions[b] (1990 to 2000)
CO_2	6091	23%	17%
CH_4	46	15%	–6%
N_2O	1.7	15%	10%
CO	145[c]	17%[c]	13%[d]
NO_x	25	22%	14%
NMVOC	29[c]	18%	–5%
SO_2	21	15%	–24%
Human population		5%	11%

[a] Derived from the EDGAR 3.2 Emission Database http://arch.rivm.nl/env/int/coredata/edgar/index.html.
[b] Derived from the Greenhouse Gas Inventory Database, maintained by the United Nations Framework Convention on Climate Change, http://ghg.unfccc.int/.
[c] Includes forest fire emissions.
[d] Does not include forest fire emissions.

hemisphere-wide distribution. These three species play important roles in global climate change and, particularly for CO_2, their budgets on continental scales are areas of active research [2]. We will not consider these three species here; rather we will focus on the others in Table 1 and the products of their photochemical decomposition, plus primary particulate emissions. The lifetimes of these species are long enough that they can be significantly redistributed from their sources across national boundaries and into marine regions that border the continents, or under favorable conditions, to other continents. In the remainder of Sect. 1 we discuss the broad source regions of NA trace gases and particulate matter emissions. In Sect. 2 we describe the major transport mechanisms that loft pollutants from the surface and export them to downwind continents, while Sects. 3 and 4 describe actual transport events from NA to Europe, and from Asia to NA. Finally, Sect. 5 reviews the potential impacts of NA emissions on tropospheric chemistry, and the implications of emissions from other continents on air quality in NA.

1.2 North American Trace Gas Emissions

A great deal of effort has been put into developing emission inventories for the countries of the world, particularly the industrialized nations. These inventories can readily be integrated to the continental scale, for NA this is simply the sum for the USA and Canada. Table 1 shows emissions estimated by the 1995 EDGAR inventory [1]. However, even for NA, there are significant uncertainties in current inventories. These uncertainties must be recognized and reduced in future work.

Carbon monoxide (CO) has a lifetime of roughly 2 months in the free troposphere, has a significant impact on photochemical O_3 production, and is important for the hydroxyl radical (OH) budget [3]. The dilution associated with long-range transport results in mixing ratios less than 1 ppmv so in these instances it is of little concern to human health. However for studies of intercontinental transport it is a useful long-lived tracer of anthropogenic emissions and forest fire plumes [4, 5]. Figure 1 in the chapter by Stohl and Eckhardt in this volume shows the estimated CO emissions across NA. Excluding wildfires, the CO emission inventory is dominated by on-road and off-road transportation sources (80% of the 1995 total in the U.S.). Thus the emission patterns largely reflect the population distribution of NA.

For CO, the comparison of time trends of estimated emissions with ambient measurements presents a puzzling contradiction. Table 1, based on data reported by the USA government to the United Nations Climate Change Secretariat in 2002, suggests an increase in CO emissions. However, a recent report from the U.S. Environmental Protection Agency (Latest Findings on National Air Quality: 2000 Status and Trends, http://www.epa.gov/air/aqtrnd00/index.html), whose analyses form the basis for all USA inventories, reports a small emission decrease of 5% from 1991–2000. In contrast to the relative small changes estimated in inventories, measured ambient concentrations of CO in the USA have decreased dramatically. That same U.S. EPA report indicates that measured concentrations in central urban areas of the USA decreased by nearly a factor of 2 from 1991–2000, and a separate study shows that rural CO levels are decreasing at a rate consistent with the urban measurement trend [6]. Furthermore, the ratio of the ambient CO to NO_x levels in urban areas has decreased by a factor approaching 3 during the decade of the 1990s [7]. From these studies it is clear that emission control efforts for CO (primarily catalytic converters on automobiles) have been much more successful than assumed in the inventory development.

Wildfires add another difficulty in defining the total emission of CO. On average, a little over one third of NA CO emissions is produced by forest fires [1], but there is great interannual variability in this source [8]. This variability is largely driven by meteorological conditions. Fires typically occur in the contiguous USA from June to September, commonly in the southeastern USA, but primarily in the western states. Boreal fires in Canada and Alaska typically occur between May and September and emit roughly 10 times more CO than fires in the contiguous USA. These fires extend east to west in a broad swath across central Canada but are more frequent in the west. The fires also reach well north of 60 degrees into the Northwest Territories, the Yukon and Alaska [9]. In addition to CO, wildfires are also important sources of NMVOCs, NO_x and particulate matter (see the chapter by Spichtinger-Rakowsky and Forster in this volume for more information on boreal forest fires).

North American SO_2 emissions are primarily the result of fossil fuel combustion, in particular from point sources such as power plants. SO_2 in the USA is mostly emitted in the eastern half of the country, with the Ohio Valley states making the greatest contribution (36% of all USA emissions). As shown in Table 1, USA SO_2 emissions are declining, but the decline is not uniform across the country. Between 1990 and 1999 the northeastern states (including the Ohio Valley) experienced a reduction in SO_2 emissions (–28%) as did the western states (–15%), but the southeastern states saw little change and emissions actually increased (15%) in the central states of Texas, New Mexico, and Colorado [10].

NO_x emissions in NA are released from both mobile and stationary point sources. Thus, the emission pattern is a superposition of that for CO (from mobile sources) and SO_2 (primarily from stationary sources.) In contrast to CO and SO_2 emissions, NO_x emissions are likely increasing as indicated in Table 1. The CO to NO_x ratio in urban areas has decreased more rapidly than CO levels have decreased [7]. This suggests that NO_x emissions from mobile sources are increasing at a rate of 2 to 3% per year, primarily from increasing emissions from heavy-duty trucks.

Anthropogenic NMVOC emissions come predominately from mobile sources, with another major contribution from solvent usage. Similar to CO, there is also a large, variable contribution from wildfires. Hence the spatial distribution for NMVOCs from these sources is similar to that for CO in Fig. 1 in the chapter by Stohl and Eckhardt in this volume. In addition, natural emissions of NMVOCs from biogenic sources likely exceed the anthropogenic and wildfire emissions listed in Table 1 [11].

1.3 North American Particulate Matter Sources

Particulate matter is a general term for an airborne mixture of solid particles and liquid droplets. Fine and coarse particles are defined as less than and greater than 2.5 μm in diameter, respectively. Particles with diameters 2.5 μm or less are typically referred to as $PM_{2.5}$ and particles with diameters 10 μm or less are referred to as PM_{10}. Particulate matter is emitted from a wide range of sources, including fossil and biofuel combustion, manufacture of chemicals, petroleum refining, metals processing, waste disposal, agricultural activities, wild and prescribed forest fires, construction, wind-blown soil, and paved and unpaved roads.

Based on 1995 emission inventories from the USA [12] and Canada [13], USA PM_{10} and $PM_{2.5}$ emissions are roughly 5 times greater than Canadian emissions. In both countries, roughly 90% of PM_{10} emissions and 70% of $PM_{2.5}$ emissions come from open sources such as agricultural activities, construction operations, forest fires and prescribed burns, natural wind erosion, and especially dust from paved and unpaved roads.

Wind generated soil dust is most common in spring and summer with a late spring or early summer maximum. The primary NA source regions are the bare soil areas of the western USA, especially in the topographical lows of the Southwest [14]. Other dust source regions are the bare soil areas of northernmost Canada and Alaska [15], and the Great Plains [16], although an assessment of satellite measurements (1980–1992) indicates that dust activity in the Great Plains is at an historical minimum [14]. These natural dust events are episodic with a high degree of interannual variability, and contribute a relatively small fraction of the total particulate emissions, for example, roughly 10% and 15% of the total USA $PM_{2.5}$ and PM_{10} emissions, respectively [12].

Due to the strong influence of agricultural activities, and to a lesser extent natural dust, particulate matter emissions are more widespread across NA than are other emissions. For example, in the USA, CO emissions are heavily weighted towards either the eastern half of the country or along the west coast. Particulate matter emissions are still prevalent in these regions but are also as important in portions of the Great Plains and the desert southwest [12].

In addition to the primary emissions, particulate matter is formed in the atmosphere as products of chemical reactions of gas-phase species. These products include sulfates, nitrates and a wide variety of organic species. Depending upon ambient conditions, this secondary particulate matter can either condense on pre-existing particles or nucleate to form new particles. The mechanisms responsible for these reactions, and the relative contribution of the primary emissions and secondary atmospheric formation to the total particulate loading are areas of active research.

1.4 North American O_3 Distribution

O_3 is a secondary pollutant, i.e. its primary source is through photochemical production in the atmosphere from NO_x, NMVOC and CO precursors that are directly emitted to the atmosphere. Ozonesonde profiles regularly measured at four sites across the USA show a clear west to east positive O_3 gradient in the lower half of the troposphere during all seasons [17]. For example, at 800 hPa the average summertime O_3 mixing ratio on the west coast is approximately 51 ppbv while it is 65 ppbv on the east coast, similarly, at 500 hPa the west coast value is 60 ppbv and the east coast value is 69 ppbv. A clear north-south O_3 gradient in the lower half of the troposphere is also present in the summer months with O_3 decreasing from the USA into Canada, for example average summertime O_3 at 800 hPa above Virginia is 60–70 ppbv but only 35–40 ppbv above northeastern Canada [18]. An analysis of O_3 measurements at 549 surface sites (mostly urban) across the USA revealed no significant positive trend in any large region of the country between 1980 and 1995, with large urban areas such as New York, Chicago and Los Angeles having negative trends due to emission controls [19]. Rural USA O_3 monitoring sites from 1980 to 1998 have a decreasing trend in high O_3 values, but an increasing trend in low O_3 values, possibly due to an increase in background O_3 mixing ratios transported into the country [20].

2 General Types of Transport Mechanisms Impacting North America

While a variety of transport pathways are responsible for trace gas export and import to NA, two major meteorological mechanisms have the greatest influence on this trace gas re-distribution: 1) deep convection, which acts on a relatively small horizontal scale, and 2) the much larger mid-latitude cyclone, whose various airstreams can, for example, vent the surface emissions of the eastern seaboard in as little as 24 h.

2.1 Deep Convection

Deep convective processes are important for venting pollutants from the NA atmospheric boundary layer (ABL) into the mid- and upper troposphere where higher wind speeds can rapidly transport the pollutants to downwind continents. Similarly, deep convection above upwind continents can loft surface emissions, which are subsequently transported to NA. Deep convection is often associated with thunderstorms, which can extend as high as the tropopause or even into the lowermost stratosphere, whereas shallow convective processes often associated with fair-weather cumulus clouds, only extend to the top of the ABL. Thunderstorms can occur within mid-latitude cyclones along the cold front and supplement the vertical transport associated with the slantwise ascent of warm conveyor belts (described below). They also occur ahead of cold fronts in the warm sectors of mid-latitude cyclones or in hot, moist unstable conditions far south of the mid-latitude storm tracks. Thunderstorms can form as individual cells or in organized patterns such as squall lines or mesoscale convective systems.

Deep convection is often associated with lightning, and recent lightning climatologies reveal the distribution of thunderstorms and deep convection across NA [21–23]. On an annual basis lightning is most frequent from eastern Texas to the east coast of Florida decreasing northwards and westwards with relatively little lightning activity above Canada and west of the Rockies. During winter, lightning activity is low across NA with the exception of the southern USA from eastern Texas to Mississippi. Lightning activity is at a maximum in summer especially in the southeastern USA but also highly frequent across the Great Plains and the Midwest. In association with the summertime peak in lightning deep convection makes its greatest impact on venting the NA ABL during this season.

Research into the importance of thunderstorms for the vertical transport of pollutants is a relatively new topic, with the first in-situ measurements study conducted in 1987 [24]. The vertical transport is rapid as shown in an analysis of an isolated thunderstorm over Colorado in which ABL air (between 0.5 and 2 km above the surface) was entrained into the storm and lofted into the anvil in a time span of about 10 min [25]. Conversely, thunderstorm downdrafts can bring trace gases from higher altitudes down to the surface [26]. Observations from a mesoscale convective complex over North Dakota show that deep convection can loft polluted ABL air into the lowermost stratosphere, blurring the chemical and dynamical distinction between the troposphere and stratosphere [27]. An early modeling study of deep convection during five consecutive Junes across the Great Plains suggests that thunderstorms are responsible for half of the ABL venting, with the rest being attributed to shallow convective processes and mid-latitude cyclones, which are weaker in summer than during winter [28].

Lightning in thunderstorms is also a major source of NO_x in the free troposphere. Estimates of lightning produced NO_x vary widely but are thought to be greater than the contribution from aircraft [29–32].

2.2 Mid-Latitude Cyclone Airstreams

Mid-latitude cyclones tracking from west to east are believed to be responsible for the bulk of the trace gas export from NA, throughout the year, even in summer when these systems are weaker [33]. And of course mid-latitude cyclones can export emissions from upwind continents. Even though these cyclones decay before they reach NA, their remnants, often laden with aged emissions, do cross the boundaries of NA. Over the past 30–40 years precipitation and clouds have been studied in terms of the component airstreams of mid-latitude cyclones [34]. Recently, both the origin and evolution of individual airstreams have been shown to influence trace gas mixing ratios and relationships in the troposphere [35–37].

The typical mid-latitude cyclone is composed of four major airstreams (Fig. 1). The warm conveyor belt (WCB), cold conveyor belt (CCB) and dry airstream (DA) produce the distinctive comma cloud of a mature mid-latitude cyclone [34, 38–40]. To a first approximation, these airstreams move along sloping isentropic surfaces. The CCB originates north of the cyclone's warm front, and relative to the cyclone center, ascends as it heads westward, with a component heading eastward at higher altitudes. The WCB is located on the eastern side of the cyclone, ahead of the surface cold front. The air originates at low altitudes in the

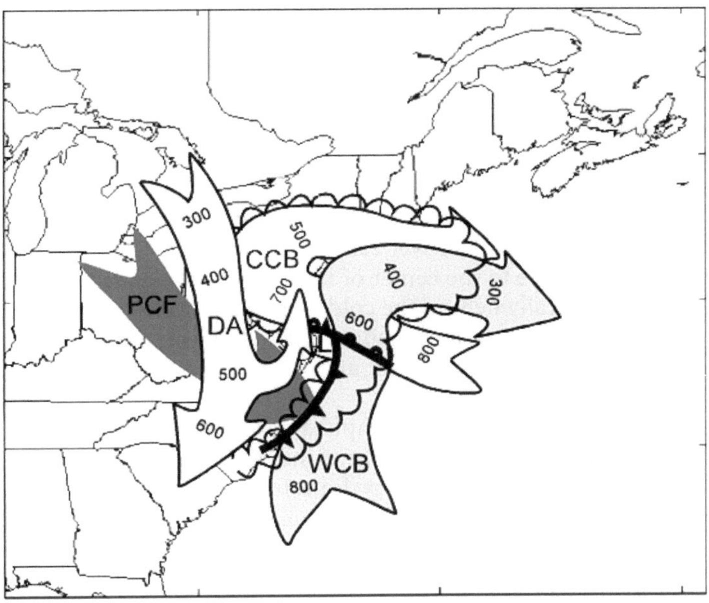

Fig. 1 Schematic diagram of a mid-latitude cyclone showing the WCB, CCB, DA and PCFA. The center of the cyclone is indicated (*L*) and the *scalloped lines* indicate the edges of the comma-cloud formed by the airstreams. The *numbers* on the WCB and CCB indicate the pressure (hPa) at the top of these airstreams, while the *numbers* on the DA indicated the pressure at the bottom of this air stream. After Fig. 3.1.24 and Fig. 3.1.27b in [40], and Fig. 2 in [51]

warm sector of the cyclone and travels northward, ascending into the mid- and upper troposphere, above the CCB. The WCB and CCB are moist airstreams and therefore their vertical motion is moist adiabatic rather than isentropic. In contrast, the DA originates at high altitudes in the upper troposphere and lowermost stratosphere on the poleward side of the cyclone and descends isentropically into the mid- and lower troposphere on the polar side of the cold front. Recently the concept of the post cold front airstream (PCFA) was introduced to explain the dry airmass in the lower and mid-troposphere that flows behind the cyclone cold front and beneath the DA [37]. The air mass within the PCFA is cold, thermally stable, and diverges laterally as the cold front elongates and the cyclone matures, preventing upwards motion within the airstream.

Figure 2 illustrates the typical motions of the four major mid-latitude cyclone airstreams. At 12 UTC, May 9, 2002, a mid-latitude cyclone was centered over the western Great Lakes. The infrared satellite image shows the high cloud tops of the WCB, with the CCB cloud emerging from beneath the WCB and advecting into the northwest quadrant of the cyclone (Fig. 2a). The cloud-free regions west of the cold front indicate the presence of the DA and the PCFA below. Using a trajectory model the domains of the airstreams were filled with thousands of trajectory particles (Fig. 2b), which were run forward in time for 48 h. After 24 h (Fig. 2c) one portion of the WCB ascended and advected into the western North Atlantic Ocean (WNAO) while the remainder ascended and wrapped cyclonically around the northwest side of the low. The lowest portion of the DA descended into the lower troposphere, west of the cold front, and mixed with the leading edge of the PCFA. The rest of the DA either wrapped into the center of the low or headed west mixing with the upper portions of the WCB. Meanwhile the CCB ascended as it wrapped around the low while the PCFA headed towards the southeast, sweeping across the eastern USA in the wake of the cold front. After 48 h, portions of the WCB and DA approached Western Europe and continued to mix over the North Atlantic. With the exception of the portion of the DA that descended to the surface behind the cold front, the remaining portions of these airstreams mixed in a large vortex over eastern Canada. The CCB did not cross the Atlantic, but stayed close to the center of the cyclone. Meanwhile the PCFA continued to diverge laterally behind the cold front staying close to the surface, and swept out much of the NA eastern seaboard. A notable feature of this simulation is the high degree of mixing between all four airstreams around the center of the low.

The most important airstream for rapid and direct intercontinental pollutant transport is the WCB. The lower portions of the WCB are located within the warm sector of the associated mid-latitude cyclone. In eastern NA, the warm sector develops on the western side of surface anticyclones where stable and often stagnant conditions allow for the accumulation of photochemically active trace gases in the lower troposphere, especially in summer [41–43]. As the cyclone moves off shore it carries the warm sector with it, exporting the polluted air mass to the WNAO [33, 44, 45]. At the same time, the air within the warm sector rises through the WCB, which forms a pathway with the potential for export to Europe [36, 46, 47].

"Climatologies" of WCB cores show that WCB inflow occurs at latitudes below 50°N, but during winter and spring the inflow only occurs below 40°N [48, 49].

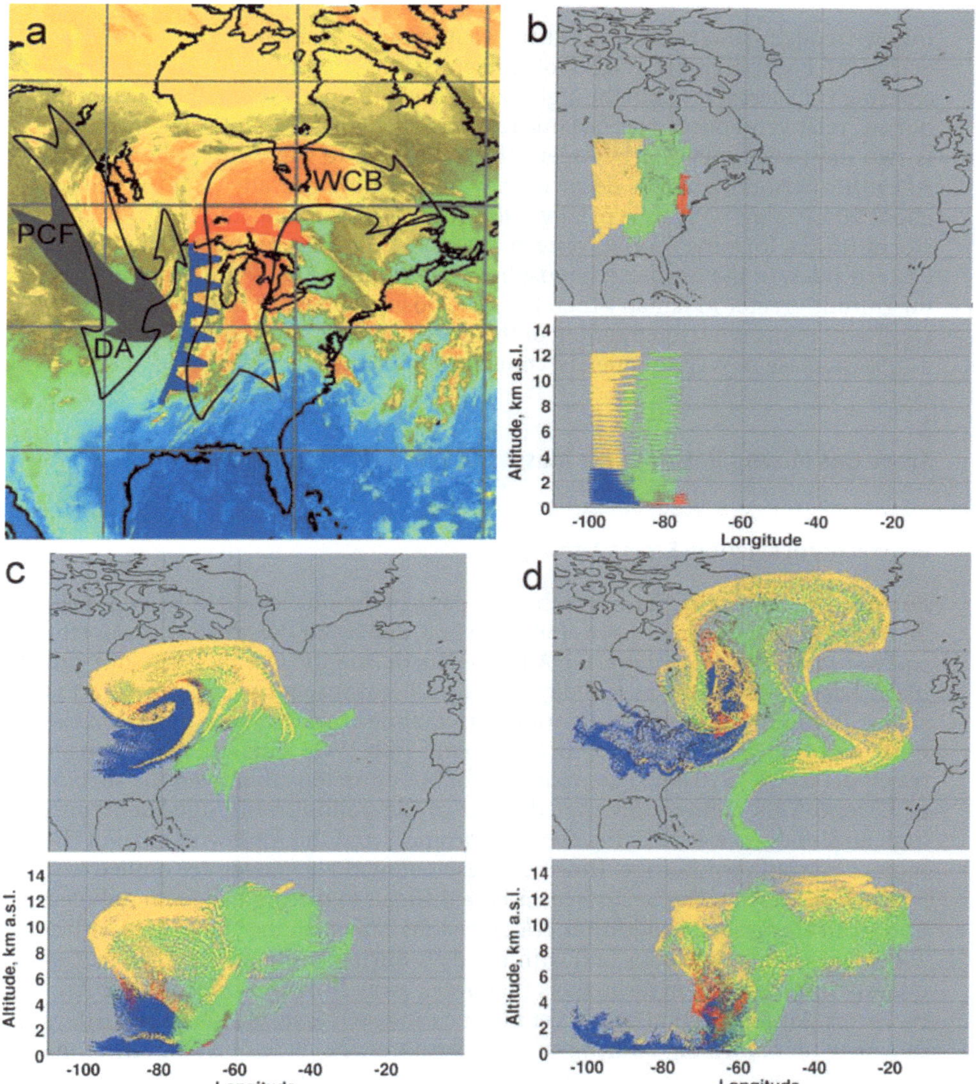

Fig. 2 a–d Infra-red GOES-East image at 12:15 UTC, May 9, 2002. *Oranges* and *yellows* indicate cold cloud tops at high altitudes, *greens* are mid-level clouds (or the cold surface of mountain ranges) and *blue* is the Earth's surface at low elevations. Also shown are the outlines of the WCB, DA and PCFA; the CCB outline has been omitted for clarity. **b** Initialization locations at 12 UTC May 9, 2002, of thousands of forward trajectory particles evenly spaced throughout the domains of the WCB (*green*), CCB (*red*), DA (*yellow*), and PCFA (*blue*) airstreams in **a**. **c** Locations of the forward trajectory particles after 24 h of transport. **d** Locations of the forward trajectory particles after 48 h of transport

Inflow frequency maxima occur at the southeastern seaboards of NA and Asia where the highest global anthropogenic emissions are located. The air masses entering WCBs traverse both the high-emission continental regions and the neighboring, relatively source-free oceanic regions. The typical lifetime of a WCB core is two days characterized by ascent from the ABL to the upper troposphere. After this period the WCBs decay as they turn anticyclonically and descend by about 20 hPa day^{-1}. WCBs typically follow the North Atlantic and North Pacific storm tracks, forming on the western sides of the oceans and decaying on the eastern sides. In the northern hemisphere the frequency of WCBs is an order of magnitude greater in January than July. The positive phase of the North Atlantic Oscillation causes a strengthening of the North Atlantic storm track resulting in a 12% increase in WCB frequency, compared to the negative phase.

3
Trace Gas Mixing Ratios in Air Masses Leaving North America

3.1
General North American Export Pathways

Focusing on transport times less than 10 days, pollutants leave NA by two main pathways during all four seasons [50]. The first, and by far the greatest, is a northeastward pathway from eastern NA to the North Atlantic Ocean. Pollutants can reach Europe in as little as two days during all seasons, but overall transport is slower in summer. These rapidly transported plumes arrive above Europe in the mid- and upper troposphere, in agreement with WCB climatologies. Longer transport times (>10 days) are required for NA emissions to impact the surface of Europe. This pathway involves descent of the pollution through the Azores high and/or downward mixing over the Pyrenees and the Alps, with most of the impact on southern Europe. However these polluted air masses are quite dilute by the time they reach the surface and are difficult to distinguish from polluted European air masses. In addition, NA emissions can reach the North Pole within about 5 days during winter and within 7 days during summer, and then subsequently return to NA in air masses that break away from the tropospheric Arctic vortex. Air masses that cross over Europe can reach eastern Asia within 6 days during winter and 8 days during summer with transport confined mainly to the mid- and upper troposphere. The second major pathway is mainly in the lower troposphere and advects pollutants from the southern and southwestern USA towards the southwest, forming a plume south of 30°N that extends into the eastern North Pacific Ocean. This plume tracks along the southern edge of the semi-permanent surface anticyclone of the eastern Pacific, reaching south of Hawaii within 5–6 days.

3.2
Trace Gas Signatures of East Coast Mid-Latitude Cyclones

Due to the importance of mid-latitude cyclone airstreams to trace gas export from NA, current research is aimed at discerning the typical trace gas mixing ratios within these airstreams as they exit NA. A recent study analyzed the chemical measurements of all cyclone airstreams sampled by aircraft during the late summer/early autumn field campaign of the 1997 North Atlantic Regional Experiment (NARE), conducted in the vicinity of the Canadian Maritimes [51]. The flight segments were classified according to the airstreams they occupied, and amalgamated into a single composite cyclone, to produce a conceptual model of the chemical and physical processing within a mid-latitude cyclone. The model separates the meteorological influences on airstream trace gas signatures from the influence of surface emissions heterogeneity. The meteorological influences on airstream trace gas composition during late summer/early autumn above the WNAO are: 1) The DA always has a stratospheric component that brings O_3 into the mid- and upper troposphere. 2) The PCFA originates to the northwest, is unaffected by wet deposition, and the sunny conditions may allow for some net photochemical O_3 production. 3) Both the CCB and WCB experience wet deposition. 4) Because of its southerly origin and association with the polluted western side of surface anticyclones the WCB draws from regions more favorable for photochemical activity than either the CCB or PCFA. 5) The CCB is generally cloudy and does not show signs of significant net photochemical production of O_3. 6) Very little NO_y is exported from the lower troposphere. 7) Airstreams in the mid-troposphere are the most susceptible to mixing with other air masses, which blurs the distinction between airstream trace gas signatures. The remainder of the chemical variation between and within airstreams is largely controlled by surface emissions heterogeneity; for example a WCB that draws from the polluted mixed layer will have higher O_3 and CO mixing ratios than one that draws from the clean marine boundary layer.

Comparison of airstream trace gas signatures between the NARE 1997 late summer/early autumn study period and the NARE 1996 springtime study period shows that the seasonal variation of photochemistry and meteorology affects the characteristic trace gas mixing ratios of the conceptual cyclone [52]. During spring, background O_3 and CO are at their maximum, cyclones track further south where continental surface emissions are greater, and transport from the stratosphere injects more O_3 into the troposphere. Figure 3 compares O_3 and CO by airstream for early spring and late summer-early autumn conditions over the WNAO. Using the positive (negative) O_3/CO slope as an indicator of photochemical O_3 production (destruction), O_3 production during late summer/early autumn is associated with the lower troposphere PCFA and all levels of the WCB, especially the lower troposphere. During early spring, significant net O_3 production is not associated with any airstream at any level, with the lower troposphere CCB associated with O_3 destruction. The negative slopes of the DA indicates that stratosphere to troposphere transport increases O_3 in the mid- and upper troposphere.

Those same NARE data sets reveal that extensive and rapid removal of NO_y occurs within the ABL as a result of wet and dry deposition during transport from the emission regions to the WNAO free troposphere [53]. During spring only 5.3% of the NO_y was estimated to reach altitudes above 3 km, during autumn this value was only 2.8%. During late summer/early autumn the lower tropospheric portions (<2 km above sea level) of the WCB, CCB and PCFA airstreams had me-

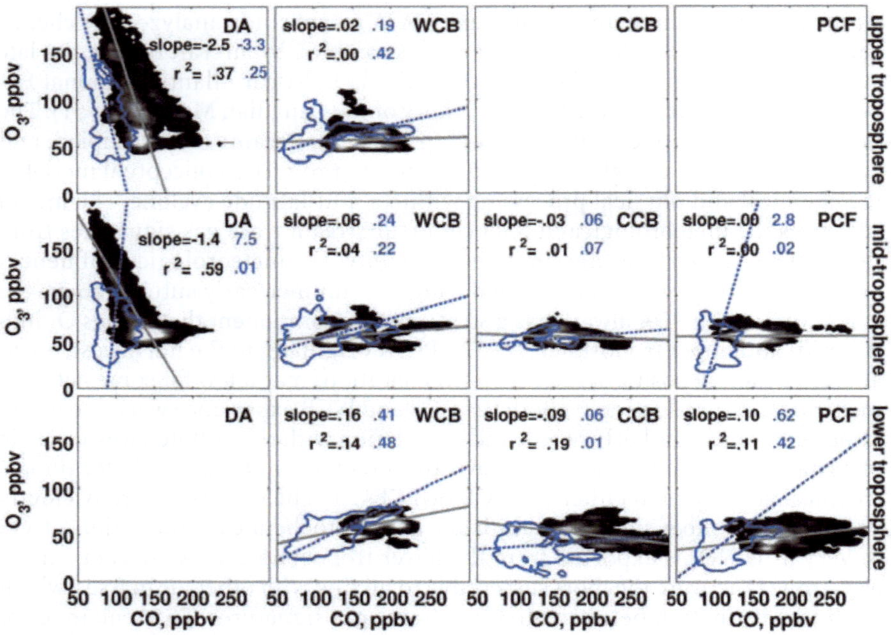

Fig. 3 Spring 1996 ozone vs. CO for all four cyclone airstreams in the lower (0–2 km), mid- (2–6 km), and upper (6–8 km) troposphere. The lighter (darker) shading corresponds to relatively higher (lower) data density. The range of the late summer-early autumn 1997 data for each air stream is outlined in *blue*. The linear regression lines for each air stream are shown for spring 1996 (*gray lines*) and summer-autumn 1997 (*blue dashed lines*), with the slope and r-squared values labeled in *black* (spring) and *blue* (summer-autumn). Note that the DA does not extend into the lower troposphere and the PCFA and CCB do not extend into the upper troposphere. After Figure 3 in [52]

dian NO_y mixing ratios of .4, .5 and .7 ppbv respectively, and 90th percentile NO_y mixing ratios of .6, .7 and 1.1 ppbv, respectively [51]. The PCFA contained the greatest values due to its continental source region and lack of precipitation events that would otherwise remove water-soluble nitrogen species. The CCB and WCB were associated with precipitation events and therefore contained less NO_y. Despite its association with the polluted warm sector of mid-latitude cyclones the WCB contained the least NO_y due to a greater influence from the clean marine boundary layer.

3.3 Transport to the North Atlantic Marine Boundary Layer

During summer, the Bermuda-Azores high pressure region extends westwards over the southeastern and Mid-Atlantic states, and mid-latitude cyclones track from west to east along the northern edge of the high. As a result the New England coast is a major export region of NA pollution during the most photochemically active time of the year. As the pollution moves offshore it can form thin layers with thickness on the order of a few tens of meters or less. The layering is formed by two major processes. First, warm polluted continental air passing over colder marine air creates stable conditions that minimizes vertical mixing. Second, the complex New England coastline produces strong shear of the horizontal wind with height, which stretches the pollution into various thin layers. Some of the pollution can become trapped in a shallow layer near the ocean surface, where conditions are so stable that little dry deposition occurs and the air mass retains the chemical characteristics of the source region. However during periods characterized by a sea breeze front or a stationary coastal front, polluted continental air is forced to rise as it moves off shore and can reach altitudes of 1–2 km before undergoing shearing processes [54–56].

Pollution episodes are detected at remote locations downwind of NA. O_3 and CO were continuously measured between 1991 and 1994 at Sable Island, a sand bar 900 km northeast of Boston. Maximum summertime O_3 and CO mixing ratios approached 100 and 300 ppbv, respectively, clearly indicating low-level export from NA. Even greater CO mixing ratios were detected in winter (up to nearly 500 ppbv) but the corresponding ozone mixing ratios were near zero indicating ozone titration due to high levels of NO and reduced sunlight. The pollution export from eastern NA to Sable Island is associated with southwesterly winds and transport within the warm sector of mid-latitude cyclones. Photochemical ozone production during summer and ozone destruction during winter appear to be typical of NA air masses that become isolated in the WNAO marine boundary layer [57].

A clear signal of NA emissions has been observed within the marine boundary layer as far away as Cape Race, Newfoundland (1300 km from New England) [58]. Pollution episodes possibly of NA origin were also detected during 1993 at the Azores which lie some 4000 km east of Washington DC [57]. Simple back trajectory analysis associated these episodes with upwind continental regions but detailed studies of the transport pathways to the Azores have not yet been conducted. Maximum CO mixing ratios at the Azores were lower than at Sable Island, reaching 220 ppbv in spring and 150 ppbv in summer. Lower values in summer were possibly due to greater isolation from the polluted layers aloft than during spring. Photochemical modeling studies indicate that on average Sable Island experiences net photochemical ozone production during summer while the central North Atlantic and the Azores experience net photochemical ozone loss [59, 60].

3.4 Trans-Atlantic Case Studies

Researchers have found relatively few instances of NA pollution impact at European surface stations. O_3 and CO mixing ratios were examined for the years 1990–1995 at Mace Head on the west coast of Ireland, an important site for establishing the background pollutant levels within air masses entering Europe from the west [61]. Only five instances of probable influence from NA emissions were found, all occurring in winter or very early spring, with a combined duration of only a few tens of hours out of the five-year record. The authors of the study concluded that either the NA pollution plume does not intersect the European coastline, or it merges into the Northern Hemisphere background during transit. A TTM study of one of these episodic events showed that a polluted air mass was exported from the USA east coast by a mid-latitude cyclone around March 17, 1994, and then followed a complex transport pathway to Mace Head, resulting in surface observations of enhanced chlorofluorocarbon mixing ratios on March 21, 1994 [62]. A separate study used a chemical transport model (CTM) and back trajectories to identify several episodes of CO and O_3 transport from NA to Mace Head during 1993–1997 [63]. When these events occur the NA CO tracer (typically less than 40 ppbv) is channeled to Mace Head between an intense Icelandic low and a surface anticyclone to the south, at altitudes below 3 km.

The strongest influence of NA emissions so far detected at Mace Head is the result of an unusually strong spate of Canadian forest fires during August 1998, resulting in maximum CO mixing ratios at Mace Head approaching 175 ppbv, roughly double the typical background value [5]. The authors of this study estimate that 58% of the CO enhancement during the month of August was the result of Canadian forest fire emissions. They suggest that the reason Canadian forest fire emissions were so much more evident than NA anthropogenic emissions at Mace Head was the result of different transport patterns. The anthropogenic emissions, originating further south, are more likely to be lofted into the mid- and upper troposphere by WCBs or deep convection. In contrast, the more northern forest fire emissions have a much lower probability of encountering WCBs and deep convection, remain in the lower troposphere for longer time periods and have a much greater likelihood of directly impacting the lower troposphere of Europe. This study also showed that Canadian forest fire emissions impact the mid-troposphere of Europe, but these and other aspects of boreal forest fire emissions are discussed in Chapter 10.

In contrast to the difficulty in identifying pollutant transport events from NA to the surface of Europe, observations of NA pollution in the free troposphere are becoming more common. Recent studies describe WCB transport of polluted ABL air from the eastern half of the NA mid-latitudes, to the mid- and upper troposphere of Europe during spring [36, 64], late summer [64] and autumn [46, 47]. One study traced the origins of a polluted layer of air detected by an instrumented aircraft on November 19, 2001 above southern Scandinavia [46]. Pollution levels were not greatly enhanced, only about 110 ppbv of CO, although one peak of 170 ppbv was detected. Using a TTM the pollution plume was tracked backwards in time to the NA ABL with the New York City region most likely making the strongest contribution to the CO measured above Europe.

While WCB transport typically takes 3–4 days from NA to Europe, explosive cyclogenesis, resulting in a meteorological "bomb" can transport pollution from the boundary layer of NA to the free troposphere of Europe in less than two days. Using satellite-based GOME measurements, a plume of NO_2 was tracked from NA to Europe within a bomb cyclone during November, 2001 [47]. NO_x mixing ratios within the plume were estimated to be nearly 2 ppbv. This form of rapid transport is four times more frequent in winter than during summer and could be important for re-distributing short-lived substances around the globe. (See Chapter 4 Huntrieser and Schlager in this volume for further discussion of trans-Atlantic transport.)

3.5 Transport into the North Pacific

Relatively few studies have addressed the impact of NA emissions on the Pacific Ocean. South of 30°N, NA low-level export occurs along the southern side of the semi-permanent surface anticyclone over the eastern North Pacific Ocean [50]. A modeling study for the spring of 2001 indicates that the export of NA pollution along this pathway contributes 10–20 ppbv of CO between 0–2 km in the remote North Pacific between Hawaii and California [65]. Trajectory analysis of total nitrate levels measured at Mauna Loa, Hawaii indicates a maximum NA influence during summer [66]. At mid-latitudes the presence of a surface anticyclone along the NA northwest coast can produce easterly offshore flow. These episodic conditions advect NA emissions to the eastern Pacific lower troposphere but their contribution to the CO budget of this region is small [65]. To date these events have not been linked to subsequent intercontinental transport.

4 Trace Gas Mixing Ratios and Particulate Matter Loadings in Air Masses Entering North America

4.1 General Import Pathways to North America

In terms of CO transport to NA, the European and Asian plumes are of greatest importance [50]. European pollution tends to travel at lower altitudes compared to NA and Asian pollution as its emission regions are usually north of the areas of warm conveyor belt formation. During winter much of the European pollution is transported at low altitudes to northern high latitudes where it can cross the North Pole and head south across NA, reaching the contiguous USA in about 7–8 days. In summer European emissions still cross the pole and reach the contiguous USA within 8–9 days but are typically located in the mid-troposphere. In contrast, the Asian plume travels at high altitudes across the Pacific and can reach NA in as little as 3 days during winter and 4 days during summer, although typical transport times are much longer. Asian pollution can descend into the semi-permanent surface anticylone of the eastern Pacific and impact NA at lower altitudes and latitudes, but this pathway requires several extra days of transport. Finally we mention the African plume, which primarily travels westward at tropical latitudes; its extension as far north as the southern USA is most common during summer.

4.2 Trans-Pacific Pollution Transport

A recent TTM study simulated the transport of Asian CO emissions over a 15-year period, and showed that transport of Asian CO across the USA west coast (125°W) has a maximum in spring and a minimum in summer [67]. During spring, the location of the CO tracer maximum is centered at an altitude of 8 km and 35°N, with relatively little eastward transport occurring north of 70°N or south of 20°N. The tracer maximum occurs in the mid- and upper troposphere due to the strong influence of WCB transport, which is rapid and generally occurs on timescales less than 10 days. For longer transport times (10–20 days) the CO tracer reaches the west coast at lower altitudes, either descending through anticyclones above the eastern North Pacific, or traveling at low altitudes all the way across the North Pacific. During summer the location of the CO tracer maximum is further north at 40°–45°N and at higher altitudes (10–11 km). WCBs are much less frequent in summer than spring [49], resulting in less efficient transport from Asia to NA and less CO reaching the west coast. Presumably summertime convective activity above Asia lofts the CO into the upper troposphere where it is advected at high altitudes to NA. Very little CO tracer intercepts the west coast at low altitudes during summer due to the presence of a strong surface anticyclone above the eastern Pacific, blocking transport from Asia.

Over the past 15 years, observations of dust and trace gas transport from Asia to NA have been reported with increasing frequency [68–75]. Table 2 shows the observed trace gas mixing ratios along the USA west coast during transport events from Asia. Measurements of 1) dimethylsulfide, sulfer dioxide, aerosol ions and radon [68], 2) hydrocarbons, O_3, and oxidized nitrogen species including PAN [70], 3) O_3, CO, PAN, hydrocarbons, aerosol scattering and many other trace gases [72, 73, 76], and 4) organochlorine pesticides in the Yukon, Canada, have been linked to transport from Asia through back trajectory analyses, but these studies give little information on the actual transport mechanisms from Asia to NA.

The transport events listed in Table 2 are often associated with elevated (up to 90 ppbv) O_3 levels. However, it is difficult to determine how much of the O_3 is the result of photochemical ozone production and how much is the result of transport from the stratosphere. As a cyclone decays portions of the WCB that lofted the Asian emissions often mix in the mid- and upper troposphere with DAs containing high levels of O_3 of stratospheric origin [77]. As a result simple correlation analysis of O_3 with CO may be misleading. Analyses to sort out these influences unambiguously are important for understanding the effect of intercontinental transport of anthropogenic emissions on tropospheric O_3 distribution.

Several case studies in the literature describe the transport pathways from Asia to NA. They can generally be classified as low altitude or high altitude transport events. In the slower, low altitude events (<3 km altitude) Asian emissions are exported at low levels by mid-latitude cyclones, and remain in the lower troposphere. The flow is channeled to the NA west coast at 45°–50°N between a surface low south of Alaska and a surface high in the eastern Pacific. This transport mechanism is associated with transport of SO_2 and CO from Asia to the coastal surface monitoring station at Cheeka Peak, Washington [71, 65]. A variation of this transport pathway is reported in a recent CTM study [65]. In early May, 2001, pollution was exported from eastern Asia at low altitudes towards the north (as far as 75°N) along the western side of a blocking anticyclone. The air mass then wrapped around the northern edge of the anticyclone and was subsequently transported to the southeast in the lower troposphere on the backside of a surface low located south of Alaska. The polluted air crossed the west coast between

Table 2 Maximum trace gas mixing ratios observed within polluted Asian air masses, above or just west of the USA west coast

Event	Altitude (km)	CO (ppbv)	PAN (pptv)	NO_y (ppbv)
1) 9 April 1999	2.8	171	622	
2) 9 April 1999	6.3	–	302	
3) 12 April 1999	4.9	–	436	
4) 28 April 1999	3.0	154	274	
5) Mar.–Apr. 1999[a]	0–2	146	91	
	2–4	152	184	
	4–6	146	343	
	6–8	135	222	
6) 14 April 2001	6.0	173		
7) 5 May 2002	5–7	313	650	0.9
8) 10 May 2002	4–6	290	670	1.1
9) 17 May 2002	2–4	245	80	0.8
10) 29 Mar 1997	CPO[b]	173	201	
11) Spring 1997, avg. of 9 events	CPO[b]	166	157	
12) Spring 1998, avg. of 9–11 events	CPO[b]	171	108	
13) March 11, 2001	CPO[b]	180–185		
14) 24–25 Apr 1985	PA[c]		70–150	0.3–0.6
15) 26 Feb. 2001	6	300		

[a] Median values are reported for aircraft measurements associated with transport from Asia and the NW Pacific, binned into 2 km layers.
[b] CPO is Cheeka Peak Observatory on the west coast of Washington State at 48.3°N, 124.6°W and 480 m above sealevel.
[c] PA is Point Arena on the westcoast of California at 38°57′N, 123°44′W and 20 m above sealevel 1–4, 6) [73]; 5) [76]; 7) [81]; 8) [77]; 9) [67]; 10) [71]; 11–12) [72]; 13) [65]; 14) [70]; 15) [78].

1 and 3 km, and was intercepted by an aircraft at 3 km with CO between 150–160 ppbv.

The faster high altitude transport events involve the lofting of Asian emissions by either deep convection or WCBs. The first detailed description of trans-Pacific transport explained the presence of anomalously high radon concentrations in the summertime upper troposphere above the eastern North Pacific Ocean [69]. The radon-rich air originated in the Asian ABL, ascended through cumulus updrafts and traveled to the NA west coast in fast-moving air on the anticyclonic side of the upper tropospheric jet. Transport times were typically 3 days, but only 48 h in one case (transport from China to a location several hundred kilometers west of San Francisco). A current study used a CTM and CO observations from the polar-orbiting MOPITT instrument to show that a pollution plume (up to 300 ppbv of CO) measured by an aircraft 700 km west of San Francisco on February 26, 2001, was exported from eastern Asia three days earlier in a WCB [78]. And another CTM study suggests that WCB transport from Asia to NA can end with the descent of the polluted air mass though a surface anticyclone in the eastern North Pacific and impact the surface of NA [65].

One important process associated with the high altitude transport events is pronounced wet deposition, or scrubbing, of the scrubbing of the transported air in deep convective cells and WCBs. This scrubbing profoundly affects the concentrations of transported emissions and their subsequent photochemistry by efficiently removing soluble species. For the three 2002 episodes in Table 2, there are relatively constant enhancement ratios for NO_y to CO, but these ratios are only 5% or less of the ratios in Asian emissions. This indicates that ≥95% of the emitted NO_y species have been removed from the air masses between emission and measurement. The remaining NO_y is generally dominated by PAN (as shown in Table 2 for the May 5 and 10, 2002 plumes). However, in the one plume that had descended to warmer, lower altitudes before measurement (May 17, 2002) the PAN had evidently thermally decomposed and been converted to nitric acid, which then accounted for most of the NO_y [79]. The aerosols were also efficiently removed in these scrubbing processes, and the subsequent occurrence of secondary aerosol formation has been observed in the May 5 and 17, 2002 plumes [80].

The long transport times and distances between Asia and NA can lead to very complex transport pathways. For example, on May 5, 2002, the NOAA WP-3D aircraft sampled a polluted air mass (150–313 ppbv of CO) between 5 and 8 km above sea level in the eastern North Pacific, 200–400 km southwest of Monterey, California. Using a TTM and geostationary satellite imagery the plume was traced back to the surface of Asia, which required more than 8 days of transport [81]. The pollution was lofted into the upper troposphere around April 28 through a WCB above central China. As the WCB decayed the plume traveled towards the southeast and descended, as is typical of WCB decay, passed over Japan and reached the mid-troposphere of the western North Pacific on May 1st and 2nd. The plume was then entrained into a second WCB that had formed a few days earlier above Japan. This second WCB transported the pollution back into the upper troposphere and across the North Pacific to Alaska, and subsequently decayed above the northeast Pacific as it rounded an upper-level ridge. The pollution plume descended again as it approached California on May 5, 2002. The majority of the WCB remnants subsequently crossed NA in the mid- and upper troposphere, with only 9% of the airstream's mass passing through the lower troposphere of the USA. This case study suggests that major pollution events from Asia do not necessarily have a direct impact on USA surface air quality.

4.3 Trans-Pacific Dust Transport

Chapter 11 discusses global dust transport, but here we discuss the effects of one recent, particularly intense event. The largest Asian dust event yet characterized in the USA ABL occurred between April 16 and 22, 2001 [82]. The dust was lofted from the desert regions of northern China and southern Mongolia by a midlatitude cyclone on April 6, 2001 [83]. The dust cloud was first observed in the free-troposphere above the west coast of NA on April 12 by TOMS satellite imagery and surface-based sun photometers [84]. Over the next few days the dust cloud continued to spread across NA in the free-troposphere and was sampled by an aircraft at altitudes greater than 4 km above the coast of Washington State on April 14 [85]. Between April 16 and 22 Asian dust was observed at surface monitoring sites across the USA, with the greatest PM_{10} (up to 60 µg m^{-3}) concentrations observed in the central west coast, the Rocky Mountains and the southeastern USA [82]. To determine how the dust was transported to the surface of the west coast on April 16, four days after it was first detected in the free troposphere a detailed trajectory analysis was conducted [82]. Thousands of forward trajectories were released from the lower troposphere of the desert source regions within the WCB and PCFA of the mid-latitude cyclone that lofted the dust on April 6. The analysis showed that the only way for the dust to be transported to the surface of the west coast on April 16 was for it to be lofted into the mid- and upper troposphere by the WCB and transported across the Pacific in the upper level winds, followed by descent into the surface anticyclone on the west coast of NA. This transport mechanism would require minimal wet-removal by precipitation which does not seem unreasonable given the extremely arid source region. The dust entrained into the PCFA traveled across the Pacific in the lower troposphere and arrived at the west coast on April 17, delivering a second plume of dust.

4.4 Trans-Atlantic: Dust from Africa, Anthropogenic Pollution from Europe

Dust is exported each summer from the desert regions of northern Africa and advected by the tropical trade winds on the southern side of the semi-permanent Bermuda-Azores High to the WNAO, the Caribbean Sea and the southern USA. A 23-year record of surface aerosol particle measurements at Miami, Florida, shows that these events typically occur from June to August, with maximum dust concentrations in July [86]. These dust events are also associated with SO_4^{2-} and NO_3^- concentrations above typical background levels. The SO_4^{2-} and NO_3^- originate in European air masses that travel southwards over Africa and entrain dust en route to NA. Analysis of several years of fine aerosol particle measurements from rural areas across the contiguous USA shows that wide-spread African dust events impact the eastern USA an average of three times per year [87]. The dust enters the USA from the south and can spread as far north as Maine and as far west as western Texas. Maximum $PM_{2.5}$ soil concentrations during these events are typically 20–30 µg m^{-3}, several times greater than average measurements in the western USA during summer, and mean $PM_{2.5}$ concentrations range from 6.0 to 12.2 µg m^{-3}. At Miami, maximum African dust concentrations range from 50 to 120 µg m^{-3}, with roughly one third to one half of these values qualifying as $PM_{2.5}$. Alone, these concentrations are not enough to exceed USA air quality standards, but in conjunction with local particulate matter emissions, could contribute to exceeding the $PM_{2.5}$ air quality standard [86].

Modeling studies indicate that the typical transport pathways from Europe to NA occur via the polar regions, tropical easterly flow on the southern side of the Bermuda-Azores high or westerly transport across Asia and the North Pacific Ocean [50, 63]. However, a modeling study for the years 1993–1997 showed that these pathways never delivered high O_3 events to NA due to the long transport times [63]. But the same study suggested that, on occasion, European emissions can make fairly large contributions to surface O_3 and CO mixing ratios at the eastern edge of NA, in one case modeled enhancements as high as 10 ppbv of O_3 and 30–50 ppbv of CO. The transport pathway occurs typically in spring and fall in the lower troposphere when an anomalous surface low becomes established for several days above the central North Atlantic at latitudes around 45°N. The European emissions can then travel westward along the northern side of the low to the east coast of NA.

5
Synthesis and Conclusions

5.1
Implications of North American Trace Gas Export

Surface measurements have shown that O_3 pollution from NA is easily detectable 1500 km downwind from the NA source region in summer [58]. And surface measurements, located 3000–4000 km downwind from the sources established that NA pollution enhances O_3 levels in the central North Atlantic during spring [57]. These observations provided the basis for the quantitative assessment of the amount of anthropogenic O_3 transported from NA. An initial study concluded that during summer, O_3 transported from NA exceeded natural O_3 from the stratosphere in the lower troposphere over the North Atlantic [58]. Further quantitative refinements of the calculations estimate that the total export flux of O_3 from NA in the summer is on the order of 6 Gmol day^{-1} [44, 63, 88]. These studies conclusively show that during the summer, the O_3 budget in the lower troposphere over the temperate WNAO is dominated by O_3 photochemically produced from precursors emitted by NA sources.

In the remote, marine troposphere, the concentrations of CO and methane are adequate to support significant photochemical O_3 formation. However, whether this photochemistry produces rather than destroys O_3 is determined by the amount of NO_x available [89]. Hence, the export of anthropogenic NO_x from the continental boundary layer can determine the amount of O_3 that is produced in the marine troposphere. Analysis of the correlation of NO_y and its component species with CO and Lagrangian trajectory analyses show that only a small fraction of the NO_x emitted in the continental boundary layer is transported to the free troposphere [53, 90]. Although transport of reactive nitrogen from the continent to the marine free troposphere is limited, continental NO_x emissions are still believed to have a significant impact on the O_3 distribution in the remote marine troposphere. Model results suggest that eventual O_3 production in the global troposphere from USA emissions is about twice as large as the direct export of O_3 from the USA boundary layer [91–93]. The amount of NO_x delivered to the lower troposphere in the WNAO during summer is sufficient to produce, on average, 1 to 4 ppbv day^{-1} of O_3 throughout the region [57, 60]. In contrast, the central North Atlantic (e.g., the Azores) is a region of photochemical O_3 destruction [57, 59, 94].

5.2 Implications of Trace Gas Import to North America

Due to rising global emissions, the impact of imported emissions and their secondary O_3 and aerosol products on NA air quality is an issue of interest and concern [95–97]. CTM studies estimate that chemical production outside of the North American boundary layer contributes an average 15–35 ppbv ozone to afternoon mixing ratios in surface air over the USA [97], and that rising Asian anthropogenic emissions between 1985 and 2010 could increase mean O_3 concentrations in the western USA by 2–6 ppbv [96]. Another CTM study estimates that major Asian pollution events are very common in the mid- and upper troposphere above the USA, but perhaps only 3–5 of these events directly impact the ABL along the USA west coast during a typical February–May period [98]. Indeed, the major Asian pollution events observed near the North American west coast have been located in the mid to upper troposphere, but with the exception of Asian dust storms, transport events from Asia have not yet been linked with large and distinct fluctuations in NA surface air quality.

Rather than dramatically impacting NA surface air quality in discrete transport events, upwind emissions are more likely to influence NA by raising background trace gas mixing ratios in well-mixed air masses. Observational evidence is beginning to support this hypothesis. Figure 4 shows background ozone levels in spring measured at five surface sites near the USA west coast. These levels have

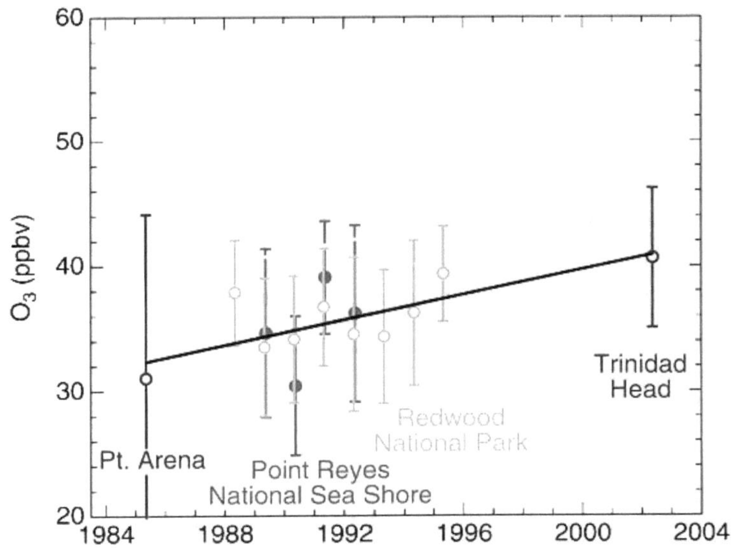

Fig. 4 Spring mean mixing ratio ±1 standard deviation for background O_3 at five sites in the western USA. The linear fit (*black line*) to the data from four coastal, sea level sites yields a slope (with 95% confidence interval) of 0.50 ± 0.36 ppbv $year^{-1}$ with an $r^2 = 0.44$. The linear fit for the inland, high altitude site (Lassen at 1.7 km) yields a slope of 0.46 ± 0.39 ppbv $year^{-1}$ with an $r^2 = 0.36$. After Figs. 1 and 2 of [99]

increased by about 10 ppbv (30%) over the past 20 years [99]. This increase correlates with growing nitrogen oxide emissions in Asia. Although a definite cause and effect relationship has not been established, spring is the season of strongest transport of Asian emissions to the Pacific so there may be a direct connection. However, this presumed Asian contribution is greater than that predicted by CTMs [95, 97, 98].

Reproducing the observed chemical composition of the air entering and leaving North America presents a significant challenge for global CTMs. Meeting this challenge will enhance our confidence in these models when they are used to predict the future state of the troposphere, especially regarding tropospheric ozone and aerosol levels. Ozone is an important greenhouse gas, and is a critical precursor for hydroxyl radical formation. This radical is the sink for other important greenhouse gases, while aerosols have direct and indirect effects on the global climate. A CTM prediction important for issues of North American air quality is the trend in ozone and aerosol levels entering the USA. The new 8-hour O_3 standard is 80 ppbv, and the increasing background O_3 that has been reported allows a progressively smaller margin for additional O_3 produced from local emissions. Likewise, imported likewise, imported particulate matter levels can constitute a significant fraction of the mandated maximum levels. Understanding these imported levels and expected trends has important implications for regional air quality control planning. While the many recent studies discussed throughout this chapter have increased our understanding of intercontinental transport and chemical transformation, accurate modeling of the present and future state of the North American and global troposphere requires a more detailed understanding of these processes than is presently available.

6 References

1. Olivier JGJ, Berdowski JJM (2001) Global emissions sources and sinks. In: Berdowski J, Guicherit R, Heij BJ (eds) The Climate System. AA Balkema Publishers/Swets & Zeitlinger Publishers, Lisse, The Netherlands, ISBN 90 5809 255 0, p 33
2. Houghton JT, Ding Y, Griggs DJ, Noguer M, van der Linden PJ, Dai X, Maskell K, Johnson CA (2001) (eds) IPCC, Climate Change 2001: The Scientific Basis, contribution of Working Group I to the Third Assessment Report of the Intergovernmental Panel on Climate Change. Cambridge University Press, Cambridge
3. Logan JA, Prather MJ, Wofsy SC, McElroy MB (1981) J Geophys Res 86:7210
4. Wotawa G, Trainer M (2000) Science 288:324
5. Forster C, Wandinger U, Wotowa G, James P, Mattis I, Althausen D, Simmonds P, O'Doherty S, Jennings SG, Kleefeld C, Schneider J, Trickl T, Kreipl S, Jäger H, Stohl A (2001) J Geophys Res 106:22887
6. Hallock-Waters KA, Doddridge, BG, Dickerson RR, Spitzer S, Ray JD (1999) Geophys Res Lett 26:2861
7. Parrish DD, Trainer M, Hereid D, Williams EJ, Olszyna KJ, Harley RA, Meagher JF, Fehsenfeld FC (2002) J Geophys Res 107:4140, doi:10.129/2001/2001JD000720
8. Wotawa G, Novelli PC, Trainer M, Granier C (2001) Geophys Res Lett 28:4575
9. Duncan BN, Martin RV, Staudt AC, Yevich R, Logan JA (2003) J Geophys Res 108:4100, doi:10.1029/2002JD002378
10. Malm WC, Schictel BA, Ames RB, Gebhardt KA (2002) J Geophys Res 107:4627, doi:10.1029/2002JD002107

11. Guenther AB, Geron C, Pierce T, Lamb B, Harley P, Fall R (2000) Atmos Environ 34:2205
12. US Environmental Protection Agency (2000) National Air Pollutant Emission Trends, 1900–1998. EPA-454/R-00-002
13. Environment Canada, http://www.ec.gc.ca/pdb/ape/cape_home_e.cfm
14. Prospero JM, Ginoux P, Torres O, Nicholson SE, Gill TE (2002) Reviews of Geophysics 40:1002, doi:10.1029/2000RG000095
15. Lunt DJ, Valdes PJ (2002) J Geophys Res 107:4669, doi:10.1029/2002JD002316
16. Gillette DA, Hanson KJ (1989) J Geophys Res 94:2197
17. Newchurch MJ, Ayoub MA, Oltmans S, Johnson B, Schmidlin FJ (2003) J Geophys Res 108:4031, doi:10.1029/2002JD002059
18. Logan JA (1999) J Geophys Res 104:16115
19. Fiore AM, Jacob DJ, Logan JA, Yin JH (1998) J Geophys Res 103:1471
20. Lin C-YC, Jacob DJ, Munger JW, Fiore AM (2000) Geophys Res Lett 27:3465
21. Zajac BA, Rutledge SA (2001) Mon Wea Rev 129:999
22. Orville RE, Huffines GR, Burrows WR, Holle RL, Cummins KL (2002) Mon Wea Rev 130:2098
23. Christian HJ, Blakeslee RJ, Boccippio DJ, Boeck WL, Buechler DE, Driscoll KT, Goodman SJ, Hall JM, Koshak WJ, Mach DM, Stewart MF (2003) J Geophys Res 108:4005, doi:10.1029/2002JD002347
24. Dickerson RR, Huffman GJ, Luke WT, Nunnermacker LJ, Pickering KE, Leslie ACD, Lindsey CG, Slinn WGN, Kelly TJ, Daum PH, Delaney AC, Greenberg JP, Zimmerman PR, Boatman JF, Ray JD, Stedman DH (1987) Science 235:460
25. Skamarock WC, Powers JG, Barth M, Dye JE, Matejka T, Bartels D, Baumann K, Stith J, Parrish DD, Hübler G (2000) J Geophys Res 105:19,973
26. Betts AK, Gatti LV, Cordova AM, Silva Dias MAF, Fuentes JD (2002) J Geophys Res 107:8046, doi:10.1029/2000JD000158
27. Poulida O, Dickerson RR, Heymsfield A (1996) J Geophys Res 101:6823
28. Thompson AM, Pickering KE, Dickerson RR, Ellis WG Jr, Jacob DJ, Scala JR, Tao W-K, McNamara DP, Simpson J (1994) J Geophys Res 99:18,703
29. Huntrieser H, Schlager H, Fiegl C, Höller H(1998) J Geophys Res 103:28247
30. Pickering KE, Wang Y, Tao W-K, Price C, Müller J-F (1998) J Geophys Res 103:31,20
31. Dye JE, Ridley BA, Skamarock W, Barth M, Venticinque M, Defer E, Blanchet P, Thery C, Laroche P, Baumann K, Hübler G, Parrish DD, Ryerson T, Trainer M, Frost G, Holloway JS, Rutledge SA, Lang T, Stith J, Zerr R (2000) J Geophys Res 105:10023
32. Huntrieser H, Fiegl C, Schlager H, Schröder F, Gerbig C, van Velthoven P, Flatoy F, Thery C, Petzold A, Höller H, Schumann U (2002) J Geophys Res 107: doi:10.1029/2002JD000209
33. Merrill JT, Moody JL (1996) J Geophys Res 101:28,903
34. Carlson TN (1998) Mid-latitude weather systems. The American Meteorological Society, Boston
35. Bethan S, Vaughan G, Gerbig C, Volz-Thomas A, Richer H, Tiddeman DA (1998) J Geophys Res 103:13413
36. Stohl A, Trickl T (1999) J Geophys Res 104:30445
37. Cooper OR, Moody JL, Parrish DD, Trainer M, Holloway JS, Ryerson TB, Hübler G, Fehsenfeld FC, Oltmans SJ, Evans MJ (2001) J Geophys Res 106:5437
38. Browning KA, Monk GA (1982) Quart J R Met Soc 108:435
39. Browning KA, Roberts NM (1994) Quart J R Met Soc 120:1537
40. Bader MJ, Forbes GS, Grant JR, Lilley RBE, Waters AJ (eds) (1995) Images in weather forecasting: A practical guide for interpreting satellite and radar imagery. University Press, Cambridge
41. Comrie AC, Yarnal B (1992) Atmos Environ 26:301
42. Comrie AC (1994) Atmos Environ 28:1601
43. Cooper OR, Moody JL (2000) J Geophys Res 105:6855
44. Berkowitz CM, Daum PH, Spicer CW, Busness KM (1996) J Geophys Res 101:28,923
45. Moody JL, Davenport JC, Merrill JT, Oltmans SJ, Parrish DD, Holloway JS, Levy H II, Forbes GL, Trainer M, Buhr M (1996) J Geophys Res 101:29213

46. Stohl A, Forster C, Eckhardt S, Spichtinger N, Huntrieser H, Heland J, Schlager H, Aufmhoff H, Arnold F, Cooper O (2003a) J Geophys Res 108:4370, doi:10.1029/2002JD002862
47. Stohl A, Huntrieser H, Richter A, Beirle S, Cooper O, Eckhardt S, Forster C, James P, Spichtinger N, Wagner T, Wenig M, Burrows J, Platt U (2003b) Atmos Chem Phys 3:969
48. Stohl A (2001) J Geophys Res 106:7263–7279
49. Eckhardt S, Stohl A, Wernli H, James P, Forster C, Spichtinger N (2003) J Climate 17:218
50. Stohl A, Eckhardt S, Forster C, James P, Spichtinger N (2002a) J Geophys Res 107:4684, doi:10.1029/2001JD001396
51. Cooper OR, Moody JL, Parrish DD, Trainer M, Ryerson TB, Holloway JS, Hübler G, Fehsenfeld FC, Evans MJ (2002a) J Geophys Res 107:10.1029/2001JD000901
52. Cooper OR, Moody JL, Parrish DD, Trainer M, Holloway JS, Hübler G, Fehsenfeld FC, Stohl A (2002b) J Geophys Res 107:10.1029/2001JD000902
53. Stohl A, Trainer M, Ryerson TB, Holloway JS, Parrish DD (2002b) J Geophys Res 107: do1:10.1029/2001JD000519
54. Angevine WM, Trainer M, McKeen SA, Berkowitz CM (1996) J Geophys Res 101:28893
55. Angevine WM, Buhr MP, Holloway JS, Trainer M, Parrish DD, MacPherson JI, Kok GL, Schillawski RD, Bowlby DH (1996) J Geophys Res 101:28935
56. Knapp KG, Balsley BB, Jensen ML, Hanson HP, Birks JW (1998) J Geophys Res 103:13399
57. Parrish DD, Trainer M, Holloway JS, Yee JE, Warshawsky MS, Fehsenfeld FC, Forbes GL, Moody JL (1998) J Geophys Res 103:13,357
58. Parrish DD, Holloway JS, Trainer M, Murphy PC, Forbes GL, Fehsenfeld FC (1993) Science 259:1436
59. Atherton CS, Grotch S, Parrish DD, Penner JE, Walton JJ (1996) Atmos Environ 30:1739
60. Duderstadt KA, Carroll MA, Sillman S, Wang T, Albercook GM, Feng L, Parrish DD, Holloway JS, Fehsenfeld FC, Blake DR, Blake NJ, Forbes G (1998) J Geophys Res 103: 13,531
61. Derwent RG, Simmonds PG, Seuring S, Dimmer C (1998) Atmos Environ 32:145
62. Ryall DB, Maryon RH, Derwent RG, Simmonds PG (1998) Quart J R Met Soc 124:417
63. Li Q, Jacob DJ, Bey I, Palmer PI, Duncan BN, Field BD, Martin RV, Fiore AM, Yantosca RM, Parrish DD, Simmonds PG, Oltmans SJ (2002) J Geophys Res 107: doi:10.1029/ 2001JD001422
64. Trickl T, Cooper OR, Eisele J, James P, Mücke R, Stohl A (2003) J Geophys Res 108:8530, doi:10.1029/2002JD002735
65. Jaeglé L, Jaffe D, Price HU, Weiss-Penzias P, Palmer PI, Evans MJ, Jacob DJ, Bey I (2003) J Geophys Res 108:8802, doi:10.1029/2002JD003121
66. Lee G, Merrill JT, Huebert BJ (1994) J Geophys Res 99:12,821
67. Forster C, Cooper O, Stohl A, Eckhardt S, James P, Dunlea E, Nicks D Jr, Holloway J, Hübler G, Parrish D, Ryerson T, Trainer M (2003) J Geophys Res (in press)
68. Andreae MO, Berresheim H, Andreae TW, Kritz MA, Bates TS, Merrill JT (1988) J Atmos Chem 6:149
69. Kritz MA, Le Roulley J-C, Danielsen EF (1990) Tellus 42B:46
70. Parrish DD, Hahn CJ, Williams EJ, Norton RB, Fehsenfeld FC, Singh HB, Shetter JD, Gandrud BW, Ridley BA (1992) J Geophys Res 97:15883
71. Jaffe D, Anderson T, Covert D, Kotchenruther R, Trost B, Danielson J, Simpson W, Berntsen T, Karlsdottir S, Blake D, Harris J, Carmichael G, Uno I (1999) Geophys Res Lett 26:711
72. Jaffe D, Anderson T, Covert D, Trost B, Danielson J, Simpson W, Blake D, Harris J, Streets D (2001) J Geophys Res 106:7449
73. Jaffe D, McKendry I, Anderson Y, Price H (2003a) Atmos Environ 37:391
74. Wilkening K, Barrie L, Engle M (2000) Science 290:65
75. Husar RB, (and 28 co-authors) (2001) J Geophys Res 106:18317
76. Kotchenruther RA, Jaffe DA, Beine HJ, Anderson TL, Bottenheim JW, Harris JM, Blake DR, Schmitt R (2001) J Geophys Res 106:7463
77. Cooper OR, Forster C, Parrish DD, Dunlea E, Hübler G, Fehsenfeld F, Holloway J, Oltmans S, Johnson B, Wimmers A, Horowitz L (2004a) J Geophys Res (in press)

78. Heald CL, Jacob DJ, Fiore AM, Emmons L, Gille JC, Sachse GW, Browell EV, Avery MA, Vay SA, Crawford JH, Westberg DJ, Blake DR, Singh HB, Sandholm ST, Talbot RW, Fuelberg HE (2003) 108:doi:10.1029/2003JD003507
79. Nowak JB et al. (2004) J Geophys Res (in press)
80. Brock CA et al. (2004) J Geophys Res (in press)
81. Cooper OR, Forster C, Parrish DD, Trainer M, Dunlea E, Ryerson T, Hübler G, Fehsenfeld F, Nicks D, Holloway J, de Gouw J, Warneke C, Roberts JM, Flocke F, Moody J (2004b) J Geophys Res (in press)
82. Jaffe D, Snow J, Cooper O (2003c) EOS 84:501,507
83. Gong SL, Zhang XY, Zhao TL, McKendry IG, Jaffe DA, Lu NM (2003) J Geophys Res 108:doi:10.1029/2002JD002633
84. Thulasiraman S, O'Neil NT, Royer A, Holben BN, Westphal DL, McArthur LJB (2002) Geophys Res Lett 29:10.1029/2001GL014188
85. Price H, Jaffe DA, Doskey P, McKendry I, Anderson T (2003) J Geophys Res 108:doi:10.1029/2002JD002930
86. Prospero JM (1999) J Geophys Res 104:15,917
87. Perry KD, Cahill TA, Eldred RA, Dutcher DD, Gill TE (1997) J Geophys Res 102:11225
88. Chin M, Jacob DJ, Munger JW, Parrish DD, Doddridge BG (1994) J Geophys Res 99:14565
89. Fehsenfeld FC, Liu SC (1993) Chapter 5, Tropospheric ozone: Distribution and sources. In: Hewitt CN. Sturges WT (eds) Global Atmospheric Chemical Change. Elsevier Applied Science, New York, p 169
90. Parrish DD, Ryerson TB, Holloway JS, Neuman JA, Williams J, Stroud CA, Frost GJ, Trainer M, Hubler G, Fehsenfeld FC, Flocke F, Weinheimer AJ (2004) J Geophys Res (in press)
91. Liang J, Horowitz LW, Jacob DJ, Wang Y, Fiore AM, Logan JA, Gardner GM, Munger JW (1998) J Geophys Res 103:13,435
92. Horowitz LW, Liang J, Jacob DJ, Gardner GM (1998) J Geophys Res 103:13,451
93. Meijer EW, van Velthoven PFJ, Thompson AM, Pfister L, Schlager H, Schulte P, Kelder H (2000) J Geophys Res 105:3833
94. Peterson MC, Honrath RE, Parrish DD, Oltmans SJ (1998) J Geophys Res 103:13,489
95. Bernsten TK, Karlsdóttir S, Jaffe DA (1999) Geophys Res Lett 26:2171
96. Jacob DJ, Logan JA, Murti PP (1999) Geophys Res Lett 26:2175
97. Fiore AM, Jacob DJ, Bey I, Yantosca RM, Field BD, Fusco AC (2002) J Geophys Res 107:10.1029/2001JD000982
98. Yienger JJ, Galanter M, Holloway TZ, Phadnis MJ, Guttikunda SK, Carmichael GR, Moxim WJ, Levy H II, (2000) J Geophys Res 105:26,931
99. Jaffe D, Price H, Parrish D, Goldstein A, Harris J (2003b) Geophys Res Lett 30:1613, doi:10.1029/2003GL017024

The Handbook of Environmental Chemistry Vol. 4, Part G (2004): 69–98
DOI 10.1007/978-3-540-40037-0

Air Pollution Export from and Import to Europe: Experimental Evidence

Heidi Huntrieser · Hans Schlager

Institut für Physik der Atmosphäre, Deutsches Zentrum für Luft- und Raumfahrt, DLR-Oberpfaffenhofen, Postfach 1116, 82230 Weßling, Germany
E-mail: Heidi Huntrieser@dlr.de; E-mail: Hans.Schlager@dlr.de

Abstract The intercontinental transport of air pollutants from and to Europe is described in detail. The major focus is on the import of air pollution from North America since a number of studies are available on this topic, especially from the North Atlantic Regional Experiment (NARE). Less attention is dedicated to the air pollution import from Asia to Europe, because very few studies exist and the importance of this contribution has only been discovered recently, e.g., during the Mediterranean Intensive Oxidant Study (MINOS). Finally, the present knowledge of the export of air pollutants from Europe is summarized. The European pollution export was investigated in great detail during the EXPORT (European Export of Precursors and Ozone by Long-range Transport) campaign. The present chapter focuses on experimental evidence of intercontinental pollution transport. Observations from different platforms – ground-based stations, aircraft, and satellites – are summarized and discussed in detail. In addition, some results from model studies are included to achieve a more general picture of the impact of intercontinental pollution transport. Finally, typical transport mechanisms and pathways of intercontinental pollution transport to and from Europe are briefly described.

Keywords Experimental measurements · Ozone production · Deep convection · Warm conveyor belts · Low level pollution transport

Abbreviations

AEROCE	Atmospheric/Ocean Chemistry Experiment
BL	Boundary layer
C_2H_2	Acetylene
CH_3OOH	Methyl hydroperoxide
CH_4	Methane
CN	Condensation nuclei
CO	Carbon monoxide
CO_2	Carbon dioxide
CONTRACE	Convective Transport of Trace Gases into the Middle and Upper Troposphere over Europe
DLR	Deutsches Zentrum für Luft- und Raumfahrt
EXPORT	European Export of Precursors of Ozone by Long-range Transport
FLEXPART	Name of particle dispersion model
FLEXTRA	Name of trajectory model
FT	Free troposphere
H_2O	Water vapor
H_2O_2	Hydrogen peroxide
LT	Lower troposphere
MT	Mid troposphere
MINOS	Mediterranean Intensive Oxidant Study
MOZAIC	Measurements of Ozone and Water Vapor by Airbus In-Service Aircraft
NA	North America
NAO	North Atlantic Oscillation
NARE	North Atlantic Regional Experiment
NMHCs	Non-methane hydrocarbons
NO	Nitrogen monoxide, nitric oxide
NO_2	Nitrogen dioxide
NO_x	Sum of nitrogen monoxide and nitrogen dioxide
NOXAR	Nitrogen Oxides and Ozone Along Air Routes
NO_y	Total reactive nitrogen
O_3	Ozone
OCTA	Oxidizing Capacity of the Tropospheric Atmosphere
PAN	Peroxyacetyl nitrate
ppbv	Parts per billion by volume
ppmv	Parts per million by volume

SO_2	Sulfur dioxide
TOR	Tropospheric Ozone Research
UKMO	United Kingdom Meteorological Office
U.S.	United States
UT	Upper troposphere
VOTALP	Vertical Ozone Transport over the Alps
WCB	Warm conveyor belt

1 Introduction

This chapter focuses on anthropogenic emissions released in Europe and their effect on other continents (see Sects. 2 and 3). Furthermore, the impact of the anthropogenic pollution imported from North America (NA) and Asia on surface air quality and free tropospheric (FT) air composition over Europe is discussed (see Sects. 4 and 5, respectively). Experimental evidence of intercontinental pollution transport from ground, air- and space-borne observations is mainly summarized, however important results from model simulations are also discussed briefly.

One of the key questions related to intercontinental pollution transport is to what extent ozone levels over Europe, NA, and Asia are affected by the import of ozone and its precursors (CO, NO_x and hydrocarbons) from upwind continents. To answer this question, it first has to be known to what extent the pollution from an upwind continent may penetrate into the boundary layer (BL) and affect surface air quality. This impact has been almost unknown up to now. The lifetime of the ozone precursor NO_x is relatively short (hours to days), however the ozone produced has a lifetime of days to months and may certainly effect downwind continents since typical transport times between continents are 5–10 days. In addition, it is important to take the trends of ozone and its precursors from upwind continents into account when official ozone standards are set, i.e., to consider regional pollution in a global context [1].

The tropospheric air composition over Europe has changed significantly since the beginning of industrialization. Volz and Kley [2] suggested that the background level of ozone has doubled since pre-industrial times due to an increase in anthropogenic emissions of ozone precursors. Long-term measurements at the TOR station Mace Head (Ireland) between 1987 and 1995 indicated an on-going positive ozone trend of +0.19 ppbv per year [3]. The general ozone increase over Europe may partly be caused by intercontinental pollution transport from NA as suggested by Parrish et al. [4], since ozone levels are slightly increasing there. Lin et al. [5] also found a positive ozone trend from measurements at rural surface sites over the United States, which they, however, partly attributed to an increase in East Asian emissions. Presently, NO_x emissions are increasing very rapidly over the Asian continent (4–5% per year) [6, 7]. The Asian emissions are discussed in more detail in the chapter by Pochanart et al. in this volume.

2 European Anthropogenic Emissions

Inventories of European anthropogenic emissions have been compiled by EMEP (Co-operative Programme for Monitoring and Evaluation of the Long-Range Transmission of Air Pollutants in Europe) [8] and CORINAIR (European Atmospheric Emission Inventory) [9]. These inventories provided input to the global inventories GEIA [10] and EDGAR [11, 12]. Table 1 summarizes the anthropogenic emissions in Europe of CO_2, CO, NO_2, SO_2, and NMHC in 1995 as given by the EDGAR 3.2 emission inventory [12].

Combustion of fossil fuel is the major source of most anthropogenic emissions. For example, road transport is the largest contributor to NO_x emissions, whereas burning of sulfur-containing coal in power plants is the main source of SO_2 emissions [13]. The emissions are localized in the highly populated and industrialized areas across Europe. The N/S ratio is about 1 for countries in western Europe and less than 1/3 for eastern Europe [8]. The main sources for anthropogenic NMHC are road transport and solvent use. Anthropogenic NMHC emissions in Europe are about 2 times higher than biogenic NMHC emissions (mostly from forests).

Emissions of particulate matter (PM) can be quantified reliably only for combustion emissions. Berdowski et al. [14] compiled the first inventory for primary anthropogenic emissions of $PM_{2.5}$ (particulate matter with an aerodynamic diameter of 2.5 μm or less) and PM_{10} (particulate matter with an aerodynamic diameter of 10 μm or less) for Europe. The highest emissions are found in densely populated regions such as the Benelux countries, western Germany and northern Italy. Emissions of PM are generally higher in Europe than in the U.S. partly due to a higher number of diesel-powered cars in Europe. For example the black carbon burden from road traffic emissions is twice as high in Europe as in the U.S. [15]. Recently, a first European inventory for selected POPs (persistent organic pollutants) has also been compiled [16].

The emissions in Europe have been relatively constant over the last decade, only in some regions have slight reductions occurred. The economical collapse

Table 1 Anthropogenic emissions in Europe for the reference year 1995 from the EDGAR 3.2 emission inventory [12]

Source	CO_2 (Pg/yr)	CO (Tg/yr)	NO_x (Tg/yr)	SO_2 (Tg/yr)	NMHC (Tg/yr)
Total (fraction of global emissions)	4.7 (18%)	61 (7%)	17 (15%)	27 (19%)	20 (12%)
Fossil fuel	4.6	45	15	20	12
Biofuel	0.01	3	0.1	0	0.4
Industrial processes	0.1	7	1	7	6
Agriculture	0.007	2	0.1	0	0.3
Biomass burning	0.001	3	0.1	0	0.4
Waste handling	0	0.3	0	0	0.6

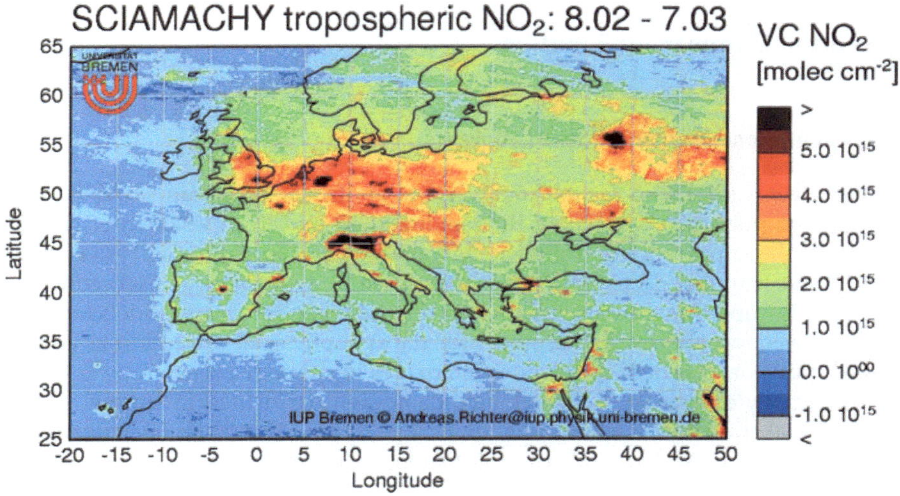

Fig. 1 Mean tropospheric NO_2 vertical column density over Europe for the period August 2002–July 2003 from SCIAMACHY on ENVISAT (provided by A. Richter and J. Burrows, University of Bremen)

in many eastern European countries has led to significant emission reductions in this region, e.g., nitrogen and sulfur emissions decreased by about 25% from 1990–1993 [17]. Since 1994 emissions have also remained relatively constant in eastern Europe. European emissions account for 9–23% of the global emissions, depending on the emitted species (see Table 1).

Satellite data offer a new approach to determine the distribution of emissions. Figure 1 shows a mean distribution of tropospheric NO_2 vertical column densities for Europe for one year derived from the SCIAMACHY (Scanning Imaging Absorption Spectrometer for Atmospheric Chartography) instrument onboard the European Environmental Satellite ENVISAT. The tropospheric NO_2 columns reflect mostly the NO_2 abundance in the atmospheric BL near the sources. The industrialized areas with major emissions can be clearly seen.

The secondary pollutant ozone is photochemically produced by hydroxyl radical oxidation of the precursors CO, CH_4, and NMHC in the presence of NO_x. In the vicinity of highly industrialized areas, the O_3 production is mainly controlled by NMHC, whereas NO_x mainly controls the O_3 abundance in remote regions. Reductions in ozone precursors may also lead to an ozone increase in some very polluted regions since high NO_x levels depress photochemical activity by reducing HO_X in the formation of HNO_3. There is evidence for a reduction in O_3 peak concentrations in recent years in some areas in Europe, e.g., the United Kingdom [18–19] and an increase in O_3 background abundance [20].

3 Export of European Air Pollution

3.1 Typical Transport Mechanisms and Pathways

Only a few studies with the focus on vertical uplifting of pollutants over Europe have been published up to now. Bethan et al. [21] and Kowol-Santen et al. [22] stressed the importance of frontal activity and WCBs for the upward transport of pollutants from the BL to the FT to explain pollution layers they observed in the FT over Europe. Simulations with mesoscale chemical transport models indicate that convection may even dominate over the synoptic scale advection as an upward transport mechanism for nitrogen oxides in summer over Europe [23].

Especially during the EXPORT field campaign performed in summer 2000, the typical transport mechanisms for European pollution were investigated in more detail [24]. Mid-latitude frontal systems associated with WCBs were found to be responsible for the uplift of pollutants into the lower troposphere (LT) and mid troposphere (MT), whereas deep convection (isolated or within frontal systems) transported the pollutants directly to the MT and upper troposphere (UT) [25, 26]. The importance of the pollution transport in thunderstorms over Europe had, in addition, been investigated previously to EXPORT [27–30]. During MINOS in summer 2001 on Crete, it was observed that BL air may even be injected into the lowermost stratosphere by deep convection [31]. However, model studies indicate that WCB lifting and convective uplifting seem to play a relatively smaller role for the export of European pollutants compared to NA and Asia, because European WCBs are more shallow than Asian and North American WCBs [32, 33]. Climatologies on the occurrence of WCBs support these results, since no pronounced maximum of WCB starting points was found over Europe [34, 35].

Instead of WCB lifting and convective uplift, the low-altitude (boundary layer) export of pollutants primarily to the north, but also to the east and south appears to dominate the European pollution transport [32, 33]. Simulations with the global climate model ECHAM and the chemistry transport model EURAD revealed the major export pathways of European pollution at low-altitudes [36, 37]. The Arctic region, Siberia, the Mediterranean area and the British Islands (with a recirculation over the Atlantic, reentering Europe further south at 35–40°N) seem to be the main receptor regions.

During winter and early spring, the pollution outflow from Europe is mainly controlled by the Siberian High [38], which transports large amounts of European surface emissions northward into the Arctic BL and partly cause the Arctic haze [39, 40]. For more details on the pollution transport into the Arctic region the reader is referred to the chapter by Bottenheim et al. in this volume. Eckhardt et al. [41] found that during high phases of the North Atlantic Oscillation (NAO) more European pollution was transported into the Arctic region. During the winter, European CO accumulates to high levels in the Arctic. Occasionally in winter and especially in spring, some of this CO is transported southward at the eastern flank of the Siberian High to the North Pacific [42].

In summer, northwesterly winds are frequent in the LT over Europe. The European pollution outflow to the south was observed during the MINOS experiment in summer 2001 in Crete [43]. Furthermore, an anticyclonic circulation over western Europe in summer can occasionally force the European pollutants to move westwards out over the Atlantic [44–47].

A further important transport mechanism for the export of European pollution is the vertical pumping of air out of deep Alpine valleys to altitudes between 2 and 4 km as studied within the framework of the VOTALP project [48–51]. During fair weather conditions, a heat low may develop over the Alps in summer, leading to additional advection of pollutants from the Po Basin and the Alpine foreland. A 2-year climatology of trajectories with starting points above the Alps indicated a main flow either to the north or to the south [51]. The northward moving trajectories ascended up to 5 km and the southward moving ones descended over the Mediterranean and northern Africa into the BL.

The specific dynamic of pollutants in the Mediterranean basin has been investigated in several projects, e.g., MECAPIP (Meso-meteorological Cycles of Air Pollution in the Iberian Peninsula) (1988–1991), RECAPMA (Regional Cycles of Air Pollution in the West Central Mediterranean Area) (1990–1991), and SECAP (South European Cycles of Air Pollution) (1992–1995) [52]. Diurnal convective cycles over extended land areas (Spanish and Turkish Plateaus) and/or along elevated mountain ridges (Alps and Apennines) may force the pollutants to be uplifted to the MT (~5 km). Further pollutant transport to the UT may be triggered by deep convection primarily over the Apennines.

3.2
Experimental Evidence

3.2.1
Observations at Surface Sites

The main transport pathway of European pollution is to the east. However, up to now, no reports from surface sites in the east quantifying this export of European pollution are available. Less frequent is the export of European pollution to the west (Atlantic), though these pollution plumes are easier to identify. In summer, it has been observed that ozone-rich air masses, which are produced by anthropogenic emissions over the European continent, can be transported episodically to remote ground-based sites in the eastern North Atlantic, e.g., to Mace Head (Ireland) and Izaña (Tenerife) [44, 53, 54]. Long-term measurements at Mace Head performed by Jennings et al. [55] indicated increasing CO and black carbon aerosol concentrations when the flow came from the United Kingdom and/or the European continent. In one of these specific pollution events, during a two-day period in May 1992, the CO concentration varied between 180–280 ppbv.

The seasonal variation of the influence of European pollution at Mace Head from 1990 to 1994 was investigated in detail by Derwent et al. [56]. Monthly average ozone and carbon monoxide concentrations were calculated for these events of pollution export from continental Europe. In winter CO increased by 60–65 ppbv and the ozone levels were depleted by 2–10 ppbv relative to background air. During summer, CO was only enhanced by 20 ppbv and in addition ozone was enhanced by 2–5 ppbv above background concentrations due to photochemical ozone production. Further, long-term measurements carried out by Simmonds et al. [3] indicated that on an annual basis, continental Europe is considered to be a net ozone sink of about 2.6 ppbv for Mace Head. The wintertime ozone destruction dominates over the summertime ozone production. Experimental evidence of the export of European pollutants out over the ocean or to other continents is rare. Though, ground-based measurements in the Arctic region indicate a strong influence of European pollution on the formation of Arctic haze as described in more detail in the chapter by Bottenheim et al. Evidence for a pronounced pollution outflow from Europe to the east is described in the chapter by Pochanart et al. Measurements at Mondy, a mountain site close to Lake Baikal, indicated elevated CO and O_3 when air masses of European origin passed the site.

3.2.2
Observations from Airborne Measurements

Airborne measurements of the European pollution export are rare as well. One major reason for this lack of observations is the difficulty of performing measurements over remote regions like Siberia, northern Africa, and the Arctic, which are the main receptor regions of the European pollution. In spite of these barriers, some evidence of pollution outflow across the European borders has been reported in the literature.

For example during the NARE 1993 field campaign, pollution layers of different origin were identified during profile measurements with the UKMO Hercules off the coast of Portugal on August 31, 1993 [46]. The lower pollution layer (up to 2.5 km), that contained elevated levels of NO_y (1 ppbv) and CO (180 ppbv), originated from continental Europe according to the trajectory analyses.

Traces of European pollution were also observed over the North Atlantic during ACE-2 (Second Aerosol Characterization Experiment) in summer 1997 when the Azores High expanded over northwestern Europe, resulting in an easterly flow south of the pressure center [47]. Especially strong SO_2 signals were recorded onboard the UKMO Hercules aircraft when the air masses originated from the west coast of Spain and Portugal, where a large number of power plants are located.

Evidence for a pronounced pollution outflow from Europe to the east has been obtained from the NOXAR measurements with a commercial airliner in the period 1995–1997 [57]. Large-scale NO_x plumes were frequently penetrated at cruising levels (UT) on flights between Europe and Siberia in summer (see Fig. 2). These plumes were observed downwind of thunderstorms and frontal systems which partly originated from continental Europe.

The first airborne field experiment with the major objective of investigating and quantifying the export of pollutants from Europe was the EXPORT campaign

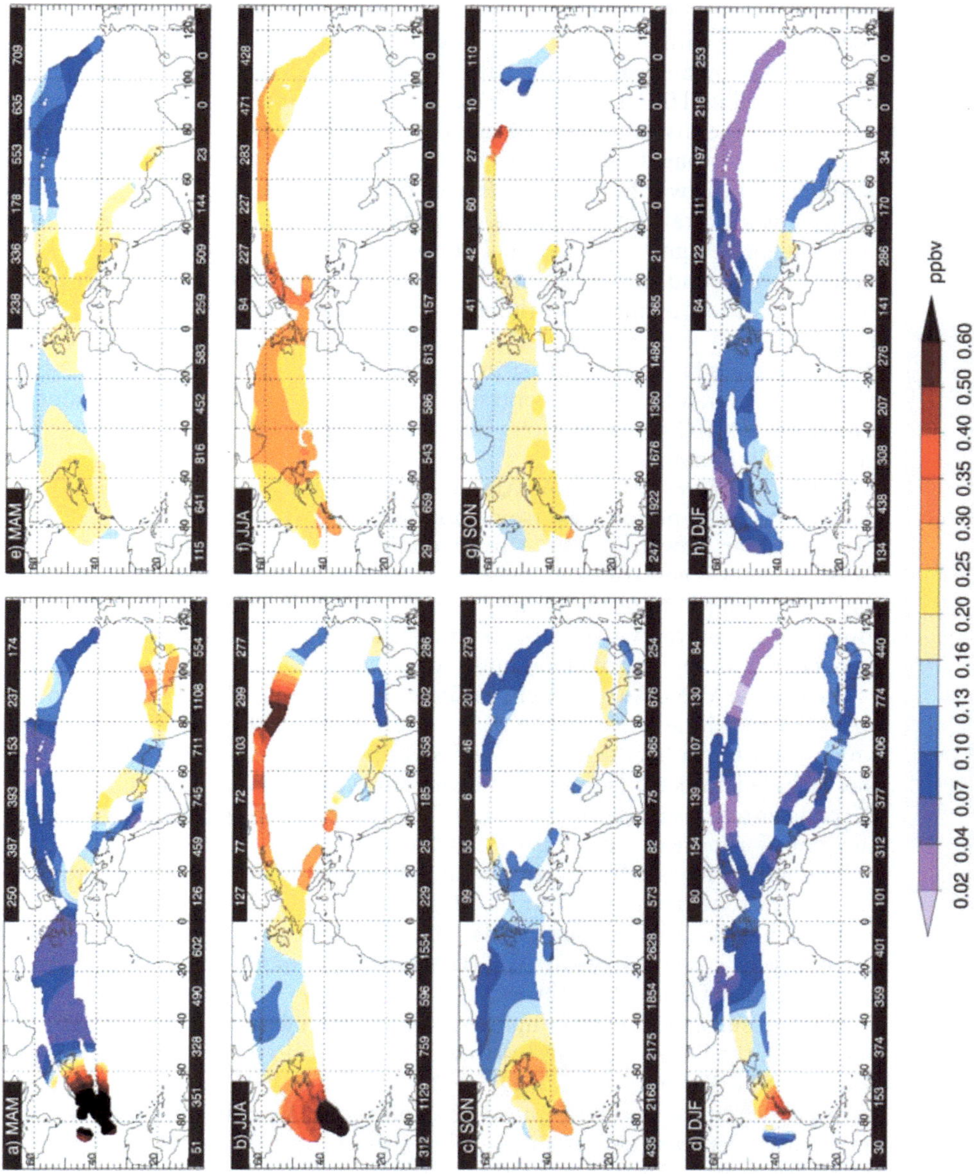

Fig. 2a–h Seasonal NO_x distribution in the upper troposphere (**a–d**) and lower stratosphere (**e–h**) for spring (MAM), summer (JJA), autumn (SON), and winter (DJF) based on the airborne NOXAR measurements [57]. (Copyright 2001, American Geophysical Union. Reproduced by permission of American Geophysical Union)

performed in summer 2000 [24]. The flights were carried out over Central and Eastern Europe with the British Hercules C-130, the French Mystère and the German DLR Falcon aircraft. During a flight over Poland on 9 August 2000, specific halocarbons like HFC-134a, and -152a, HCFC-141b and -142b (all markers of anthropogenic emissions) were elevated in a layer between 1 and 2 km. This pollution layer originated over western Europe according to trajectory analyses and was advected eastwards at low levels over Poland. Furthermore, the Falcon and the Hercules aircraft investigated a complex pollution situation in the FT over Germany on 2 August 2000 [26]. Pollution from western Europe was slowly advected upwards and eastwards with the passage of a cold front and the associated WCB. In addition, a number of thunderstorms were triggered ahead of the cold frontal line. As a result elevated CO mixing ratios of 100–140 ppbv were measured in several layers in the MT and UT in the thunderstorm outflow and WCB region (see Fig. 3). Even if the CO concentrations were similar in all observed pollution layers, the NMHC concentrations were distinctly higher in the fresh outflow from the thunderstorms [58].

During the MAP (Mesoscale Alpine Program) field phase in autumn 1999, elevated layers of polluted air from continental Europe were observed with an airborne lidar leeward of the Alps extending over the Adriatic Sea [59]. On 24 September, an aerosol layer spread over >200 km at an altitude of 4 km which was attributed to mountain venting over the Alps followed by advection to the south. Furthermore, evidence for new particle formation was found in the layer associated with regions of cloud outflow.

During the MINOS experiment in Crete in summer 2001 elevated ozone was observed over large areas in the BL by aircraft [43]. CO tracer simulations, in which the CO contribution from each continent of the northern hemisphere was included, indicated a major influence from western and eastern Europe on CO concentrations in the BL over Crete. The reader is referred to the chapter by Lawrence for more details on the MINOS results.

Polluted air masses from continental Europe may even contribute to the tropospheric ozone maximum over the Middle East region in summer as observed by airborne MOZAIC measurements [60]. Simulations with a global 3-D model of tropospheric chemistry (GEOS-CHEM) revealed that the main flow in the LT actually comes from the northwest.

More recently, the impact of uplifted polluted air masses on the trace gas composition in the FT over Europe was investigated within the CONTRACE project with the DLR Falcon aircraft [61]. On 22 November 2001 European emissions were lifted into the lower FT ahead of a cold front. A pollution plume was observed over Germany between 2 and 4 km altitude heading southeastwards. The plume contained elevated CO (150 ppbv) and NO_y (6 ppbv). Ozone was low (20 ppbv) due to the titration by NO in late fall.

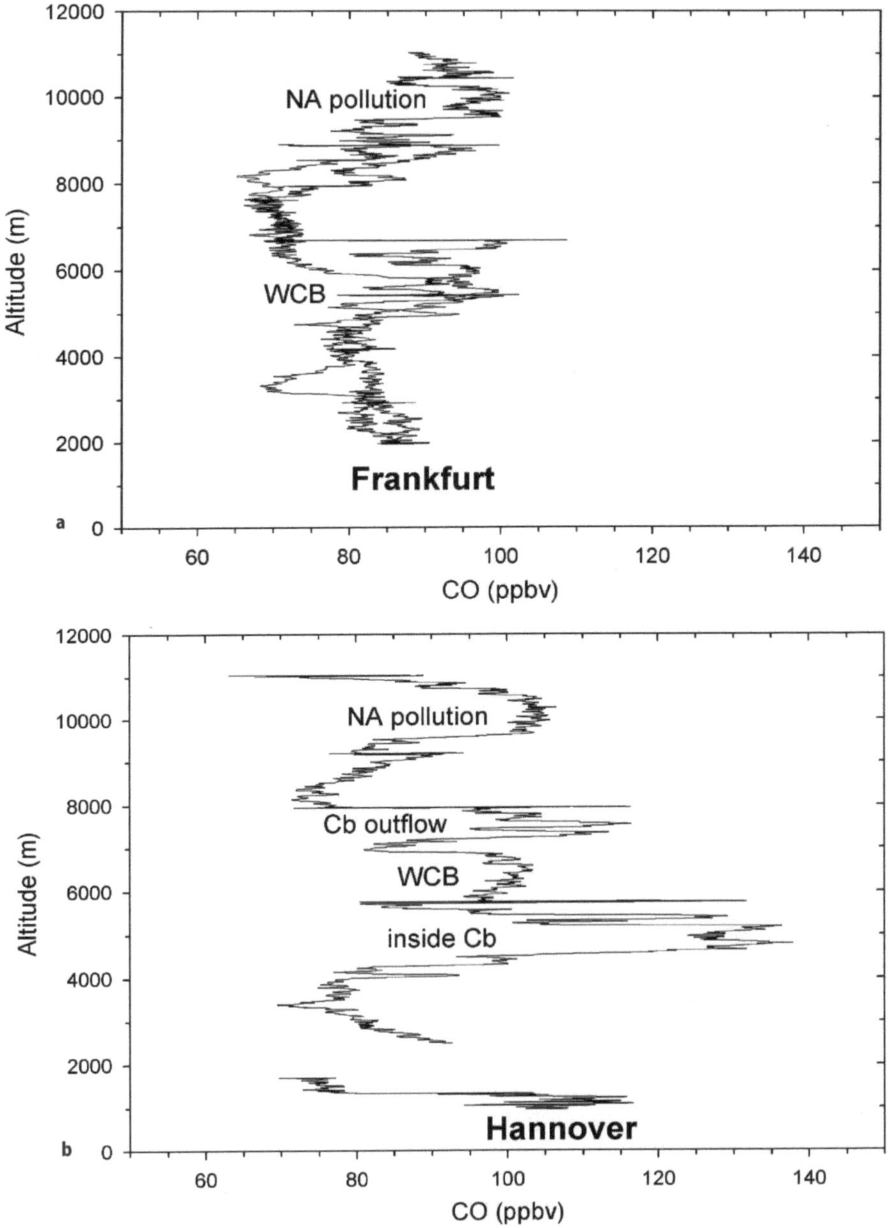

Fig. 3 Vertical CO profiles measured on the DLR Falcon on 2 August 2000 as part of EXPORT. A warm conveyor belt (WCB) with enhanced CO mixing ratios was penetrated in the lower and mid troposphere over Frankfurt **a**, and Hannover **b**. In the uppermost troposphere a layer with enhanced CO advected from North America (NA) was observed. At Hannover also nearby thunderstorms (Cb) embedded in/ahead of the WCB were penetrated [26]. (Copyright 2003, GSF-Forschungszentrum für Umwelt und Gesundheit. Reproduced by permission of GSF)

4 Import of Air Pollution from North America

4.1 Typical Transport Mechanisms

In this section, the typical transport mechanisms for the North American pollution heading east are briefly described. For a further description the reader is referred to the chapter by Cooper and Parrish.

Data analyses from the NARE 1993 summer intensive by Angevine et al. [62] showed that lofting due to a sea breeze front is a common transport process for the export of North American pollutants into the North Atlantic Ocean. Thereby, the polluted air mass becomes stable stratified and spreads out in a thin layer 1–2 km above the ground. Depending on the meteorological situation (e.g., existence of a Icelandic low pressure system), this low-level pollution layer can be transported all the way to Europe [63].

The major processes responsible for vertical uplifting of polluted air masses to higher altitudes over the United States are synoptic-scale WCBs associated with frontal systems [34, 35, 64–68], and meso-scale deep convection [69–74]. Depending on the lightning activity in thunderstorms large amounts of NO may be produced by lightning and injected into the UT and lower stratosphere region. Such large-scale NO_x and NO_y plumes have frequently been observed over the North Atlantic at cruising altitude by commercial airliners equipped with scientific instrumentation [57, 75–77]. Further, Newell et al. [78] and Thouret et al. [77] observed the polluted layers over the North Atlantic most frequently between 5 and 7 km altitude. The mean thickness of the layers was estimated to be 0.5–1 km. Furthermore, in the case where the pollutants are transported all the way up to the UT, they can easily enter into the jet stream and be rapidly transported to Europe. An airstream climatology carried out by Stohl [34] showed that Europe frequently receives the outflow from WCBs that originate on the seaboards of eastern NA.

Airborne measurements during the 1996 AEROCE experiment over the North Atlantic Ocean indicated that polluted air masses can be uplifted by prefrontal convection over the North American continent and then transported eastwards out over the North Atlantic Ocean into another subsiding postfrontal air mass with enhanced ozone [79]. This mechanism can explain why in some cases high-ozone episodes over the North Atlantic can be associated with low humidity and high CO. Stohl and Trickl [65] reported on a similar case where the vertical ozone distribution over Europe was inverted. The North American pollution layer was found in the UT at the same time as ozone of stratospheric origin was observed in the LT. It was shown that the pollution outflow from a WCB can override the dry intrusion airstream of a preceding frontal system. Further, Parrish et al. [66] described a NARE case where ascending tropospheric air masses and descending stratospheric air masses mixed. They observed that the region of the stratospheric intrusion caught up with the polluted air lifted in the frontal zone.

4.2 Typical Transport Pathways and Timescales

The pathway of the outflow of anthropogenic pollution from NA into the North Atlantic and the further transport to Europe is mainly controlled by a semi-permanent cyclone over Iceland and a subtropical anticyclone located over the Azores [80]. This semi-permanent situation causes confluence and a westerly flow between the two pressure centers. The strength of the flow depends on the pressure gradient between the Iceland low and the high pressure system over the Azores. If a strong jet stream develops, which is the case especially in the winter season, the air can be channeled over to Europe very rapidly within a few days. Over Europe typically the transport path is anticyclonic which causes subsidence of the polluted North American air masses.

Stohl et al. [33] used the Lagrangian particle dispersion model FLEXPART to study typical pathways and timescales of intercontinental pollution transport during a 1-year period. In general, the cross-Atlantic transport of a CO tracer took about 4 days. Typically, the North American pollution layer first arrived over northwestern Europe in the MT or UT. Thereafter, there was a clear southward and downward transport of the North American CO tracer into the Mediterranean region.

A 15-year climatology of WCBs carried out by Eckhardt et al. [35] further showed that the North Atlantic Oscillation (NAO) strongly influenced the pathway of polluted air masses transported over the Atlantic. During high NAO index phases the WCB outflow was located more northerly (about 10°) and easterly (about 20°), and lifted to higher altitudes. The typical timescale for the uplift associated with a WCB, the ascent from the BL to the UT, was two days. Thereafter, the WCB trajectories typically moved anticyclonically and descended. After 4–6 days travel over the North Atlantic the WCB trajectories were most frequently found over the Mediterranean area.

In a case study carried out by Wild et al. [46] during the NARE 1993 campaign, similar patterns for the cross-Atlantic transport of North American pollutants were found, as described above from climatology and model studies. Ten-days forward trajectories were calculated from a region at the North American east coast (Gulf of Maine) where the King Air aircraft had intercepted a pollution plume on August 20. The trajectory path led over Nova Scotia, southern Greenland, north of Iceland, west of Ireland and finally over the west coast of Portugal where further aircraft measurements were performed to investigate the aged polluted air mass. During the first three days of the cross-Atlantic transport the polluted air mass was uplifted rapidly from 840 hPa to 260 hPa due to a cyclonic flow. As the flow changed to anticyclonic (Greenland-Iceland area) the polluted air mass started to descend slowly. After 10 days, the initial altitude was almost reached again (800 hPa). In this case study of a single pollution event, the transport across the Atlantic to the British Islands took 3.5 days.

4.3 Experimental Evidence

Experimental evidence for intercontinental pollution transport from NA to Europe is provided by in situ and remote (lidar) measurements at surface sites, and observations from aircraft and satellites. The findings from these observations are summarized next.

4.3.1 *Observations at Surface Sites*

Signatures of intercontinental pollution transport from NA to Europe have been observed since the end of the 1980s at remote stations in the central and eastern North Atlantic and along the European west coast. Preferred ground-based stations for these kind of observations are: (1) Santa Bárbara (38.7°N, 27.3°W, 1010 m), Terceira, on the Azores in the central North Atlantic, (2) the mountain station Izaña (28.3°N, 16.5°W, 2370 m) on Tenerife, Canary Islands, (3) Mace Head (53.3°N, 9.9°W) on the Irish west coast, and (4) Porspoder (48.0°N, 4.8°W) in Brittany on the French west coast. At two of these stations (Izaña and Mace Head) long-term measurements have been carried out since the beginning of the program AEROCE in 1987. In 1988, the TOR program was established as part of a collaborative European effort to investigate the ozone budget over Europe.

Several observations at the TOR station Izaña and on the island of Terceira in the Azores have indicated that North American pollution may reach these sites [45, 81–85], however, clear evidence is still missing. Therefore, the findings will not be discussed here in more detail.

At the Irish station Mace Head, where long-term records of CO and O_3 are available, mainly European but also a few North American pollution events have been observed [44, 53, 86–87]. During a five-year period (1990–1994) Derwent et al. [56] identified only five possible events of long-range transport from NA. The mean contribution from NA was almost negligible (1–3 ppbv excess O_3 during winter and spring, and less than 5 ppbv CO during summer and autumn). Derwent et al. concluded that the pronounced plume of enhanced ozone and CO frequently observed in the western North Atlantic [4, 88] probably almost never intersects with Mace Head or the plume merges with the background before it reaches the site.

One particular pollution event observed at Mace Head has been described in more detail by Jennings et al. [55]. Simultaneous measurements of CO and CN over the time period August 1991 and August 1992 indicated possible pollution transport from NA to Mace Head. During these events, a distinct positive correlation between CO and CN number density was observed. The average CO concentration measured during one event in April 1992 was 174 ppbv (range 165–185 ppbv) and the CN concentration varied between 500 and 1000 particles cm^{-3}. Trajectory analysis indicated lifting of these air masses over the North American continent. Before the air masses reached Mace Head they descended from the MT (5 km). Furthermore, measurements of inert CFCs (chlorofluorocarbons) carried out at Mace Head indicated some traces of North American pollution

[87, 89]. In comparison to European pollution plumes the CFC peaks caused by North American pollution events were one order of magnitude smaller.

The TOR station in Porspoder on the French west coast is frequently exposed to aged European emissions which makes it difficult to quantify the contribution from North American pollution. Long-term records of O_3, PAN, NMHCs and NO_x in combination with back-trajectories have been used to study the long-range transport to Porspoder [90, 91]. In the first studies no clear signatures from North American pollution plumes could be observed. However, Fenneteaux et al. found that the influence of North American emissions on the European west coast was strongest in spring and reached a minimum in summer. The meteorological situation in spring favors the transport of freshly polluted North American plumes to Europe. During these events high levels of ozone and PAN have been observed at Porspoder which suggests that the original composition of the plumes remained during the transatlantic transport, and the effect of deposition and mixing was only minor.

Discussion A 1-year simulation with the particle dispersion model FLEXPART by Stohl [34] may explain why observations of downward transport of North American pollution plumes to European surface sites are rare. For fresh North American tracers (e.g., 4–6 days) the concentrations are highest in the MT and UT (5–8 km) over Europe and north of 60°N. Older tracers are transported southward and downward over the Alps and Pyrenees, and finally reach the Mediterranean region, where only few measurement stations are located (none of the mentioned ones). Recent studies of one selected North American pollution event (19–22 November 2001) showed the typical descent of the pollution plume over the Alps [92–93]. Routine measurements from the GAW (Global Atmospheric Watch) mountain station Zugspitze (47.4°N, 11.0°E, 2962 m) in the Alps indicated a pronounced influence of the North American plume during two days (20–22 November 2001). The measurements in the FT clearly showed a positive O_3-CO correlation and up to 52 ppbv O_3 and 170 ppbv CO (see Fig. 4). In comparison, typical background concentrations in November 2001 were 40 ppbv O_3 and 120 ppbv CO. The records from a second GAW station (Hohenpeißenberg, 47.8°N, 11.0°E, 976 m), located close to the Zugspitze but in the BL, also showed signatures from the North American plume. The effect of the North American pollution plume on the concentration of air pollutants in the BL was quite different from the Zugspitze. Due to mixing with the BL air the positive O_3-CO correlation could no longer be observed so clearly. In comparison to the typical background concentrations at Hohenpeißenberg, the North American pollution plume appeared rather "clean". Though, enhancements in O_3 by 15% (corresponding to 4 ppbv) and SO_2 (also by 15%) in comparison to the average mixing ratios measured in November 2001 were observed. For CO and NO_y reversed conditions were observed – a decrease by 15% and 65%, respectively, in comparison to the monthly means.

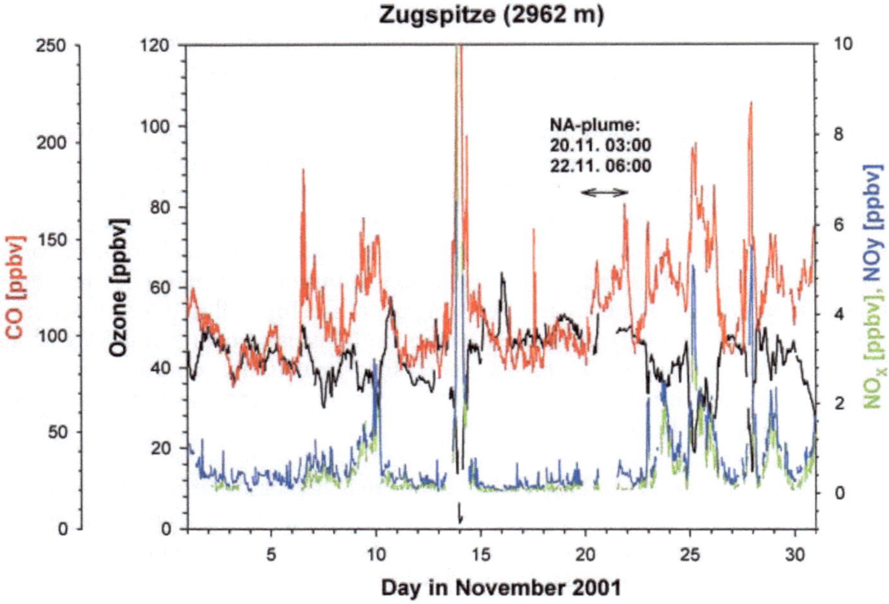

Fig. 4 Trace gas measurements from the mountain site Zugspitze (2962 m) in Germany during November 2001: mixing ratios of CO (*red*), O_3 (*black*), NO_x (*green*), and NO_y (*blue*) are shown. In the period 20–22 November a pollution plume from North America (NA) reached the site (provided by H. Huntrieser, DLR-Oberpfaffenhofen, based on data from E. Scheel, Institut für Meteorologie und Klimaforschung, Forschungszentrum Karlsruhe)

4.3.2
Observations with Ground-Based Lidar

Within the EC-project VOTALP, measurements with an ozone lidar were carried out in 1996–1997 in Garmisch-Partenkirchen (47.5°N, 11.1°E, 740 m), Germany to study the stratosphere-troposphere exchange of ozone. On 29 May 1996, an event was monitored with the ozone lidar where the stratospheric intrusion penetrated down to the ground [94]. During the whole event, which lasted from May 28 to June 1, secondary ozone tongues were also observed in the MT and UT. Results from trajectory calculations indicated that some of these descending ozone pockets could be related to the transport of polluted air masses from the North American BL [95].

A second important tropopause fold event over Garmisch-Partenkirchen was observed in the period 28–29 May 1997. Detailed studies with trajectories (FLEXTRA), a particle dispersion model (FLEXPART), and potential vorticity analyses confirmed the stratospheric origin [65]. In the same period secondary ozone tongues (80–100 ppbv) were observed in the UT behind the fold. Trajectory analyses indicated that these ozone rich air masses originated from the polluted BL over eastern NA during a high ozone episode with photochemical ozone production. The low potential vorticity and high water vapor content also supported these findings. To confirm the North American origin, FLEXPART model runs with two passive tracers, a stratospheric ozone tracer, and a tracer for North American emissions, were performed. The results clearly indicated a strong North American ozone source which was also supported by several MOZAIC flights passing the uplifted pollution plume over the central North Atlantic. This textbook example is one of the first studies published which without any doubt identifies imported ozone from NA over Europe.

Discussion Stohl and Trickl [65] concluded that long-range transport of pollutants (ozone) from NA to Europe may be more important than assumed at this time. However, since ozone measurements in the UT are limited to a few lidar, ozonesondes, and airborne measurements, monitoring of North American ozone layers over Europe is difficult. Further the use of trajectories and transport models is essential to distinguish between transport from the stratosphere and transport from the polluted BL. In many cases it can be difficult to distinguish the two ozone sources since the behavior over Europe is similar (descending air masses from the UT).

4.3.3 *Observations from Airborne Measurements*

During the NARE 1993 summer intensive, the American activities were complemented by measurements made for the European project OCTA [96]. Flights were performed along the European west coast and across the Atlantic with the UKMO Hercules C-130 aircraft. Wild et al. [46] examined the origin of tropospheric layers with enhanced ozone observed on August 31, 1993 about 150 km west of Portugal (39.0°N, 11.2°W). The ozone layer located around 6 km originated from the North American BL (New York/Boston area) 7 days earlier, according to back trajectory calculations. Ozone increased from 40 to 57 ppbv, CO from 80 to 105 ppbv, NO_y from 350 to 600 pptv, and the aerosol loading doubled.

In 1997 NARE continued and within the European framework ACSOE (Atmospheric Chemistry Studies in the Oceanic Environment) the Hercules C-130 aircraft operated in the region of the Azores [97]. Widespread layers of polluted air were observed during all flights, especially in the MT between 3 and 8 km altitude. On 14 September 1997, a flight was made south of the Azores in the outflow from the Hurricane Erika. Elevated concentrations of typical anthropogenic tracers like O_3 (100 ppbv), CO (120 ppbv), and NO_y (1 ppbv) were seen in a layer between 5 and 7 km including highly positive correlations between these trace gases. Back trajectory calculations indicated that these polluted air masses were uplifted from the BL 5–6 days earlier by a frontal system moving out over the southeastern U.S. and then transported into the path of the hurricane.

On October 14, 1993 during the POLINAT (Pollution in the North Atlantic Flight Corridor) field experiment, a pollution layer was observed more by coincidence in the UT north-west of Ireland with the German DLR Falcon [98]. During the flight section in the polluted air masses (56°N, 10°W, at 8.8–9.5 km altitude) enhanced SO_2 (3 ppbv), acetone (3 ppbv), and CN ($\geq$6 nm) concentrations (1500 cm^{-3}) were observed. It was suggested that acetone originated directly from the emission source and/or was produced by the photooxidation of emitted hydrocarbons (e.g., propane). Further, CO_2 increased by about 6 ppmv from 358 to 364 ppmv. Trajectory analysis indicated a clear influence from the BL over north-eastern U.S., where the polluted air masses were uplifted 4–5 days prior to the measurements.

Scientific long-term measurements onboard commercial aircraft traveling between Europe and the U.S. (e.g., CARIBIC, MOZAIC, and NOXAR) provide further evidence that occasionally large-scale pollution plumes move out from the North American continent eastward and impact the trace gas composition in the FT significantly. The horizontal distribution of CO sampled along the CARIBIC (Civil Aircraft for Regular Investigation of the Atmosphere Based on an Instrument Container) flights routes clearly indicate a CO maximum downwind of NA [99]. Measurements within the framework of NOXAR give further evidence that occasionally large-scale NO_x plumes move out from the North American east coast [57, 75, 76]. Especially in summer, ozone is also elevated in the outflow air masses. Climatological studies of the MOZAIC data set show similar results for NO_y as described above for NO_x. An outflow region with elevated NO_y concentrations has been observed in the UT region over the western North Atlantic by, e.g., Volz-Thomas et al. [100].

Results from SONEX (Subsonic Assessment Ozone and Nitrogen Oxide Experiment) with NASA's (National Aeronautics and Space Administration) DC-8 aircraft in autumn 1997 revealed the importance and advantage of high-resolution model-derived winds in comparison to traditional trajectory analysis to interpret chemical signatures along the flight track [73]. On October 29, 1997 a flight starting from the Azores (39°N, 27°W) heading southwest showed a variety of trace gas signatures. Layers with enhanced NO (1.5 ppbv), NO_y (2 ppbv) and CN (30,000 cm^{-3}) in the UT could be clearly attributed to air masses that were influenced by lightning 1–2 days earlier over the eastern Gulf of Mexico and off the southeast coast of the United States. In the MT a layer containing enhanced CO (>90 ppbv), ozone (>75 ppbv), PAN, H_2O, C_2H_2, H_2O_2, CH_3OOH was traced back to the heavily polluted east coast of the U.S. where convective clouds (without lightning) lofted the surface pollution. In-between a mixed layer with signatures from both sources was encountered.

During the EXPORT experiment carried out in July/August 2000 over Central Europe airborne measurements were performed with three aircraft (the French Mystère, the UKMO C-130, and the German DLR Falcon). On 9 August, 2000 halocarbon concentrations (e.g., HFC-134a and -152a, HCFC-141b and -142b which are typical markers of anthropogenic pollution) were enhanced both in the LT and UT on a flight over Poland [24]. The lower layer could be attributed to European outflow and the upper layer in 8 km to North American emissions. Further, on 10 August, 2000 an enhanced ozone layer (70–90 ppbv) was observed in 5–6 km altitude by the lidar on the French Mystère aircraft between Munich and Prague [101]. The North American origin of the layer was confirmed by back trajectories [24]. A MOZAIC flight descending over Frankfurt on the same day also measured elevated ozone concentrations (70–90 ppbv) in a layer between 5 and 7 km [101]. Further evidence that signatures of North American emissions may reach Europe was found on 2 August 2000 (see Fig. 3). A European WCB was investigated over Germany and above the uplifted European pollution, a pollution layer containing North American emissions was observed with the Falcon aircraft [26]. The layer was found in the uppermost troposphere (9.5–11 km) and, e.g., CO was elevated (~110 ppbv).

The impact of long-range transport on the trace gas composition in the eastern Mediterranean area was investigated in great detail during the MINOS experiment in August 2001 including flights with the DLR Falcon [43]. Average backward-trajectories for the MINOS field intensive period indicated a major flow from the west in the MT and UT (4–12 km). Occasionally these air masses advected from the west contained elevated concentrations of CO indicating an origin from anthropogenic sources (e.g., NA).

Recently, during the CONTRACE field experiment in November 2001, it was possible to guide the DLR Falcon aircraft very precisely into aged North American pollution plumes over Europe during several flights [61, 102]. The flight planning was based on two different chemical weather forecasts (MATCH-MPIC and FLEXPART) as documented by Lawrence et al. [103] and Stohl et al. [93]. On 19th November several vertical profiles were flown between Oslo and Stockholm indicating a polluted CO-layer located in the LT and MT with CO mixing ratios reaching up to 170 ppbv. Surprisingly for the winter season, a positive O_3/CO correlation was observed in the plume and the O_3 mixing ratio increased from 43 to 53 ppbv ($\Delta O_3 = +10$ ppbv). Comparisons with O_3 measurements from a mountain station in eastern NA, where the passage of the lifted pollution plume was observed, showed enhanced O_3 mixing ratios in the same range (50–60 ppbv). Most likely O_3 had already been produced photochemically in the polluted BL over eastern NA and not in transit over the North Atlantic. On 22 November a more complex chemical weather situation was predicted for Central Europe with lifting of European emissions to the LT ahead of an approaching cold front and simultaneously, the advection of a pollution plume from eastern NA in mid tropospheric layers. Similar CO mixing ratios were observed in both plumes making it difficult to distinguish the two plumes without additional information from other trace gases such as NO_y and O_3 (see Fig. 5). The European pollution plume was characterized by large enhancements in the CO (150 ppbv) and NO_y (6 ppbv) mixing ratios. The NO_y/CO ratio was 0.135 which is a typical value for fresh emissions. In comparison, the NO_y/CO ratio for the North American pollution plume was 0.010 indicating a tracer age of 4 days according to Stohl et al. [102]. The observed CO and NO_y mixing ratios in this plume were 160 ppbv and 1 ppbv, respectively. The two plumes were also characterized by very different O_3 to CO relationships (see Fig. 5). In the plume from NA, a positive O_3/CO slope was observed similar to the values as described above for the 19 November plume. The European plume showed a strong negative O_3/CO relationship with O_3 mixing ratios reaching down to 20 ppbv. In addition, the SO_2, and acetone mixing ratios were especially enhanced in the pollution plume from NA (2–3 ppbv SO_2 and 5–6 ppbv acetone).

Discussion Airborne measurements clearly show a distinct outflow of pollutants from NA to the east. The rather thin pollution layers (~1 km) extend over several 100 km and have been most frequently observed in the MT. Most airborne

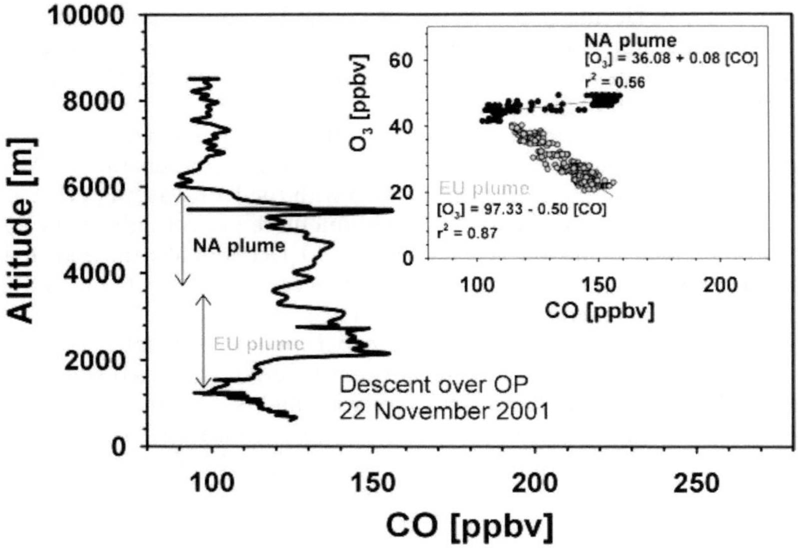

Fig. 5 Aircraft profile measurements over Oberpfaffenhofen (OP) in southern Germany on 22 November 2001. Two polluted layers with different origin (European/North American) were penetrated with the DLR Falcon aircraft as part of CONTRACE. In the European (EU) and North American (NA) pollution plume negative and positive O_3/CO correlation was observed, respectively, as shown in the upper right [102]. (Copyright 2003, GSF-Forschungszentrum für Umwelt und Gesundheit. Reproduced by permission of GSF)

observations arise from measurements over the North Atlantic Ocean. In comparison, measurements over the European continent are rare. Over the continent, the well-defined structure of the plumes, as maintained during the transport over the Atlantic in the jetstream, dissolves when the plumes approach new frontal systems and mix with uplifted European pollution. For the detailed prediction and analysis of the fine structure of the North American pollution plumes entering Europe, high resolution back trajectories and tracer forecasts are essential tools.

4.3.4
Observations from Satellites

Measurements from space with satellites are an excellent tool for studying the large-scale spreading of air pollution between continents. In comparison to airborne, ground-based and lidar measurements the monitoring of intercontinental pollution transport from satellites is a rather new topic in atmospheric science. Since the first estimates of the global tropospheric ozone distribution from space by using a combination of TOMS (Total Ozone Mapping Spectrometer) and SAGE (Stratospheric Aerosols and Gases Experiment) data in the beginning of the 1990s it was obvious that on average a large scale ozone plume extends from eastern NA into the central North Atlantic [105]. The cause of the ozone plume could partly be attributed to ozone production from industrial emissions [106]. A pronounced tropospheric ozone plume over eastern NA has also been observed in a recent study by Fishman et al. [107] which combines TOMS and SBUV (Solar Backscatter Ultraviolet instruments) measurements. Creilson et al. [108] used the same technique to estimate variations in tropospheric ozone and found an increase in ozone over Europe during positive phases of the NAO (North Atlantic Oscillation) when a direct pathway for the pollution outflow from North America to Europe is prominent. Further, data from the MOPITT (Measurements of Pollution in the Troposphere) instrument on the Terra satellite, which measures CO, give further evidence of the anthropogenic origin of the large-scale plume observed over the North Atlantic [109].

Recently, one rare event of NO_2 transport from NA to Europe observed in the period 7–11 November 2001 by GOME (Global Ozone Monitoring Experiment) has been reported by Stohl et al. [110]. The average lifetime of NO_x in the LT is less than 1 day which explains why on a climatological basis the NO_2 distribution is mainly located around the urban centers and almost no advection with the main wind direction is visible [111]. Since intercontinental pollution transport from NA to Europe normally takes about a week, it was not expected to observe this transport in tropospheric GOME-NO_2 data. However, Stohl et al. [110] reported on an extremely fast event where the pollution transport across the

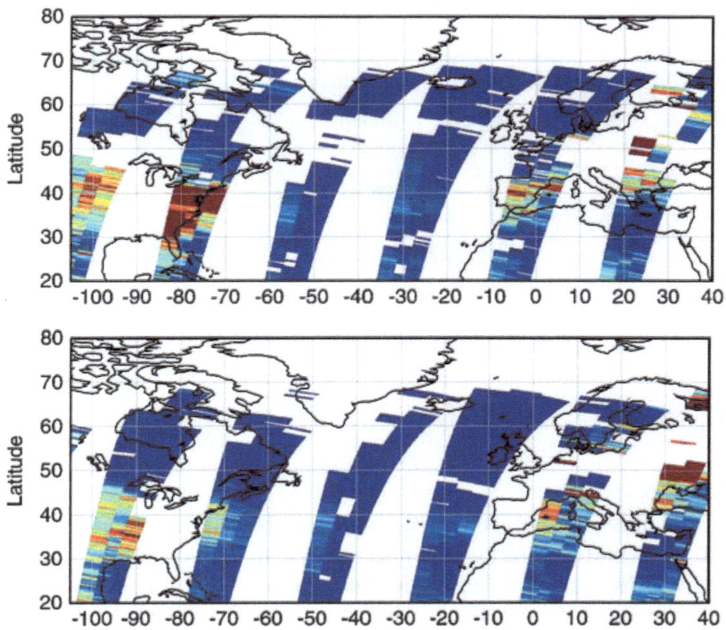

Fig. 6a–e Tropospheric vertical columns of NO_2 (in 10^{15} molecules cm^{-2}), retrieved from GOME satellite data on **a** 7, **b** 8, **c** 9, **d** 10, and **e** 11 November 2001. In white regions data are missing, mainly because of >50% cloud cover [110]. (Copyright 2003, European Geosciences Union. Reproduced by permission of European Geosciences Union)

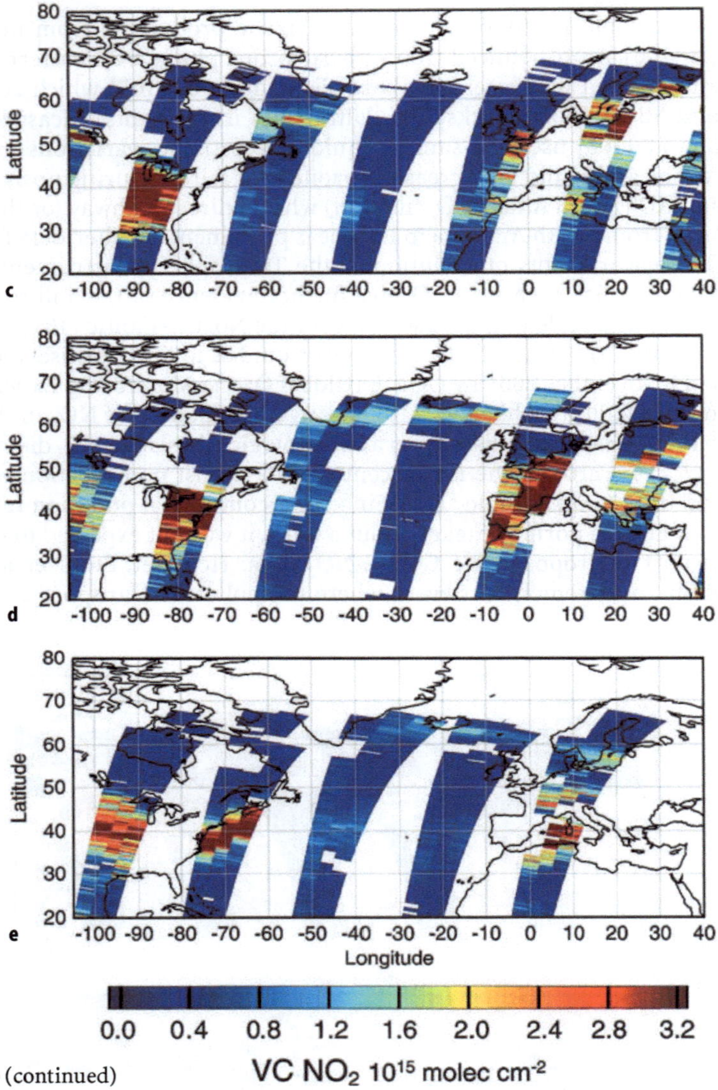

Fig. 6c–e (continued)

Atlantic took only about one day. This explains why this event could be monitored by GOME. The fast transport was generated by a so-called "bomb", a low pressure system which deepens and intensifies very rapidly [112]. The bomb created an "express highway" for the pollution which was uplifted by a WCB over the North American east coast on 7 November 2001. In Fig. 6 the temporal evolution of the NO_2 plume is shown by observations from GOME. Further, an excellent agreement with the tracer transport model FLEXPART is visible (Fig. 7) which also confirms the anthropogenic origin of the NO_2 plume. On 9 November, a GOME

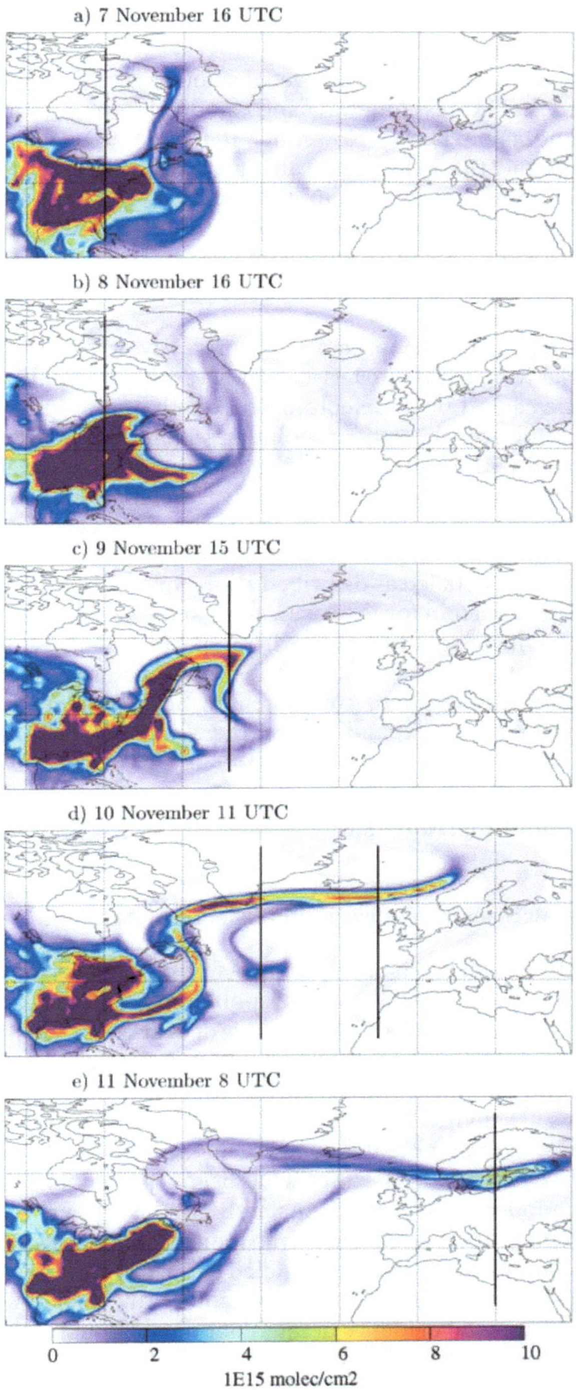

Fig. 7a–e Total vertical columns of the FLEXPART NO_x tracer (in 10^{15} molecules cm^{-2}) on **a** 7, **b** 8, **c** 9, **d** 10, and **e** 11 November 2001 [110]. (Copyright 2003, European Geosciences Union. Reproduced by permission of European Geosciences Union)

signal of enhanced NO_2 was observed over Newfoundland. One day later, the NO_2 filament entered the "express highway" and extended from Greenland across the Atlantic to Scandinavia. On 11 November the main NO_2 plume was located between Iceland and the Baltic Sea. During the transport across the Atlantic the pollution plume descended from 4–6 km to 2–4 km. Due to the general descending motion of the air mass, clouds evaporated and exposed the plume to the GOME instrument. This situation favored the measurements because GOME cannot measure through thicker clouds.

Discussion GOME satellite observations impressively show that short-lived trace species like NO_2, emitted over eastern NA, can survive the fast transport over the Atlantic caused by a meteorological bomb and even be detected over eastern Europe. However, the uplift of pollutants preferably takes place in cloudy regions associated with frontal systems which in most cases makes it difficult to detect this process from space. Reliable measurements can only be obtained for clear skies or when thin clouds, e.g., cirrus cover the sky. Therefore, in most cases it is probably only possible to detect fragments of the pollution plumes from space and not as detailed as described above for an exceptional event. Tracer models are an essential tool for the verification and interpretation of the patchy measurements of single pollution events from space. In contrast to the problems with the detection of single pollution events, average pictures of pollution plumes have successfully been derived from satellite measurements, e.g., the North American pollution plume expanding over the western North Atlantic.

5
Import of Air Pollution from Asia

5.1
Typical Transport Mechanisms, Pathways, and Experimental Evidence

One important transport mechanism for the Asian pollution is the uplift in WCBs as described already for North American pollutants in Sect. 4.1. A 15-year climatology study of WCBs carried out by Eckhardt et al. [35] showed that the most frequent WCB starting points are located between 25°N and 45°N, and one of the maxima was found over the Northern Pacific just east of the Asian coast line. The Asian pollutants are uplifted more rapidly within WCBs than North American pollutants. However, then more than one WCB is needed to transport Asian pollution across the whole Pacific. Further, it has been reported that deep convection can transport radon-rich air from the Asian BL to the UT very rapidly, where it can be injected into the jet stream and transported all the way to the west coast of the United States [113]. For a more detailed description of the Asian pollution transport to NA the reader is referred to Chapter 3 by Cooper and Parrish. In comparison experimental evidence of intercontinental pollution transport from Asia to Europe is very sparse. The main reason is the long distance between these two continents if we assume a westerly flow which is the general situation in the midlatitudes. Most signatures of Asian pollution plumes diminish before the air masses reach Europe. However, it has been reported in the literature that anthropogenic pollution from Asia can be transported in the opposite direction to the west in an easterly tropical jet stream (20°–30°N).

Global 3-D chemistry transport model simulations by Li et al. [60] predicted a tropospheric ozone maximum over the Middle East in summer. Measurements of vertical ozone profiles during MOZAIC flights in that region confirmed the predicted ozone maximum. Further, the model was used to estimate the origin of the ozone maximum. It was found that the Asian influence from the east dominated in the southern part of the maximum. Large amounts of NO_x were transported into the UT due to strong convection over the polluted East Asian region and Indian monsoon lightning. The pollution transport to the Middle East region and the prevailing large-scale subsidence over that region favored the production of ozone.

The MINOS field study in summer 2001 in Crete indicated a strong influence by Asian pollutants on the Mediterranean air composition [43, 103]. Especially in the MT and UT, where the main flow comes from the west, the contribution to background CO was dominated by Asian pollutants as calculated from model tracer analyses. Further, more pronounced pollution events were observed when the Asian pollution was transported from the east. The pollution was uplifted to the UT with the strong convection in the Asian monsoon. An extended upper-level high-pressure ridge over the Middle East transported the pollution westward over Africa and subsequently northward towards Crete. Airborne measurements during MINOS indicated a pronounced CO pollution layer (~100 ppbv) in the UT on average, which was attributed to the Asian pollution arriving from the east. The reader is referred to the chapter by Lawrence in this volume for more details on the MINOS results and to the chapter by Pochanart, Wild and Akimoto for more information on the Asian pollution outflow in general.

6
Summary and Conclusions

The present knowledge of the export and import of anthropogenic pollutants from and to Europe has been summarized in this chapter. The influence of North American emissions on the North Atlantic Region and Europe has been documented reasonably well. In contrast, little is known about the influence of Asian emissions, and the impact of European emissions on downstream continents. Experimental evidence of the impact is difficult to quantify and model studies are required to obtain a more complete picture of the effects. However, with new observations from advanced satellite instruments from space, it can be expected that the understanding of these processes based on experimental data will improve in the coming years.

The European pollution export was first studied in more detail during the EXPORT field campaign carried out in summer 2000 over Central Europe [24]. Airborne measurements were performed in WCBs and thunderstorms that lifted the pollutants to the MT and UT. A low-level (boundary layer) pollution export mainly to the east was also observed. In addition, model studies indicate that the European pollution is mainly exported at low levels [32, 33]. The four receptor regions of the European pollution is the Arctic region (primarily), Siberia, the

Mediterranean area and the British Isles (with recirculation over the Atlantic, reentering Europe further south) [36, 37]. Within the framework of VOTALP in 1996 and 1998 a further transport mechanism for European pollutants was investigated in great detail. Deep valleys in the Alps may pump up pollutants to the LT and MT during fair weather conditions [48].

The NARE and OCTA field campaigns carried out in summer 1993 between the North American east coast and the European west coast aroused the scientific interest in the transport of intercontinental pollution from NA to Europe [81, 82, 96, 114]. Due to the relatively short distance between these two continents, the pronounced westerly flow, and the fact that the North American east coast belongs to the major anthropogenic emission source regions in the northern hemisphere, it was expected that Europe is most affected of all the continents by intercontinental pollution transport, as confirmed later by model studies [32]. The major transport mechanism of the pollutants emitted along the east coast of the United States is the uplift in WCBs ahead of cold fronts, which preferably develops off the coastline close to the main emission source region [34]. Once in the MT/UT the pollutants then may easily enter the jet stream, propagating along a northerly route passing Greenland and Iceland, and then traveling to Europe [65].

Simulations with global 3-D chemical transport models indicate that the Asian emissions probably have the strongest impact on the global tropospheric ozone distribution of all continents [32]. However, signatures of Asian pollution layers mostly diminish before they reach Europe since the layers mix with the background air over the long distance (for transport with the prevailing westerlies). Only recently, during the MINOS field experiment in summer 2001 in Crete, the important influence of Asian anthropogenic emissions transported from the east (tropical jet) on the tropospheric air composition in the eastern Mediterranean area was recognized [43]. Besides the strong uplift of Asian pollutants by deep convection, the NO_x produced by Indian monsoon lightning was also found to play a major role for the pollution transport and ozone formation.

Acknowledgements The authors are grateful for access to the trace gas data from the Zugspitze and Hohenpeissenberg provided by S. Gilge (Deutscher Wetterdienst) and E. Scheel (Institut für Meteorologie und Klimaforschung, Forschungszentrum Karlsruhe).

7
References

1. Collins WJ, Stevenson DS, Johnson CE, Derwent RG (2000) Atmos Environ 24:255
2. Volz A, Kley D (1988) Nature 332:240
3. Simmonds PG, Seuring S, Nickless G, Derwent RG (1997) Atmos Chem 28:45
4. Parrish DD, Holloway JS, Trainer M, Murphy PC, Forbes GL, Fehsenfeld FC (1993) Science 259:1436
5. Lin CY, Jacob DJ, Munger JW, Fiore AM (2000) Geophys Res Lett 27:3465
6. van Aardenne JA, Carmichael GR, Levy H, Streets D, Hordijk L (1999) Atmos Environ 33:633
7. Streets DG, Waldhoff ST (2000) Atmos Environ 34:363
8. EMEP (2000) DNMI research note 37. Det Norske Meteorologiske Institutt

9. CORINAIR (2002) In: EMEP/CORINAIR Emission Inventory Guidebook, 3nd Edition, Technical Report No 30 of the European Environment Agency
10. Olivier JGJ, Berdowski JJM (2001) Global emissions sources and sinks. In: Berdowski J, Guicherit R, Heij BJ (eds) The Climate System. A.A. Balkema Publishers/Swets & Zeitlinger Publishers, Lisse, The Netherlands, ISBN 90 5809 255 0, pp 33–78
11. Greadel TF et al. (1993) Global Biochem Cycles 7:1
12. Olivier JGJ, Berdowski JJM (2001) Global emissions sources and sinks. In: Berdowski J, Guicherit R, Heij BJ (eds) The Climate System. A.A. Balkema Publishers/Swets & Zeitlinger Publishers, Lisse, The Netherlands, ISBN 90 5809 255 0, pp 33–78
13. Benkovitz CM (1996) J Geophys Res 101:29239
14. Berdowski JJM, Mulder W, Veldt C, Visschedijk AJH, Zandveld PYJ (1997) TNO Institute of Environmental Science, Energy Research and Process Innovation. Apeldoom, The Netherlands
15. Köhler I, Dameris M, Ackermann I, Haas H (2001) J Geophys Res 106:17997
16. Berdowski JJM, Baas J, Bloos JPJ, Visschedijk AJH, Zandveld PYJ (1997) The European Emission Inventory of Heavy Metals and Persistent Organic Pollutants, Report 104 02 672/03. TNO, Apeldoom, The Netherlands
17. Tuovinen JP, Barrett K, Styve H (1994) Transboundary Acidifying Pollution in Europe: Calculated Fields and Budgets 1985–1993, ECE Cooperative Programme for Monitoring and Evaluation of Long-Range Transmission of Air Pollutants in Europe, Oslo
18. Fowler D et al. (2001) Transboundary air pollution, acidification, eutrophication, and ground-level ozone in the UK. UK Department of Energy, Food and Rural Affairs, London
19. Roemer M (2001) Trends of ozone and precursors in Europe, TOR-2 Status Report, TNO Report R 2001/244
20. Simmonds PG, O'Doherty S, Spain TG (2001) Advanced global atmospheric gases experiment, Mace Head, Ireland, Final Report to DETR, INSCON, Ringwood, UK
21. Bethan S, Vaughan G, Gerbig C, Volz-Thomas A, Richer H, Tiddeman DA (1998) J Geophys Res 103:13413
22. Kowol-Santen J, Beekmann M, Schmitgen S, Dewey K (2001) Geophys Res Lett 28:2907
23. Hov Ø, Flatøy F (1997) J Atmos Chem 28:319
24. Penkett SA (2002) EUROTRAC Newsletter 24:4
25. Reeves CE (2003) In: Penkett SA, Law KS, Platt U, Volz-Thomas A (eds) EXPORT-E2 Final Report. EUROTRAC-2 ISS, GSF-National Research Center for Environment and Health, Munich, Germany, p 147
26. Schlager H, Huntrieser H, Heland J, Ziereis H (2003) In: Penkett SA, Law KS, Platt U, Volz-Thomas A (eds) EXPORT-E2 Final Report. EUROTRAC-2 ISS, GSF-National Research Center for Environment and Health, Munich, Germany, p 151
27. Hauf T, Schulte P, Alheit R, Schlager H (1995) J Geophys Res 100:22957
28. Huntrieser H, Schlager H, Feigl C, Höller H (1998) J Geophys Res 103:28247
29. Ström J, Fischer H, Lelieveld J, Schröder F (1999) J Geophys Res 104:12221
30. Huntrieser H et al. (2002) J Geophys Res 107: ACH5(10.1029/2000JD000209)
31. Fischer H et al. (2002) Atmos Chem Phys Discuss 2:2003
32. Wild O, Akimoto H (2001) J Geophys Res 106:27729
33. Stohl A, Eckhardt S, Forster C, James P, Spichtinger N (2002) J Geophys Res 107: 4684(10.1029/2001JD001396)
34. Stohl A (2001) J Geophys Res 106:7263
35. Eckhardt S, Stohl A, Wernli H, James P, Forster C, Spichtinger N (2003) J Clim (accepted)
36. Bey I, Schultz M (2003) In: Penkett SA, Law KS, Platt U, Volz-Thomas A (eds) EXPORT-E2 Final Report. EUROTRAC-2 ISS, GSF-National Research Center for Environment and Health, Munich, Germany, p 37
37. Feldmann H, Ebel A, Jakobs HJ, Memmesheimer M, Kerschgens M (2003) In: Penkett SA, Law KS, Platt U, Volz-Thomas A (eds) EXPORT-E2 Final Report. EUROTRAC-2 ISS, GSF-National Research Center for Environment and Health, Munich, Germany, p 67
38. Raatz WE (1989) Atmos Environ 23:2501
39. Christensen JH (1997) Atmos Environ 31:4169

40. Novelli PC, Masarie KA, Lang PM (1998) J Geophys Res 103:19015
41. Eckhardt S, Stohl A, Beirle S, Spichtinger N, James P, Forster C, Junker C, Wagner T, Platt U, Jennings SG (2003) Atmos Chem Phys Discuss 3:3222
42. Staudt AC, Jacob DJ, Logan JA, Bachiochi D, Krishnamurti TN, Sachse GW (2001) J Geophys Res 106:32571
43. Lelieveld J et al. (2002) Science 298:794
44. Doddridge BG, Dirmeyer PA, Merrill JT, Oltmans SJ, Dickerson RR (1994) J Geophys Res 99:22923
45. Prospero JM, Schmitt R, Cuevas E, Savoie DL, Graustein WC, Tarekian KK, Volz-Thomas A, Diaz A, Oltmans SJ, Levy II H (1995) Geophys Res Lett 22:2925
46. Wild O, Law KS, McKenna DS, Bandy BJ, Penkett SA, Pyle JA (1996) J Geophys Res 101:29269
47. Andreae MO, Elbert W, Gabriel R, Johnson DW, Osborne S, Wood R (2000) Tellus 52B:1066
48. Furger M et al. (2000) Atmos Environ 34:1395
49. Prévôt ASH, Dommen J, Bäumle M (2000) Atmos Environ 34:4719
50. Prévôt ASH, Weber RO, Furger M, Dommen J, Neininger B (2002) In: Midgley PM, Reuther M (eds) Proc EUROTRAC-2 Symposium. Margraf Verlag, Weikersheim, Germany
51. Dommen J, Prévôt ASH, Henne S, Nyeki S, Weingartner E, Baltensperger U (2003) In: Penkett SA, Law KS, Platt U, Volz-Thomas A (eds) EXPORT-E2 Final Report. EUROTRAC-2 ISS, GSF-National Research Center for Environment and Health, Munich, Germany, p 61
52. Millán MM, Salvador R, Mantilla E (1997) J Geophys Res 102:8811
53. Simmonds PG, Derwent RG (1991) Atmos Environ 25A:1795
54. Schmitt R (1994) In: Borrell P, Borrell PM, Seiler W (eds) Proceedings of EUROTRAC Symposium '94: Transport and Transformation of Pollutants in the Troposphere. SPB Academic Publishing, The Hague, Netherlands, p 313
55. Jennings SG, Spain TG, Doddridge BG, Maring H, Kelly BP, Hansen ADA (1996) J Geophys Res 101:19447
56. Derwent RG, Simmonds PG, Seuring S, Dimmer C (1998) Atmos Environ 32:145
57. Brunner D, Stähelin J, Jeker D, Wernli H, Schumann U (2001) J Geophys Res 106:27673
58. Purvis RM, Lewis AC, McQuaid JB, Barjat H, Dewey K, Kent J (2002) In: Midgley PM, Reuther M (eds) Proc EUROTRAC-2 Symposium. Margraf Verlag, Weikersheim, Germany
59. Nyeki S, Eleftheriadis K, Baltensperger U, Colbeck I, Fiebig M, Fix A, Kiemle C, Lazaridis M, Petzold A (2002) Geophys Res Lett 29:33(10.1029/2002GL014897)
60. Li Q et al. (2001) Geophys Res Lett 28:3235
61. Huntrieser H et al. (2002) In: Midgley PM, Reuther M (eds) Proc EUROTRAC-2 Symposium. Margraf Verlag, Weikersheim, Germany
62. Angevine WM, Trainer M, McKeen SA, Berkowitz CM (1996) J Geophys Res 101:28893
63. Li Q et al. (2002) J Geophys Res 107:ACH4(10.1029/2001JD001422)
64. Moody JL et al. (1996) J Geophys Res 101:29213
65. Stohl A, Trickl T (1999) J Geophys Res 104:30445
66. Parrish DD, Holloway JS, Jakoubek R, Trainer M, Ryerson TB, Hübler G, Fehsenfeld FC, Moody JL, Cooper OR (2000) J Geophys Res 105:24363
67. Cooper OR, Moody JL, Parrish DD, Trainer M, Holloway JS, Ryerson TB, Hübler G, Fehsenfeld FC, Oltmans SJ, Evans MJ (2001) J Geophys Res 106:5437
68. Cooper OR, Moody JL, Parrish DD, Trainer M, Ryerson TB, Holloway, JS, Hübler G, Fehsenfeld FC, Evans MJ (2002) J Geophys Res 107:ACH1(10.1029/2001JD000901)
69. Dickerson RR et al. (1987) Science 235:460
70. Thompson AM et al. (1994) J Geophys Res 99:18703
71. Liu SC et al. (1999) Geophys Res Lett 26:2441
72. de Reus M, Ström J, Curtius J, Pirjola L, Vignati E, Arnold F, Hansson HC, Kulmala M, Lelieveld J, Raes F (2000) J Geophys Res 105:24751
73. Hannan JR et al. (2000) J Geophys Res 105:3807
74. Milne PJ, Prados AI, Dickerson RR, Doddridge BG, Riemer DD, Zika RG, Merrill JT, Moody JL (2000) J Geophys Res 105:9981
75. Brunner D, Stähelin J, Jeker D (1998) Science 282:1305

76. Jeker DP, Pfister L, Thompson AM, Brunner D, Boccippio DJ, Pickering KE, Wernli H, Kondo Y, Stähelin J (2000) J Geophys Res 105:3679
77. Thouret V, Cho JYN, Newell RE, Marenco A, Smit HGJ (2000) J Geophys Res 105:17379
78. Newell RE, Thouret V, Cho JYN, Stoller P, Marence A, Smit HG (1999) Science 398:316
79. Prados AI, Dickerson RR, Doddridge BG, Milne PA, Moody JL, Merrill JT (1999) J Geophys Res 104:26219
80. Tucker GB, Barry RG (1984) In: Van Loon H (ed) World Survey of Climatology, vol 15, Climates of the Oceans. Elsevier Sci, New York, p 193
81. Fehsenfeld FC, Daum P, Leaitch WR, Trainer M, Parrish DD, Hübler G (1996) J Geophys Res 101:28877
82. Fehsenfeld FC, Trainer M, Parrish DD, Volz-Thomas A, Penkett S (1996) J Geophys Res 101:28869
83. Schmitt R, Volz-Thomas A (1997) J Atmos Chem 28:245
84. Parrish, DD, Trainer M, Holloway JS, Yee JE, Warshawsky MS, Fehsenfeld FC Forbes GL, Moody JL (1998) J Geophys Res 103:13357
85. Schultz M, Schmitt R, Thomas K, Volz-Thomas A (1998) J Geophys Res 103:13477
86. Bandy BJ, Penkett SA, Simmonds PG, Spain G, Merrill JT (1994) In: Borrell P, Borrell PM, Seiler W (eds) Proc EUROTRAC Symposium '94: Transport and Transformation of Pollutants in the Troposphere. SPB Academic Publishing, The Hague, Netherlands, p 221
87. Ryall DB, Maryon RH, Derwent RG, Simmonds PG (1998) Q J R Meteorol Soc 124:417
88. Fehsenfeld FC (1994) In: Borrell P, Borrell PM, Seiler W (eds) Proc EUROTRAC Symposium '94: Transport and Transformation of Pollutants in the Troposphere. SPB Academic Publishing, The Hague, Netherlands, p 56
89. Evans MJ et al. (2000) Atmos Environ 34:3843
90. Dutot AL, Colin P, Etienne A, Perros PE, Toupance G (1994) In: Borrell P, Borrell PM, Seiler W (eds) Proc EUROTRAC Symposium '94: Transport and Transformation of Pollutants in the Troposphere. SPB Academic Publishing, The Hague, Netherlands, p 239
91. Fenneteaux I, Colin P, Etienne A, Boudries H, Dutot AL, Perros PE, Toupance G (1999) Atmos Chem 32:233
92. Huntrieser H et al. (2003) J Geophys Res (submitted 2004)
93. Stohl A, Forster C, Eckhardt S, Huntrieser H, Heland J, Schlager H, Aufmhoff H, Arnold F, Cooper O (2003) J Geophys Res 108:4370 (doi:10.1029/2002JD002862)
94. Eisele H, Scheel HE, Sladkovic R, Trickl T (1999) J Atmos Sci 56:319
95. Trickl T, Cooper OR, Eisele H, James P, Mücke R, Stohl A (2003) J Geophys Res 108:8530 (doi: 10.1029/2002JD002735)
96. Volz-Thomas A, Schultz M, McKenna DS, Schmitt R (1994) EOS Trans AGU, 75(44), Fall Meet Suppl, p 113
97. Law KS et al. (2001) IGACtivities Newsletter 24:17
98. Arnold F, Schneider J, Gollinger K, Schlager H, Schulte P, Hagen DE, Whitefeld PD, van Velthoven P (1997) Geophys Res Lett 24:57
99. Zahn A, Brenninkmeijer CAM, Slemr F, Mühle J, Hermann M, Heintzenberg J, Ziereis H, Schlager H, van Velthoven PFJ (2003) In: Penkett SA, Law KS, Platt U, Volz-Thomas A (eds) EXPORT-E2 Final Report. EUROTRAC-2 ISS, GSF-National Research Center for Environment and Health, Munich, Germany, p 175
100. Volz-Thomas A, Pätz HW, Houben N (2003) In: Penkett SA, Law KS, Platt U, Volz-Thomas A (eds) EXPORT-E2 Final Report. EUROTRAC-2 ISS, GSF-National Research Center for Environment and Health, Munich, Germany, p 161
101. Ravetta F, Kubicki A, Ancellet G (2002) In: Midgley PM, Reuther M (eds) Proc EUROTRAC-2 Symposium. Margraf Verlag, Weikersheim, Germany
102. Huntrieser H et al. (2003) In: Penkett SA, Law KS, Platt U, Volz-Thomas A (eds) EXPORT-E2 Final Report. EUROTRAC-2 ISS, GSF-National Research Center for Environment and Health, Munich, Germany, p 95
103. Lawrence MG et al. (2003) Atmos Chem Phys 3:267
104. Stohl A, Trainer M, Ryerson TB, Holloway JS, Parrish DD (2002) J Geophys Res 107:ACH11(10.1029/2001JD000519)

105. Fishman J, Fakhruzzaman K, Cros B, Nganga D (1991) Science 252:1693
106. Thompson AM (1992) Science 256:1157
107. Fishman J, Wozniak AE, Creilson JK (2003) Atmos Chem Phys 3:893
108. Creilson JK, Fishman J, Wozniak AE (2003) Atmos Chem Phys Discuss 3:4431
109. Remedios JJ, Richards NAD (2002) In: Midgley PM, Reuther M (eds) Proc EUROTRAC-2 Symposium. Margraf Verlag, Weikersheim, Germany
110. Stohl A et al. (2003) Atmos Chem Phys 3:969
111. Leue C, Wenig M, Wagner T, Klimm O, Platt U, Jähne B (2001) J Geophys Res 106:5493
112. Sanders F, Gyakum JR (1980) Mon Wea Rev 108:1589
113. Kritz MA, le Roulley JC, Danielsen EF (1990) Tellus 42B:46
114. Penkett SA, Fehsenfeld FC, Prospero JM (1994) In: Prinn RG (ed) Global Atmospheric-Biospheric Chemistry: the 1st IGAC Scientific Conf, OHOLO Conference Series Books. Plenum Publishing, New York

The Handbook of Environmental Chemistry Vol. 4, Part G (2004): 99–130
DOI 10.1007/978-3-540-40037-0

Air Pollution Import to and Export from East Asia

Pakpong Pochanart · Oliver Wild · Hajime Akimoto

Atmospheric Composition Research Program, Frontier Research System for Global Change, 3173-25 Showa-machi, Kanazawa-ku, Yokohama 236-0001, Japan
E-mail: pakpong@jamstec.go.jp; oliver@jamstec.go.jp; akimoto@jamstec.go.jp

Abstract Continued rapid industrialization is expected to make East Asia the largest source region for air pollution in the coming decades. This chapter reviews the import and export of air pollutants, mainly carbon monoxide and ozone, by long-range transport to and from East Asia. Air pollutant sources and transport characteristics dominated by the monsoon regime over Northeast and Southeast Asia are summarized. Over Northeast Asia both import and export of air pollution contribute significantly to the intercontinental transport of pollution. Trans-Eurasian transport of air pollution by westerly flows from Europe to Northeast Asia, and trans-Pacific transport of Northeast Asian pollution to the Pacific and North America have been verified by both observations and model studies. The influence of Siberian forest fires on pollutants over Northeast Asia has also been clearly identified. Over continental Southeast Asia, import of air pollution occurs mainly from source regions over the Indian subcontinent during the summer monsoon and from Northeast Asia during the early winter monsoon. Biomass burning within the region, however, strongly dominates regional pollution. The transoceanic export of air pollution from Southeast Asian biomass burning is known to be one of the most significant sources of ozone and CO over the North Pacific in the springtime.

Keywords Anthropogenic emissions · Biomass burning · East Asia · Intercontinental transport · Trans-Eurasia

1 Introduction

1.1 The Increasing Asian Contribution to Global Emissions

East Asia covers a large area at the eastern end of the Eurasian land mass, from the tropics to the polar regions of Siberia. At present, nearly one third of the world's population, ~2 billion people, reside in this region. For climatological purposes we subdivide East Asia into Northeast Asia (comprising China, Taiwan, Japan, North and South Korea, Mongolia, and Eastern Siberia) and Southeast Asia (comprising Myanmar, Thailand, Vietnam, Laos, Cambodia, the Philippines, Malaysia, Singapore, Indonesia, Brunei, and East Timor). Rapid population and economic growth have characterized the entire region in the last few decades, and have led to a rapid increase in anthropogenic emissions of air pollutants and greenhouse gases. Figure 1a shows how NO_x emissions from the continents of North America, Europe and Asia have developed over the last 30 years [1–6]. Asian emissions were far smaller than those from North America or Europe in the 1970s, but a rapid increase in emissions during the last three decades has brought the contributions from Asia to a level rivaling those of the other two continents in the 1990s. Since the mid-1990s, the total emissions from Asia have exceeded those from North America and Europe and Asia is now a major contributor to air pollutant emissions on a global scale. Since the increase in Asian emissions is expected to continue for the next couple of decades, Asian contributions are likely to dominate global emissions in the earlier decades of the twenty-first century.

Figure 1b shows a similar plot of the emission trend focusing on Northeast Asia and comparing it with equivalent regions of Europe and the United States. These three regions lie at mid-latitudes in the Northern Hemisphere, where westerly winds dominate, and hence emissions are expected to have a substantial influence on the intercontinental transport of ozone and other air pollutants. Although Fig. 1b resembles Fig. 1a, it should be noted that East Asian

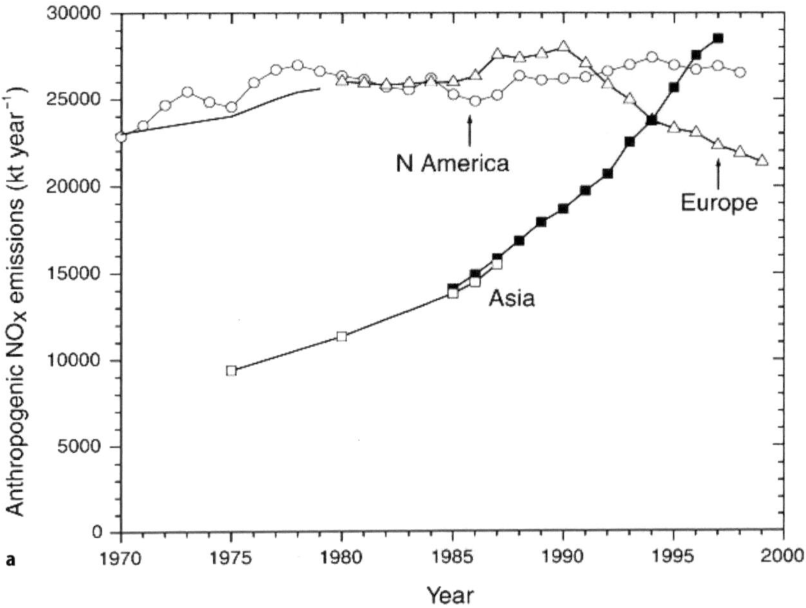

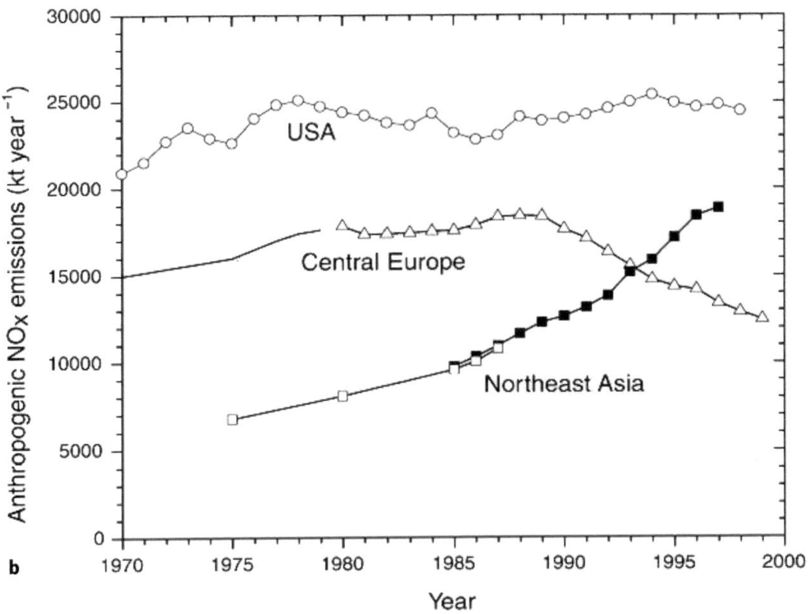

Fig. 1a,b Trends in emissions of NO_x over 30 years over: **a** the continents of North America (USA and Canada), Europe (including European part of Russia), and Asia (East Asia and South Asia); **b** the USA, Central Europe (excluding Scandinavia and Russia), and Northeast Asia (China, Taiwan, Japan, North and South Korea) [1–6]

emissions exceeded those of central Europe in the mid-1990s, and are currently approaching those of the United States. It is therefore evident that while North American and European emissions have changed little since the 1980s, and are likely to have had relatively stable intercontinental impacts on inflow to East Asia, the outflow of pollutants from East Asia and its consequent intercontinental impacts on other regions are increasing rapidly. The differing trends make the East Asian region a particularly important region to study, as a good understanding of pollutant emissions and their trends is required to interpret correctly the current and future intercontinental impacts of air pollution from the region on other parts of the Northern Hemisphere.

1.2 Trans-Eurasian Transport vs Trans-Pacific Transport

There has been considerable interest in the outflow of pollutants from Asia over the last decade, driven largely by a realization that they may have a substantial impact on the Pacific and beyond, and a number of large international measurement campaigns have focused on this region, including PEM-West A and B [7, 8], TRACE-P [9], ACE-Asia [10], and the PEACE campaigns [11, 12]. Increases in ozone in Asian continental outflow have been extensively studied with both ground-based observations [13–15] and ozone sounding data analysis [16, 17]. Intercontinental transport of ozone, CO and dust from East Asia across the Pacific Ocean has been examined in a number of model and observational studies [18–24] (see Chap. 3).

In contrast to the outflow from East Asia, European outflow and intercontinental transport of pollutants across the Eurasian continent to East Asia have been little studied. A study of backward trajectories from points along the 100°E meridian by Newell and Evans [25] indicated that a substantial fraction of air masses arriving over East Asia cross Europe in winter and early spring, and may therefore be exposed to pollutant emissions. Wild and Akimoto [26] presented the first global CTM model analysis of the intercontinental transport of ozone between the three continents in the Northern Hemisphere, and considered the characteristics of transport from Europe to Asia. This chapter focuses principally on the inflow and outflow of air pollution by long-range transport to and from Northeast Asia. The observational evidence for trans-Eurasian and trans-Pacific transport of air pollution, mainly of ozone and CO, are discussed and summarized. The chapter also introduces the effect of Siberian forest fires on air quality in Northeast Asia as they make a substantial contribution to the inflow of polluted air from over the Eurasian continent [27, 28], although this is covered in greater detail in Chap. 10.

1.3 Regional Studies in Southeast Asia

A further focus of this chapter is the import and export of air pollution to and from Southeast Asia. Although this region lies in the tropics and is thus removed from the prevailing westerly flow at mid-latitudes, monsoonal flow and the associated movement of the Inter-tropical convergence zone (ITCZ) and more active photochemistry may contribute to substantial intercontinental impacts. Southeast Asia is experiencing a similar rapid economic growth to that in Northeast Asia, but is also a region where large-scale biomass burning occurs on a regular basis. This biomass burning has a very different influence on the atmospheric pollutants in the region from the boreal fires affecting Northeast Asia, as reported by Pochanart et al. [29, 30] who observed strong seasonal variations in ozone and CO characterized by biomass burning in the dry season and clean air in the wet season. However, few atmospheric chemistry studies have been carried out in this area, and transport of air pollutants to and from the region over continental and intercontinental scales has yet to be investigated in detail.

2 Inflow of Air Pollution to Northeast Asia

2.1 Transport Mechanisms and Patterns

The climatology of the Northeast Asian region is principally governed by the seasonal exchange between air from two quite different origins: continental air from Eurasia brought by westerly winds and marine air from the Pacific brought by the East Asian monsoon (Fig. 2) [14, 31, 32]. In summertime, surface high-pressure systems associated with the Pacific anticyclone cause an inflow of clean oceanic air to Northeast Asia [14, 15]. For the rest of the year the meteorology of the region is influenced by continental outflow, particularly during wintertime when the East Asian winter monsoon is dominant [33–35]. Westerly flow at mid-latitudes transports air from the western regions of the Eurasian land mass, and is more pronounced in the free troposphere [25]. The East Asian winter monsoon, which is associated with a strong anticyclone over the continent and active cold surges, is one of the most energetic monsoon circulation systems and exerts a strong impact on both extra-tropical and tropical circulation on a global scale [33]. The dominant surface feature of this monsoon flow is a huge thermal anticyclone known as the Siberian High, which typically has a central pressure in excess of 1030 hPa. The Siberian High extends over a large part of northern Asia, and produces northwesterly flow over much of East Asia. At higher elevations, the flow patterns are linked to the coastal trough (~500 hPa) and the jet stream (~200 hPa). The very strong jet stream is associated with large vertical wind shear, and strong cold advection [36]. The surface Siberian High, 500 hPa trough, and 200 hPa jet characterize the three- dimensional large-scale monsoon circulation and are intense north of 30°N. At lower latitudes, the Siberian High and winter monsoon are shallower due to the influences of the trade winds [33].

The dominant periods for the inflow of air pollution to Northeast Asia are winter and early spring. Summertime is much less important, despite being the period of strongest photochemical activity, because the East Asian monsoon brings clean air masses from the Pacific Ocean (Fig. 2). The mixing ratios of moderately long-lived air pollutants suitable for identifying intercontinental

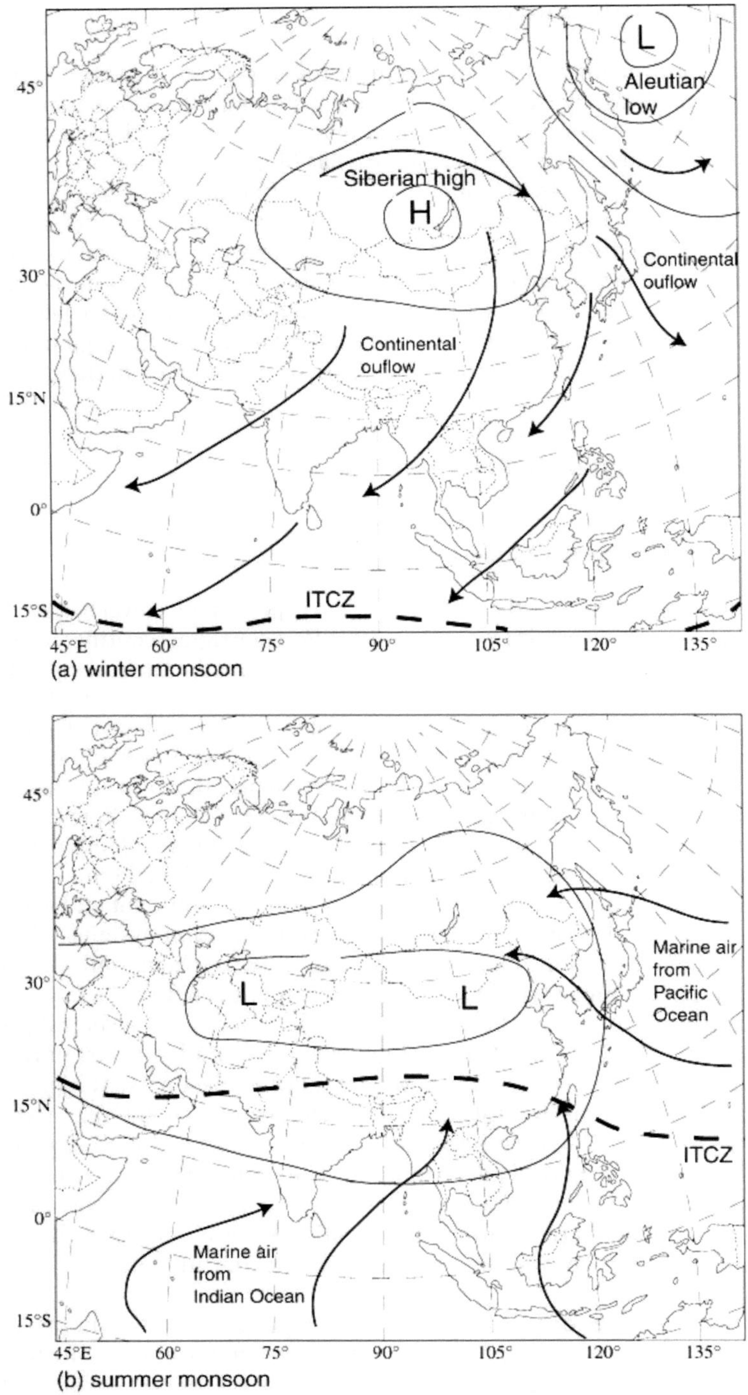

Fig. 2a,b Pressure distribution and surface wind-flow pattern during: **a** the winter monsoon; **b** the summer monsoon. The *heavy dashed line* represents the position of Inter-tropical convergence zone (ITCZ)

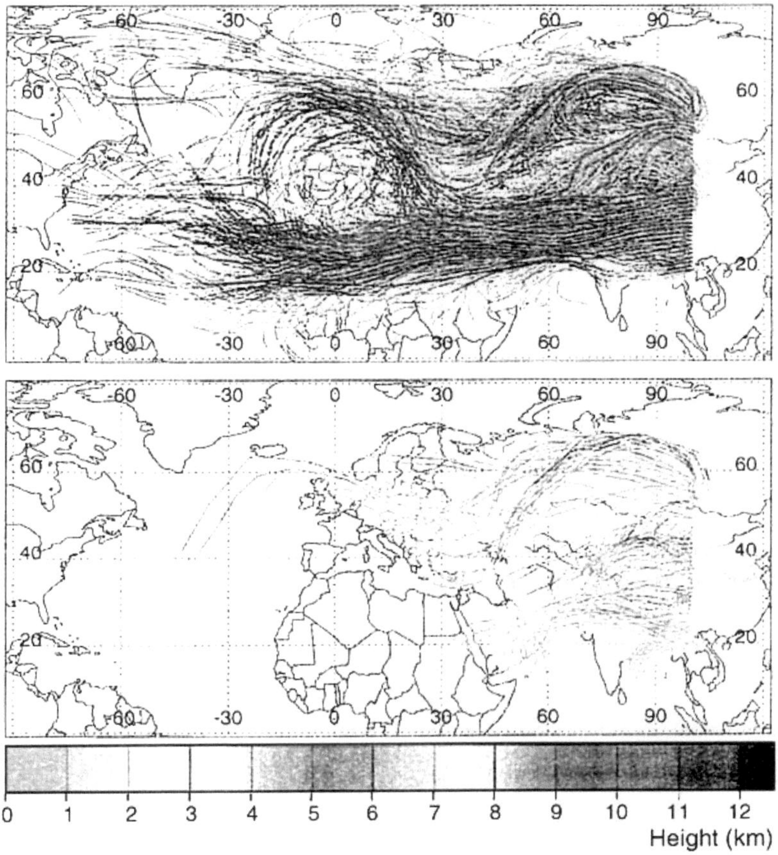

Fig. 3 Five-day backward trajectories calculated from a wall at 100°E and at pressures equal to 0.9, 0.8, ..., 0.3 of the surface pressure for 17–22 March 1997. The upper panel shows trajectories starting at heights above 2.5 km, and the lower panel those starting at heights below 2.5 km (after Newell and Evans [25])

transport (e.g., CO and ozone [37], see also Chap. 3) in coastal regions of East Asia are therefore generally low during summer [14, 15, 27, 38, 39]. However, near the intense emission regions and in the inner regions of East Asia, relatively high mixing ratios of air pollutants may be found even during summer and fall [40–43] as a result of regional-scale pollution transport within East Asia. In the winter, the jet stream in the upper troposphere carries pollution from further west, potentially from Europe and North America, while the Siberian High may bring surface pollution on a more northerly route via the Arctic [44]. Figure 3 shows the paths of air masses arriving at the eastern boundary of East Asia along the 100°E meridian between 20 and 60°N in March, illustrating the typical transport patterns in spring. Air masses in the upper troposphere arrive from Europe, northern Africa, or from as far away as the east coast of North America. In the lower troposphere, air masses arriving at higher latitudes are transported mainly from Europe while those arriving at lower latitudes are transported from the Middle East and northeastern Africa.

2.2 Transport of Anthropogenic Air Pollution from Europe to Northeast Asia

Observed mixing ratios of ozone and CO at remote sites in East Asia typically have a maximum in early spring. The spring maximum of ozone and CO is a well-recognized feature of the remote Northern Hemisphere [45–52] and is thought to be governed principally by chemical processes [53], but may also be partly attributed to the intercontinental transport of ozone and CO [26]. However, while this seasonality agrees well with the trajectory studies at the eastern boundary of East Asia [25], more detailed analysis of intercontinental transport over Eurasia has yet to be made. We address here the unique characteristics of intercontinental transport over Eurasia, and the evidence for it based on observations and model studies.

2.2.1 *Characteristics of Intercontinental Transport from Europe to Northeast Asia*

Transport of air pollution typically occurs in distinct meteorological episodes which lead to 'pollution events' detectable at downwind sites. However, over intercontinental distances atmospheric turbulence and mixing may lead to the disappearance of these episodes and to a general enhancement in background concentrations, particularly for transport occurring in the lower troposphere. Whereas pollution events detected at observation sites may be readily attributable to intercontinental transport, contributions to well-mixed background concentrations are not easy to identify. Transport of pollutants over the Pacific and Atlantic Oceans, covered in Chaps. 3 and 4, typically occurs by convection or frontal lifting followed by westerly flow in the free troposphere [54]. In contrast, transport over Eurasia generally occurs at higher latitudes and at lower altitudes [55]. Lower tropospheric flow over the continent, where boundary layer mixing is typically more vigorous, is likely to lead to a smaller contribution from distinct pollution events than is the case for trans-oceanic transport, and hence identification of the effects of intercontinental transport from observations over central Asia is more difficult. Higher NO_x emission intensities over Europe and low insolation lead to substantial depletion of ozone in the regional boundary layer in winter [56], and hence export of pollutant CO in this season may be accompanied by suppressed levels of ozone. In addition, pollution from sources over Eurasia may be mixed in with pollution from Europe during transport, complicating identification still further; trans-oceanic transport, in contrast, encounters little additional pollution except that from shipping [57]. These unique features of trans-Eurasian transport, together with a lack of observational studies over the inner parts of Eurasia, present a particularly difficult challenge to understanding the impacts of Europe on East Asia, one that is only starting to be tackled at the present time.

2.2.2
Evidence of Trans-Eurasian Transport from Observations

There have been few observational studies of air mass transport over the remote parts of Eurasia, although more observations are planned in the region in the near future. Most of the long-term or campaign-based observations over the continent have been carried out either in western Europe or in the coastal region of East Asia, close to the major regional pollution sources. The East Asian coastal sites range from sub-tropical locations in Hong Kong and Taiwan to higher latitude sites in China, Korea and Japan, and provide an extensive database of measurements of ozone, CO and other trace gases (Fig. 4) [14, 15, 27, 39, 40–43, 58–63]. However, identification of trans-Eurasian transport at these sites is difficult due to the intense anthropogenic emissions from East Asia itself, which swamp the signal of upwind sources. Characterization of regional inflow thus requires measurements upwind of Asian sources.

Long-term ground-based monitoring of pollutants has been performed at Mondy, a mountain site in eastern Siberia southwest of Lake Baikal. Analysis of ozone and CO data obtained during 1997–1998 indicates that air masses trans-

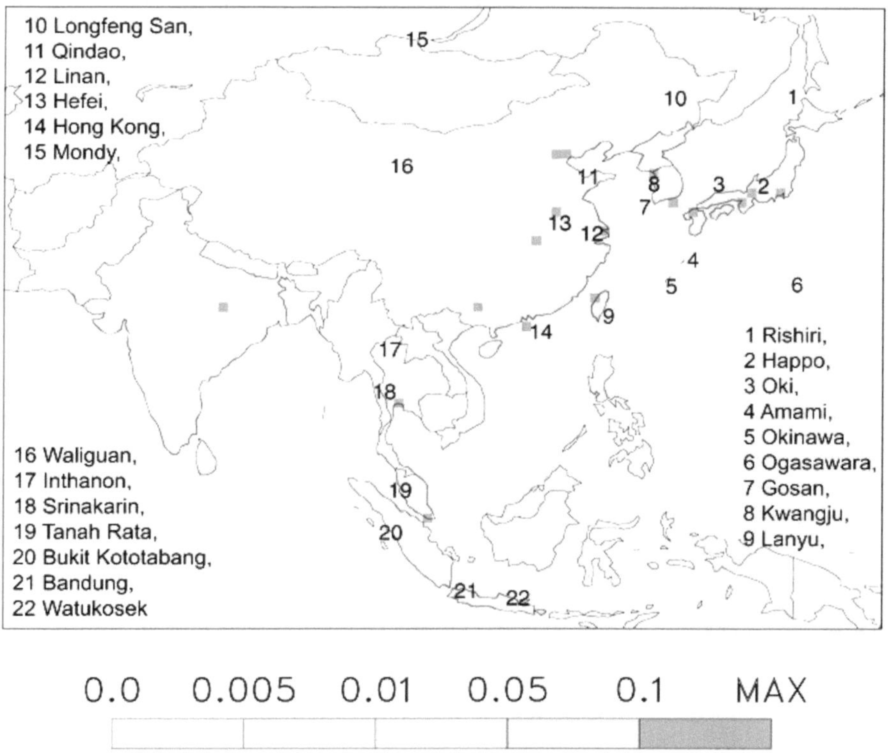

Fig. 4 Locations of existing representative atmospheric measurement sites in East Asia, shown together with annual NO_x emissions for 1990 [63] (units of Tg NO_2 m^{-2} $year^{-1}$)

ported from Europe contain an average of 2–3 parts per billion by volume (ppbv) higher ozone and 6–14 ppbv higher CO than those arriving from other regions [34]. Analysis of the transit time for these air masses from Europe to Mondy reveals that CO mixing ratios decrease significantly with increasing transport time (Fig. 5). The slope of CO against transport time shows a mean decrease of ~6 ppbv day^{-1} of CO in summer and ~2 ppbv day^{-1} in winter during air flow from Europe to east Siberia. This emphasizes the dilution and chemical destruction occurring in pollutant-rich air masses from Europe as they are transported over Eurasia. The winter season sees higher CO mixing ratios and a much shallower slope, reflecting longer chemical lifetimes and dilution into a higher-level background. At present, these results are probably the only observations confirming the intercontinental transport of air pollution from Europe to Asia. Long-term observational data since 1994 are also available from Mt. Waliguan in remote China, lying at the same longitude as Mondy but 15° further south (see Fig. 4) [41]. This WMO/GAW (World Meteorological Organization/Global Atmosphere Watch) baseline site provides valuable data on ozone and CO although it lies south of the major transport pathways for European pollution over Eurasia. Waliguan also experiences significant westward flow from East Asian sources in summertime.

Surface measurements of trace gases have been made along the Trans-Siberian Railroad between Moscow and Vladivostok during a number of campaigns [64–66]. Measurements of CO during TROICA 3 (Trans-Siberian Observational Into the Chemistry of the Atmosphere) in April 1997 show decreasing mixing ratios eastwards along the railroad from the European to the Asian parts of Russia [66]. CO mixing ratios decrease from ~230 ppbv at 45–65°E to relatively constant values below 200 ppbv eastward of 80°E. It was concluded that CO in western Siberia is considerably influenced by large scale advection of CO-rich air from Europe, a similar conclusion to that drawn at Mondy. Upper tropospheric measurements over Eurasia are available from measurement programs using commercial airliners, including the NOXAR project (Nitrogen Oxides and Ozone Measurements along Air Routes) [67, 68] and MOZAIC (Measurement of Ozone and Water Vapor by Airbus In-Service Aircraft) [69, 70]. While stratosphere-troposphere exchange is a major source of ozone at cruising altitudes, export from the continental boundary layer over Eurasia was also found to be an important source of ozone in the upper troposphere [68, 70]. The scarcity of observations over the inner parts of Eurasia is hindering attempts to explore trans-Eurasian transport in more detail. Large observational campaigns, integrating data from aircraft flights, ground-based sites, and satellite observations, have proved very useful in probing the transport and chemical transformation of pollutants over the Pacific and Atlantic Oceans [7, 8, 71]. Such large international campaigns have yet to target the more complex interactions occurring over Eurasia, but observational aircraft campaigns have been proposed and are expected to be executed in the near future [72].

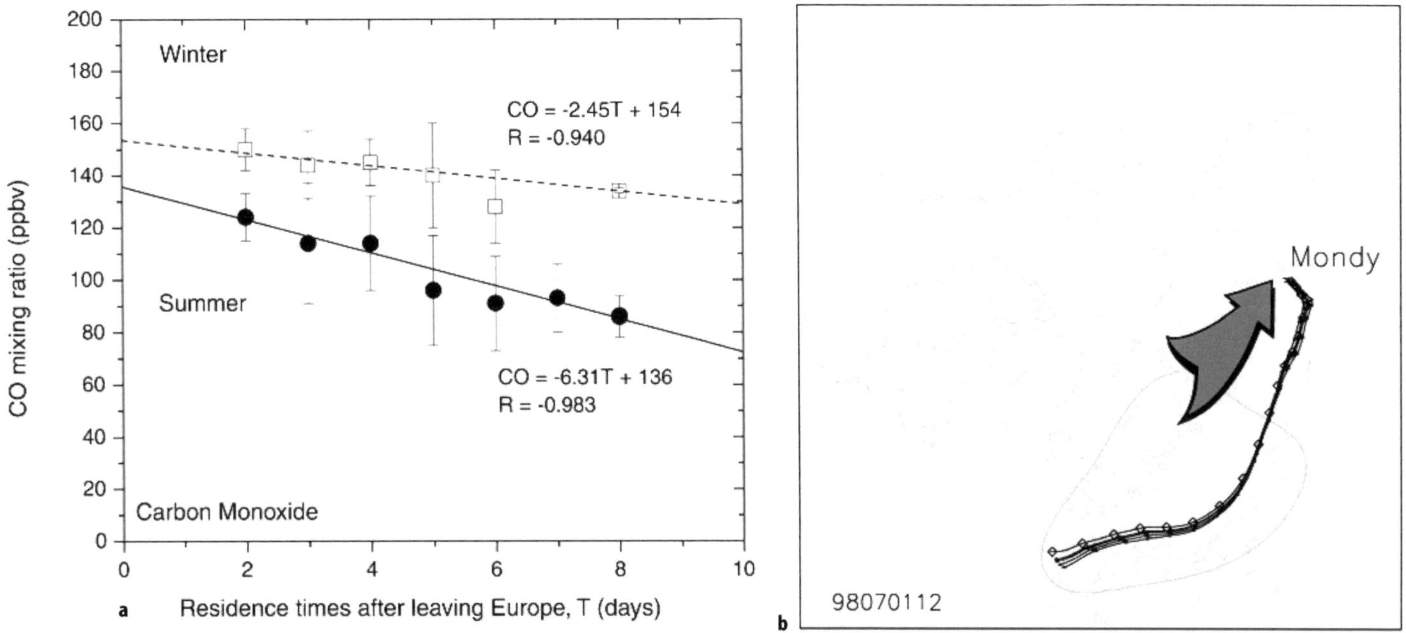

Fig. 5. **a** Relationship between the transport times of European air masses arriving at Mondy (i.e., the transit time over Eurasia) and the corresponding CO mixing ratios in summer (June–August) and winter (December–February). The symbols are *solid circles* for summer and *open squares* for winter, and *error bars* represent the average value ± one standard deviation. *Solid line* and *dotted line* show correlation from a simple linear regression analysis. **b** An example of ten-day backward trajectories calculated for July 1, 1998 12 UT showing a five-day transit time between leaving Europe and arriving at Mondy. *Open diamonds* indicate 12 h intervals

2.2.3
Evidence of Trans-Eurasian Transport from Model Studies

Few global chemical transport model studies have yet focused on the inflow of pollutants to East Asia except in analysis of observations made over the Pacific. Bey et al. [73] analyzed data from the PEM-West B campaign held in February–March 1994, and concluded that 15% of the outflow of CO from Asia came from upwind sources over Europe, and 9% from North America. A similar analysis of PEM-Tropics B data concluded that 12% of the tropospheric CO column over the northeastern Pacific came from European fossil fuel sources in March–April 1999, and 7% from North American sources [74]. It was also reported that European sources are responsible for an average of 16% of CO, about 25 ppbv, over the northeastern Pacific in spring 2001 [75]. For the TRACE-P campaign held in February–April 2001, it was found that European sources made a significant contribution to East Asian outflow, up to almost 50 ppbv during outflow events, but that this was typically masked by much larger Asian contributions, except to the north of Japan [76]. However, distinct transport episodes bringing enhanced levels of European CO were identified over southern Japan during the PEACE-A campaign in winter 2001, demonstrating that European impacts are not restricted to small enhancements of the continental background. At a maximum, the estimated CO from European anthropogenic surface emission was estimated to contribute as much as 40% of the total CO around Japan [77].

CTM studies of intercontinental transport suggest that contributions from European fossil fuel sources to surface ozone over East Asia vary between 1.8 ppbv in April and 0.2 ppbv in July on a monthly-mean basis, with an annual mean of 1.1 ppbv [26]. The corresponding North American impacts are 1.5 ppbv to 0.3 ppbv, with an annual mean of 0.9 ppbv. In both cases impacts in the lower free troposphere are 10–20% larger. While these enhancements appear small, it should be noted that there is a strong latitudinal variation over the region, with very little European impact over Southeast Asia, but a rather larger impact on Japan. Analyzing ozonesonde data from Japan and Hong Kong, Liu et al. [78] estimate a mean daytime surface ozone enhancement of 3 ppbv over East Asia. They also note that in the middle and upper troposphere, North American sources have a larger effect on ozone over East Asia than those from Europe, reflecting greater photochemical activity and different transport mechanisms.

Model analysis of CO and ozone at Mondy, updated from simulations presented in Wild and Akimoto [26], shows the spring maximum and summer minimum characteristic of remote sites and reflecting chemical lifetimes and transport patterns (Fig. 6). European influence dominates for both species, except for ozone in wintertime, where chemical destruction over Europe and slow chemical time scales at high latitudes lead to suppressed impacts, and North American sources dominate. The variations in the European influence seen here, 10–30 ppbv for CO and 1.0–3.5 ppbv for ozone, are broadly consistent with the trajectory analysis of Pochanart et al. [34] once the impact of Europe on the continental background, neglected in that study, are considered.

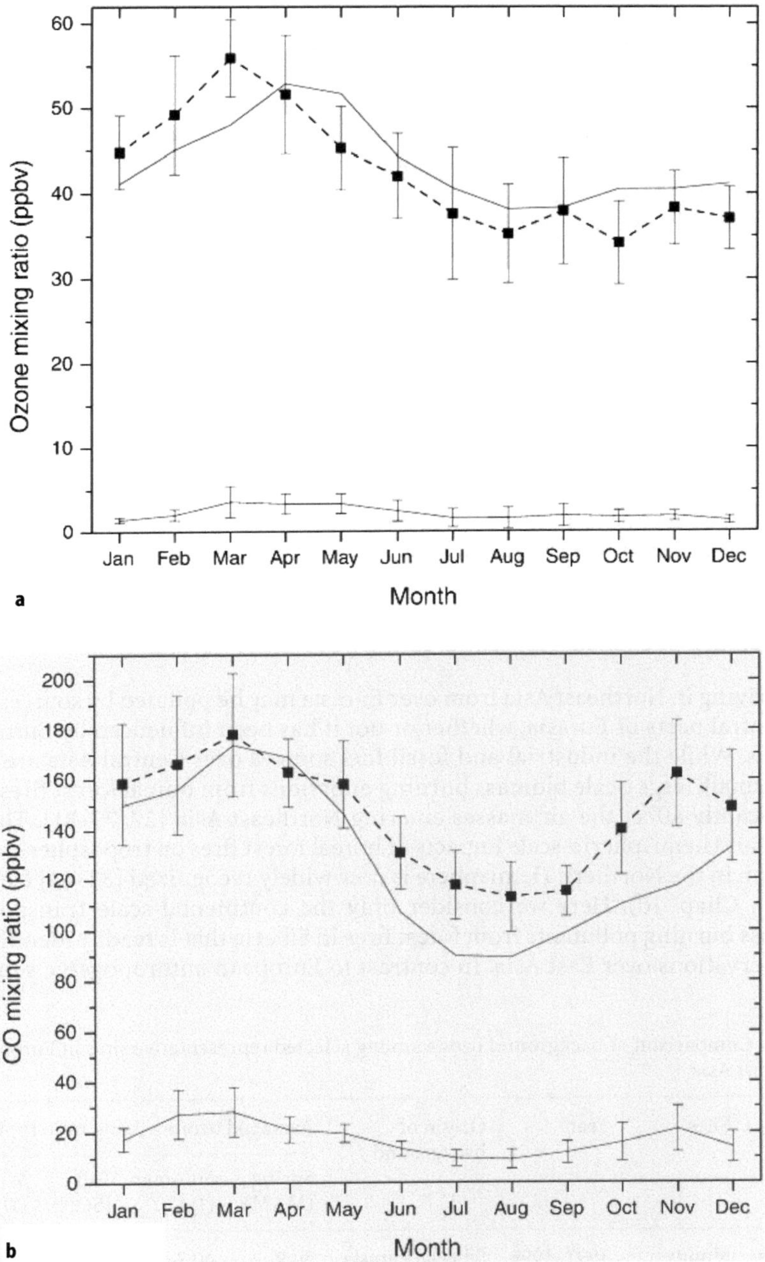

Fig. 6 Ozone and CO mixing ratios at Mondy from a global 3-D CTM [26] compared with observations, together with the contributions to ozone and CO in the model from European pollution sources. Solid line and shade represent observed values and their variability. Solid squares and error bar represent values from model calculation. European influences are shown as thin solid line with error bars

2.2.4
Comparison of Background Air Pollution Across Eurasia

Comparison of regional background mixing ratios of air pollutants over Europe and Northeast Asia may provide further indirect evidence of trans-Eurasian transport. Background mixing ratios of ozone may be estimated from long-term observational data using trajectory techniques to filter out possible influence from regional pollution sources. Analysis of this type suggests that air arriving in Europe has significantly less ozone than that arriving in Northeast Asia. Table 1 compares ozone mixing ratios in background air at sites in Western Europe [47, 48] and in East Asia [34, 35]. Air arriving in Europe originates from the Atlantic, while that arriving in Northeast Asia picks up emissions from over Eurasia. The difference between the two largely reflects photochemical production from sources over the continent, with a small contribution from differences in stratospheric influence over the regions. Background ozone increase in East Asia due to these sources is ~9 ppbv on average. While chemical time scales, deposition, and stratospheric influence prevent more detailed analysis based on observations alone, it is clear that Asian emissions build on higher background mixing ratios of ozone than are present over Europe.

2.3
Transport of Air Pollution from Central Eurasian Sources to Northeast Asia

Air arriving in Northeast Asia from over Eurasia may be polluted by sources over the central parts of Eurasia, whether or not it has been influenced by European sources. While the industrial and fossil fuel sources over Central Asia are relatively small, large-scale biomass burning emissions from boreal forest fires may significantly affect the air masses entering Northeast Asia [27, 79–81]. The regional and hemispheric-scale impacts of boreal forest fires on tropospheric composition in the Northern Hemisphere is now widely recognized [82–84] (see details in Chap. 10). Here we consider only the continental-scale transport of biomass burning pollutants from forest fires in Siberia that is readily identifiable in observations over East Asia. In contrast to European anthropogenic sources,

Table 1 Comparison of background ozone among selected representative sites in Europe and Northeast Asia

Region	Sites	Year	Origin of background	Averaged ozone mixing ratio (ppbv)			
				Spring (MAM)	Summer (JJA)	Fall (SON)	Winter (DJF)
NE Asia	Mondy	1997–1999	Siberia/Eurasia	50.9	40.7	39.8	42.4
	Happo	1996–1998	Siberia/Eurasia	51.0	41.3	40.8	42.7
Europe	Mace Head	1990–1994	Atlantic Ocean	40.0	30.6	33.5	35.6
	Arosa	1996–1997	Atlantic Ocean	40.9	31.8	31.6	36.6
Difference between Europe and Northeast Asia				10.5	10.0	7.8	6.4

which are relatively constant from year to year, wild fires in Siberia vary greatly in location, season, and coverage from year to year dependent on regional climate [85, 86]. The fires may occur over a large region between western Siberia and the coast of the Siberian Far East. Warm, dry conditions and low precipitation, as in the spring and summer of 1987 and 1998, provide good conditions for fires and lead to years with exceptionally large emissions from burning extended into the fall [79, 87, 88].

Background mixing ratios of pollutants arriving in East Asia may be strongly perturbed by boreal forest fire emissions, especially in years with intense burning, and these may largely hide the signal from intercontinental transport over Eurasia. Enhancements in trace gas mixing ratios, especially in CO, were observed in Russia and Siberia from ground-based measurement at Mondy [34], from isotopic CO measurement during the TROICA 2 and 3 projects [64, 66], and from spectroscopic measurements of CO total column at Zvenigorod [89, 90]. In particular the intense Siberian fires in 1998 significantly enhanced background CO levels at most observation sites in the remote Northern Hemisphere [83, 84]. As shown in Fig. 7a, increases in CO mixing ratios of ~40 ppbv were clearly observed at Mondy in eastern Siberia during fall 1998. Similar CO enhancements were also observed at high latitudes [84]. Uplift of these pollution plumes to the free troposphere, enhanced by convection stimulated by the fires themselves, brings pollution into the prevailing westerly flow, and leads to direct impacts on East Asia. The impacts of Siberian fires on air pollutant mixing ratios over East Asia have been clearly detected in northern Japan [27, 91, 92], and may extend to the lower latitudes of central Japan [81]. Figure 7b clearly shows enhancement of CO mixing ratios at Rishiri Island in northern Japan due to the transport of forest fire plumes from far eastern Siberia during the summer and fall of 1998 [27]. Spectroscopic column measurements of CO, C_2H_6, and HCN at Moshiri and Rikubetsu in Hokkaido Island, northern Japan also indicate enhancements in 1998, with a maximum change of 54–65% relative to seasonal mean values [92]. At Happo in central Japan, trajectory analysis of CO data during May-September 1998 indicates that Siberian air masses that pass forest fires contain ~29 and ~5 ppbv higher CO and ozone mixing ratios, respectively, than those that do not during their transport to Japan.

The downwind impact of Siberian fires on East Asia has also been examined in model studies. Simulations for May 1987 indicate that Siberian fires could increase ozone by more than 25% in parts of Siberia and northern Japan [80]. There is even a suggestion that the influence of Siberian forest fires may be detected as far away as continental Southeast Asia, where CO mixing ratios in the continental air masses transported from northern China/Siberia during September and October show ~45–75 ppbv increases in 1998 compared with those from other years between 1997 and 2000 at Srinakarin, Thailand [30]. Forest fires may therefore contribute significantly to the pollutant inflow to East Asia, and greatly complicate the attribution of pollutant enhancements to upwind, intercontinental sources.

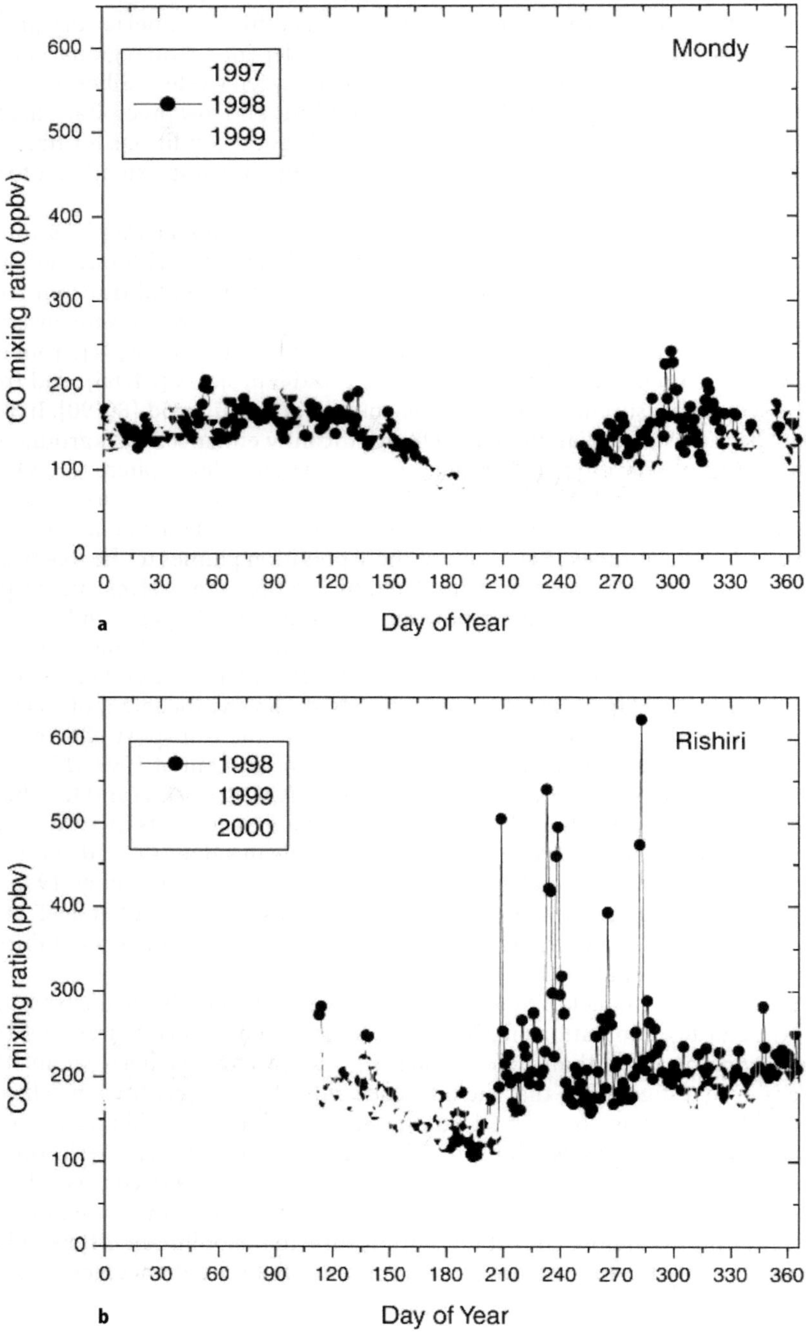

Fig. 7a,b Daily averaged plots of CO mixing ratios: **a** at Mondy in eastern Siberia; **b** at Rishiri Island in northern Japan

3 East Asian Pollution Outflow

The importance of pollutant outflow from East Asia on climate [93] and air quality [19, 20] is well established, and continuing increases in emissions are likely to lead to greater impacts in the future [94, 95]. In addition, East Asia has been shown to have a larger global impact on oxidants such as ozone than North American or European sources due to its more southerly location and more efficient lifting of pollutants [26]. Outflow of pollution from East Asia has been the focus of long-term measurements along the coast and a number of large international aircraft measurement campaigns over the western and central Pacific, and is much better characterized than the inflow to East Asia. Here we outline the pathways for export of pollution, and summarize the extensive measurements over the western Pacific region.

3.1 Export Mechanisms and Pathways

As described previously, the climatological conditions over East Asia are characterized by northwesterly flow at the surface for much of the year, governed by the Siberian High and carrying East Asian pollution to the western Pacific, and by southeasterly flow in summertime when the Pacific High brings cleaner marine air to the region. Westerly flow prevails in the middle and upper troposphere associated with the jet stream which is particularly strong in spring and autumn. The confluence of cold dry air from Siberia and warmer air from the Pacific causes active cyclogenesis over coastal regions of East Asia, leading to a succession of cyclones which move off the coast and track east or northeastwards over the Pacific.

The passage of these cyclones and their associated fronts provides the dominant pathway for pollutant export from East Asia [73]. The transport pathways are similar to those over the North Atlantic characterized by Cooper et al. [96], and are described in detail in Chap. 3. Warm, humid air from southerly latitudes is drawn northeastwards into these systems and is lifted into the free troposphere in 'warm conveyor belts' (WCBs), which bring air into the westerly flow which may carry it across the Pacific [54, 73]. These warm conveyor belts may draw boundary layer air from East Asia or the Western Pacific region between about 10 and 40°N, and deposit it at higher altitudes along the North Pacific storm track between about 30 and 60°N [54]. Where these WCBs draw air from East Asia, this pathway provides a mechanism for rapid export of pollution (e.g., [9]). However, the WCBs may also draw in clean marine air, leading to dilution of pollutant outflow [97, 98]. The variability in evolution of these cyclonic systems leads to high variability in pollutant transport [99], due to both the flow characteristics of the individual cyclone and the footprint region from which the WCB is drawn. In addition, the intermittent nature of this export leads to considerable variability in pollutant transport over the Pacific, and contributes to the episodic nature of air reaching North America [24].

A second important transport mechanism associated with these traveling cyclones is low-level outflow in air following the cold front. Pollutants from the more heavily-populated regions of continental northeast Asia may accumulate during anticyclonic periods in winter and spring and periodically may be exported in the boundary layer and lower troposphere behind cold fronts [100]. This low-level northwesterly flow is known to bring high levels of pollutants to Japan [101] and to Taiwan [102]. The dry, descending air stream behind the cold front may also draw high levels of stratospheric ozone down into the lower troposphere [103, 104], which may mix in with this polluted air. While this air is generally confined to the lower troposphere over the western Pacific, it may be entrained into the pre-frontal ascending flow in the following cyclone [99], thus lifting it to higher altitudes where it may be rapidly transported eastwards.

Deep convection provides a more rapid route for lifting pollution into the upper troposphere, where it may be transported northeastwards over the Pacific. This is a major export pathway for pollutants from biomass burning over Southeast Asia in the springtime [105]. Cold temperatures associated with the Siberian High suppress convection as a significant route for export of industrial pollutants from northeast Asia in the wintertime, but its region of influence migrates slowly northwards during the spring [106], and it provides efficient lifting to the upper troposphere during the summer [107]. The monsoon flow prevailing in this season typically confines continental outflow at lower altitudes to regions north of 40°N [8]. However, while mid-level westerly flow is weaker during this season, pollution lifted into the upper troposphere by convection may be caught up in southward transport around the Tibetan anticyclone. This leads to pollution from East Asian sources entering the tropical easterlies, where it may be transported westwards in the upper troposphere to the Middle East [108] and as far as the Mediterranean [109] (see also Chap. 6).

During the winter, anticyclonic conditions with subsiding air over the continent trap pollutants in the boundary layer, and continental outflow to the western Pacific principally occurs close to the surface [110]. In addition, there is significant transport by northerly winds into the tropics, leading to substantial impacts from Northeast Asian sources on Hong Kong [62, 111] and on Southeast Asia [29, 30]. Convective lifting of this pollution into the upper troposphere in the tropics may provide a route for transport into the mid-latitude westerlies and hence for intercontinental transport [112]. As cyclone formation becomes stronger and more frequent towards spring, westward transport dominates, leading to enhanced impacts on the Western Pacific north of about 20°N [8]. Westward transport is two to three times faster during this early spring period than in the summer [113], and transport across the North Pacific is also known to be strongest during spring [114].

3.2
Observations of Pollutant Export

Transport from East Asian sources is observed most clearly for soil dust, which is readily seen in SeaWIFS satellite images and in the aerosol index derived from the TOMS satellite instrument, as well as in reduced regional visibility and strong deposition of yellow sand ("Kosa") over Korea, Japan, and the North Pacific [115]. Dust is periodically lofted from the deserts of central Asia and western China, typically by strong surface winds associated with frontal systems [116], and is carried at relatively low altitudes, generally below 3 km, behind the cold front across China. Although episodes of dust transport vary greatly, many are lifted sufficiently high into the lower troposphere (2–7 km) to be transported out over the Pacific [117]. While Korea and Japan experience a number of Kosa events each spring, particularly large transport events may periodically carry dust to North America. Two recent events in April 1998 and in April 2001 could clearly be followed across the Pacific in SeaWIFS images, and contributed to a widespread reduction in visibility over much of western and central North America [116–118]. Results from the Asian Pacific Regional Aerosol Characterization Experiment (ACE-Asia) in spring 2001 indicate that dust from East Asia carries many pollutants with it, and that its properties are modified by black carbon and SO_2 from fossil fuel sources in the region [10].

Ground-based measurements of atmospheric oxidants have been made extensively along the coastal region of East Asia, and provide a useful gauge of the contributions from the continental region to the Pacific. As illustrated in Fig. 8, ozone and CO measurements at rural and remote sites in Japan clearly show higher mixing ratios in air coming from the continent than in less-polluted background air masses from Siberia or the Pacific [13–15, 35, 59, 119]. At Happo and Oki, Japan (see Fig. 4), the averaged annual increases of ozone and CO due to East Asian anthropogenic emission are about 11 ppb and 54 ppb, respectively [14, 35]. Long-term ozonesonde measurements over Japan and Hong Kong have provided further information, with pollution layers from biomass burning, fossil fuel and stratospheric sources highlighting the importance of different outflow mechanisms [78, 104].

Export of oxidants from East Asia has been the focus of several large aircraft campaigns over the Western Pacific. The NASA Global Tropospheric Experiment (GTE) Pacific Exploratory Missions PEM-West A and PEM-West B sampled the region in September-October 1991 and February-March 1994, respectively, close to the expected minimum and peak periods for export. In general, higher concentrations of ozone precursors were found in continental outflow during PEM-West B than PEM-West A, consistent with slower chemical processing in spring and more efficient transport, and they had a wider region of influence, north of 20°N rather than north of 40°N during PEM-West A [8]. Net ozone production was found to occur over much of the Western Pacific during spring, with continental impacts in the marine boundary layer extending out as far as 2000 km into the Pacific [120]. The Transport and Chemical Evolution over the Pacific (TRACE-P) campaign explored the springtime outflow in greater detail in February–April 2001 [9]. Aircraft measurements found very high levels of ozone (140 ppb), CO (1.1 ppm), SO_2 (31 ppb), and other pollutants over the Yellow Sea in fresh boundary layer outflow from Shanghai [121]. However, significant outflow in the upper and middle troposphere was clearly found as far east as 150°E, with hydrocarbon ratios suggesting combustion emissions only 2–3 days old [121]. Intercontinental transport was clearly revealed by aircraft measurements and MOPITT CO columns over the eastern North Pacific, with elevated ozone

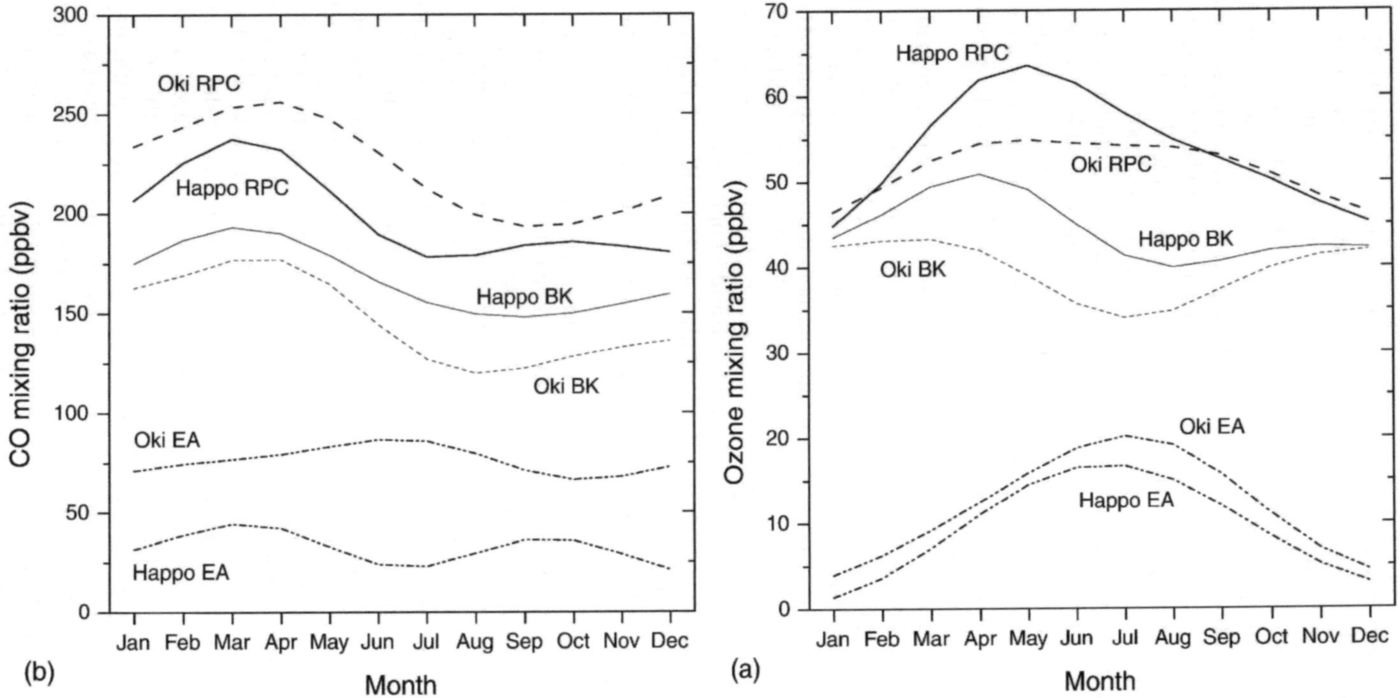

Fig. 8a,b Seasonal variation of: **a** ozone; **b** CO in different air mass categories at two representative sites in northeast Asia; Happo (36°41N, 137°48E, 1840 m asl) and Oki (36°17N, 133°11E, 90 m asl). The figure shows regionally polluted continental (RPC) outflow from Siberia that passes over the intense anthropogenic emission regions of East Asia before arriving at the sites, background continental (BC) outflow from Siberia that is not perturbed by intense anthropogenic emissions, and the difference between these categories representing the East Asian influence (EA)

(70 ppb) and CO (200 ppb) over a wide region (about 2500 km) at 2–4 km altitude that was attributable to East Asian sources [122]. Aircraft measurements suggest that more than half the NO_y and SO_x emitted over East Asia are lost over the continent, with half again lost by wet removal during transport into the free troposphere, but that the net flux out to the Pacific is about 20% of the emissions [123], mostly in the form of PAN and sulfate. More recent smaller-scale measurement campaigns such as the Pacific Exploration of Asian Continental Emission (PEACE), which sampled air over the western Pacific in both winter and spring 2002, are expected to provide substantial additional information on this outflow.

Increasing emissions of precursors over East Asia [94, 124] suggest that future impacts on oxidants at a hemispheric scale will be greater. Increases in boundary layer and lower tropospheric ozone at ozonesonde stations in Japan between 1969 and 1990 are about 2% and 1% per year, respectively [16], about 8 ppbv per decade in the boundary layer. Surface measurements at Okinawa between 1989 and 1997 suggest a trend of 2.5% per year in continental outflow from East Asia [17]. Observations over the west coast of the USA over 18 years indicate an increase of about 10 ppbv, about 30%, some which may be attributed to East Asian sources [125]. However, the global nature of impacts on ozone makes attribution to specific sources difficult, and comparison of trends in other species is valuable. Comparison of measurements from PEM-West B and TRACE-P, separated by seven years, shows a decrease in halocarbons consistent with a decline in anthropogenic emission [126], but trends in ozone cannot clearly be separated from the effects of more active photochemistry due to the three-week time lag between campaigns [127] and to greater stratosphere-troposphere exchange [128]. The large variability in outflow pathways and uncertainty regarding the representativeness of the sampling make extraction of trends very difficult for many species, and further observations will be required.

4 Air Pollution Transport over Continental Southeast Asia

4.1 Southeast Asia, Climatology, and Air Pollution

Southeast Asia lies between about ~20°N and ~10°S latitude in the tropics and comprises two distinct regions: the continental regions of the Indochina peninsula and the island countries separating the Indian and Pacific Oceans. Southeast Asia is a large source of several air pollutants, and may make a significant contribution to regional and global pollution due to (1) increasing anthropogenic emissions associated with rapid development and industrialization [5, 94, 129], (2) biogenic emissions from large tropical forests [130], (3) wide-spread biomass burning [131–136], and (4) the greater oxidizing capacity in tropical regions due to higher UV intensity and humidity [137, 138].

Biomass burning is a major source of atmospheric pollutants in the tropics [30, 131, 136, 139]. Unlike boreal forest fires, tropical biomass burning in Southeast Asia is strongly influenced by anthropogenic post-agricultural waste burning, and occurs consistently each year [131, 133, 140, 141]. The global impacts of tropical biomass burning are described in Chap. 9; in this section, we summarize the import and export of air pollution related to both anthropogenic emissions and biomass burning over continental Southeast Asia, principally in the Northern Hemisphere.

The climatology of continental Southeast Asia is dominated by a strong monsoon (Fig. 2) [29, 30], with meteorological features similar to those of South Asia (see Chap. 6) [142–144]. The Southeast Asian monsoon and the associated movement of the ITCZ define the climatology and air mass transport in this region [30, 142]. Since most of continental Southeast Asia is in the tropics, seasons are not as distinct as in more temperate zones, and typically only wet and dry seasons can be clearly distinguished. When the ITCZ moves northward across continental Southeast Asia into the Northern Hemisphere, the summer monsoon brings marine air masses from the Indian Ocean to the region. This region experiences a wet season for about six months (May–October) before the ITCZ moves back to the Southern Hemisphere and long-range transport of continental air masses from northern Asia in the winter monsoon prevails during the subsequent dry season (November–April). Biomass burning occurs during the dry season, typically with a peak in February–March in continental regions. However, it should be noted that because of the different circulations in the Northern and Southern Hemispheres, wet and dry seasons in continental Southeast Asia and in insular Southeast Asia are at opposite times of the year, and consequently biomass burning also occurs in different periods. The movement of the ITCZ which normally separates the Northern and Southern Hemisphere may also cause air mass exchange between the two Hemispheres.

4.2 Air Pollution Import to Continental Southeast Asia

There is no clear observational evidence that the intercontinental transport of air pollution exerts a strong impact on continental Southeast Asia, partly due to the very few studies conducted in this region, and partly to the transport characteristics of the region. Import of air pollution to continental Southeast Asia occurs mainly as a result of long-range transport of regional-scale pollution originating within the Asian continent, particularly from anthropogenic emissions from Northeast and South Asia [29, 30]. Transport of pollution from other parts of Asia does not normally occur, but may be observed occasionally in episodic events. Model studies, however, suggest that intercontinental transport of air pollution could be observable in continental Southeast Asia. Results from a global model of tropospheric chemistry during the PEM-West B campaign in spring 1994 show that biomass burning from Africa may contribute significantly to outflow from continental Southeast Asia in the upper troposphere [73].

During the summer monsoon (wet season), continental Southeast Asia is dominated by marine air masses from the low and middle latitudes of the Southern Hemispheric Indian Ocean. These marine air masses bring small to moderate amounts of air pollution to continental Southeast Asia. Comparing observed surface CO mixing ratios at a continental site in rural Thailand with three remote sites in the Indian ocean, there is a clear increase in CO when oceanic air is transported to continental Southeast Asia (Fig. 9) [30, 52, 142, 145, 146]. Observed CO mixing ratios are lowest in the pristine marine environment of Amsterdam Island, increase at the Seychelles and Kaashidhoo, and are highest in the continental environment at Srinakarin. Trajectory analysis at Srinakarin indicates that the increasing CO is due to large pollution sources over eastern Africa and the Indian subcontinent, in addition to local/sub-regional sources within continental Southeast Asia. The ~20 ppbv higher CO mixing ratios in the Seychelles than at Amsterdam Island has mainly been attributed to pollution outflow from the African continent and Madagascar [142]. The further increases in CO of ~10 ppbv and ~45 ppbv on reaching Kaashidhoo and Srinakarin, respectively, are mainly due to pollution sources in the Indian subcontinent [30, 142], a large source of air pollutants [143, 147–149]. This increase in air pollutants towards the continent due to large anthropogenic sources over South Asia has been confirmed during INDOEX ship cruise campaigns of the last decade [149, 150] (see also Chap. 6).

It is interesting that, while observations in remote South Asia indicate the potential influences of intercontinental transport of pollutants from biomass burning in Africa and Australia [142], such effects are not clearly observed in continental Southeast Asia, probably due to (1) the limited observational data in this region and (2) the stronger signal of air pollutants emitted from the Indian subcontinent and within continental Southeast Asia. The only significant evidence ever found in the region is a ~60 ppbv enhancement of CO at Srinakarin over typical levels in air from the Indian Ocean, associated with marine air masses influenced by biomass burning in Indonesia and northern Australia during the El Niño year of 1997 [30].

During the winter monsoon (dry season), continental Southeast Asia is influenced by the transport of regionally polluted continental air masses from the northeast and from western Asia [25, 29, 30]. In the early months of the winter monsoon, the Siberian high and the associated continental outflow dominate most regions of East Asia. Continental outflow from Siberia and Northeast Asia passes over the heavily polluted regions of East Asia and picks up significant amounts of pollution during transport to Southeast Asia. Together with strong subsidence in the early dry season, significant increases in air pollutant levels are observed in continental Southeast Asia. Compared with oceanic air masses containing ~100 ppbv of CO at Srinakarin, more than 300 ppbv of CO is typically observed in continental outflow from Northeast Asia [30]. In the later months of the dry season, long-range transport of air masses from the Middle East, western Asia, and from as far away as Europe may be found [25, 29, 30]. However, biomass burning also occurs during these months and contributes to the observed annual maximum in air pollutants levels over continental Southeast Asia. The strong impact of regional biomass burning makes it extremely difficult to detect any evidence of the long-range transport of air pollution from other regions to continental Southeast Asia.

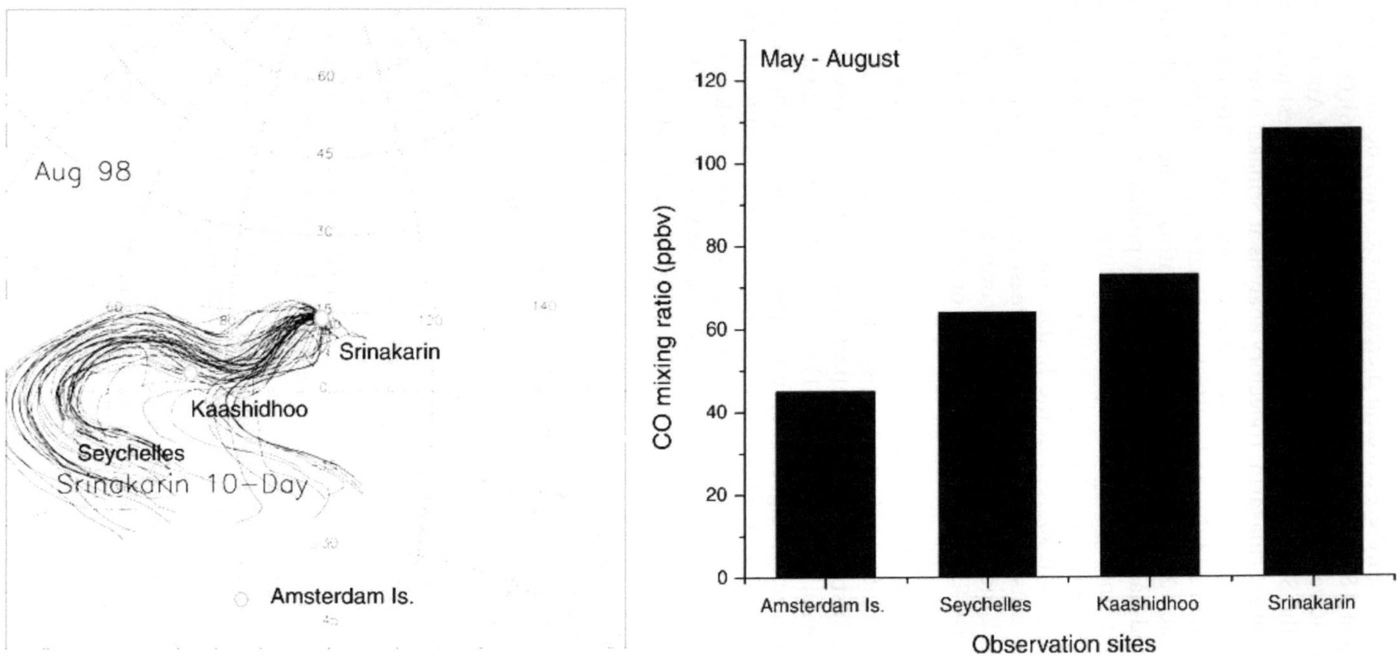

Fig. 9 Ten-day backward air mass trajectories arriving at Srinakarin, Thailand during the wet season of 1998 (August). Three other representative sites in the Indian Ocean are shown – Kaashidhoo, Seychelles, and Amsterdam Island. The *right panel* compares the CO mixing ratios at these four sites during the early period of the summer monsoon (May–August). Data at Srinakarin, Kaashidhoo, the Seychelles, and Amsterdam Island are from Pochanart et al. [30], Lobert and Harris [142], Holloway et al. [145], and Gros et al. [146], respectively

4.3 Air Pollution Outflow from Continental Southeast Asia

The export of air pollution from continental Southeast Asia is most significant during the burning period in the late dry season. Outside the burning season, the export of air pollution from continental Southeast Asia is relatively small. During the wet season, air masses from the Indian Ocean influenced by anthropogenic emissions over the Indian subcontinent and continental Southeast Asia are transported to the northeast. Climatological studies at Mt. Fuji in central Japan reveal evidence of the transport of air masses from Southeast Asia, but enhancements of ozone and CO associated with these air masses are not observed [151]. Conversely, in the early dry season air pollutants from continental Southeast Asia may be exported to the Indian Ocean. Trajectory categorization of CO data at Kaashidhoo indicates that the highest CO mixing ratios (>140 ppbv) are found in air masses arriving from continental Southeast Asia [142].

During the late dry season between January and April, biomass burning is a large source of atmospheric pollution in continental Southeast Asia. The seasonal variations over continental Southeast Asia are revealed by the number of hot spots detected by the Along Track Scanning Radiometer (ATSR) satellite instrument, shown in Fig. 10 [152, 153]. Aside from the intense Indonesian fires in 1997, a larger number of fires are typically found in continental Southeast Asia than in India or Indonesia. During spring 1998 and 1999, ATSR hot spots in continental Southeast Asia make up as much as 30–35% of the Northern Hemispheric total. The global distribution of CO emissions calculated for the PEM-Tropics B campaign (March–April 1999) indicates that the CO emissions from biomass burning in Asia (mainly from India and continental Southeast Asia) account for 40% of global CO emissions and are almost four times higher than CO from fossil fuel sources over the whole of Asia [74]. A good correlation of ATSR hot spot data with CO, ozone and POLDER aerosol index have been found in Thailand [30, 153]. In February–March, the most intense burning period of the year, the maximum CO and ozone mixing ratios observed in rural Thailand are a factor of four to six times higher than their levels during the wet season [30]. A similar feature is also observed in South Asia [142, 144, 148]. Compared with the relatively low mixing ratios during the summer monsoon (see Fig. 9), the month-averaged maxima of CO at Seychelles, Kaashidhoo, and Srinakarin are 150, 240, and 450 ppbv, respectively [30, 142].

Convergence and updrafts lift biomass burning plumes into the middle and upper troposphere where they may be caught up by the westerly flow at subtropical latitudes and transported to the east. This export of air pollutants associated with Southeast Asian outflow has been identified over the Pacific from both model studies and observations [73, 74, 154]. Analyses of tropospheric ozone soundings over Hong Kong indicate the influence of Southeast Asian biomass burning over southeastern China [78, 155–159]. Using six years of ozonesonde data between 1993 and 1999, Chan et al. [158] found that 90% of ozone enhancement events occurred during the Southeast Asian biomass burning season and that 77% of these ozone-rich layers lay between 2 and 6 km altitude. The excess ozone in these layers contributed 12% of the tropospheric ozone

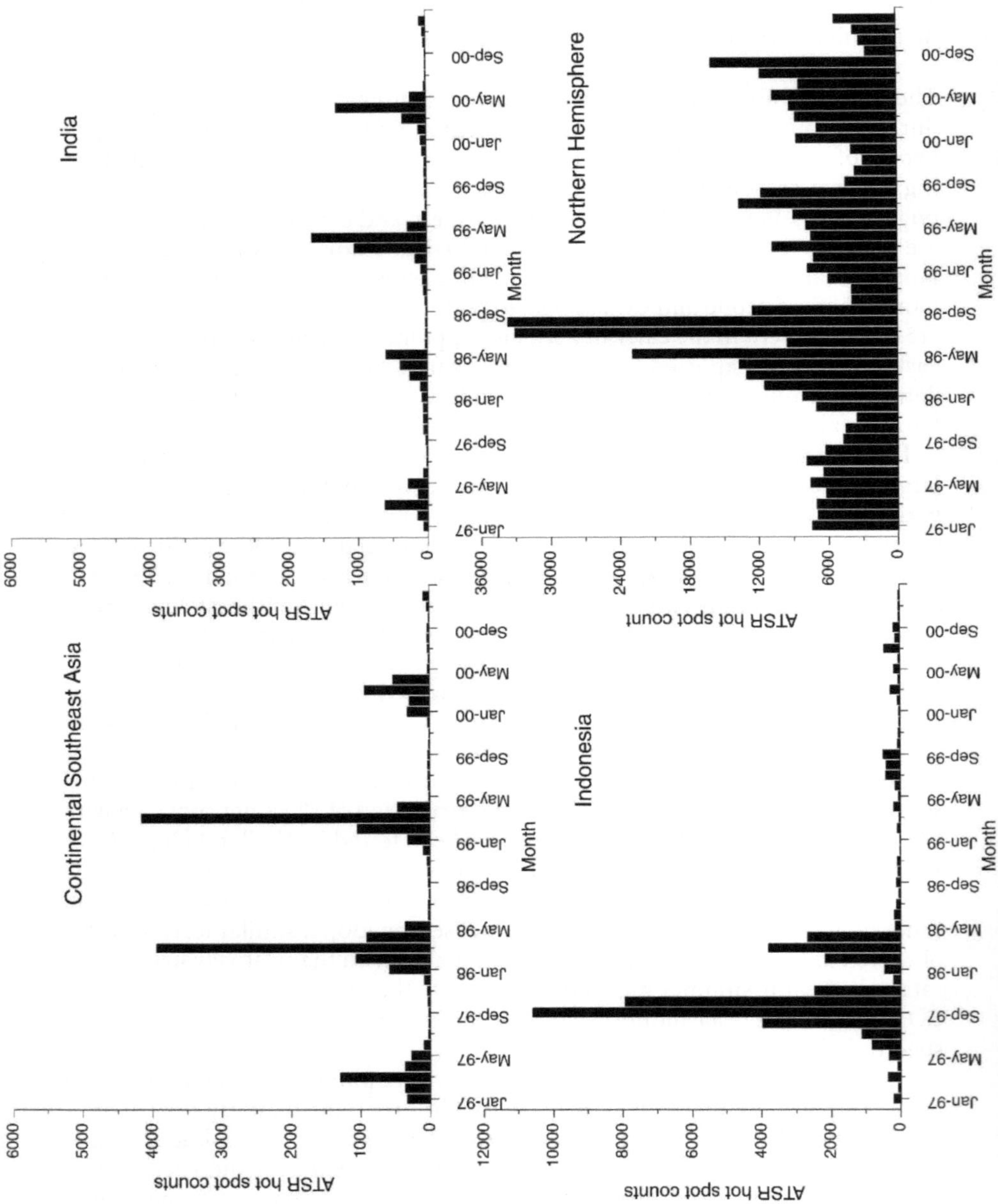

Fig. 10 ATSR hot spot counts in continental Southeast Asia (5–25°N, 90–110°E), the Indian subcontinent (5–30°N, 70–90°E), Indonesia (10°S–5°N, 95–130°E) and those of the entire Northern hemisphere (ATSR data use with permission, copyright ESA 1999, ESA/ESRIN, ATSR World Fire Atlas Project, with support from the IGBP-DIS Office)

column. Enhancements in ozonesonde data due to biomass burning are observed as far away as Hilo in the eastern Pacific. During the PEM-Tropics campaign in spring 1999, most of the CO enhancement events observed in the free troposphere over the northern tropical Pacific, and 20% of the tropospheric column, were attributed to transoceanic transport of air pollution from biomass burning in Southeast Asia [74].

5 Summary

This chapter has reviewed the import and export of air pollution by long-range transport to and from East Asia, focusing on carbon monoxide and ozone. We have summarized the sources and transport of air pollutants over Northeast and Southeast Asia, both of which are important regions for air pollutant emissions. Rapid industrialization is likely to make Northeast Asia the largest source region globally in the coming decades. A similar rapid industrialization over Southeast Asia is supplemented by wide-spread biomass burning over the region. Based on the available evidence from observations and model studies, the following conclusions are drawn:

1. The monsoon regimes provide the main climatological controls on the import and export of air pollution over East Asia. During the winter monsoon, long-range transport of air pollution associated with strong continental outflow is dominant, particularly in Northeast Asia. During the summer monsoon, the impact of air pollution transport is less significant as both Northeast Asia and Southeast Asia are under the influence of relatively clean air masses transported from the Pacific and Indian Oceans, respectively.
2. Intercontinental transport of air pollution is more pronounced in Northeast Asia, which lies below the prevailing mid-latitude westerly flow. However, the import of air pollution from Europe across the Eurasian continent is not as evident as the export of air pollution from East Asia across the Pacific Ocean. Trans-Pacific transport of air pollution from Northeast Asia occurs by strong convection or frontal lifting to the free troposphere followed by the rapid westerly flow, while trans-Eurasian transport from Europe occurs at higher latitudes and at lower altitude over the Eurasian continent, making identification of distinct 'pollution events' difficult. Trans-Eurasian transport may also encounter additional pollution sources over Eurasia, complicating identification of the signal from European sources.
3. The lack of observational studies over the inner parts of Eurasia, upwind of the anthropogenic emissions from East Asia, makes identification of trans-Eurasian transport difficult. Most of the observational studies in Northeast Asia have been carried out in the coastal regions, close to and perturbed by regional pollution sources. However, limited data from long-term ground-based observations at Mondy in southeastern Siberia and from trans-Siberian railroad-based TROICA campaigns across Eurasia provide some evidence of trans-Eurasian air pollution transport. These studies found higher CO and ozone mixing ratios in air masses transported from Europe, and a decrease in

CO mixing ratios across Eurasia with increasing distance from Europe, suggesting the advection and mixing of CO-rich air from Europe into the background troposphere. Further observational studies are needed to characterize the inflow to East Asia in greater detail.
4. Large-scale boreal forest fires in central regions of northern Eurasia are a significant source of air pollution affecting Northeast Asia. Enhancements of CO and aerosol associated with forest fires over Eastern Siberia have been clearly detected over the northern parts of Northeast Asia, and are particularly strong in years with intense fires.
5. Trans-Pacific transport of air pollution has been studied in greater detail than trans-Eurasian transport. The principal export mechanisms involve the passage of frontal systems, which play a major role in spring and autumn, and deep convection, which dominates export in summer. Major international aircraft campaigns over the western and central Pacific have greatly helped in characterizing the outflow from east Asia, and have clearly identified the substantial impacts that continental outflow can have over much of the North Pacific in springtime, with impacts reaching as far as North America.
6. Over continental Southeast Asia, the import of air pollution related to intercontinental transport is relatively small. During the wet season, when the summer monsoon and marine air masses from the Indian Ocean prevail, a moderate amount of air pollution from the emission sources over the Indian subcontinent may reach continental Southeast Asia. The signals are stronger in the early dry season, when the winter monsoon brings continental outflow from the heavily-polluted regions of Northeast Asia to continental Southeast Asia.
7. Export of air pollution from continental Southeast Asia, mainly from biomass burning within the region, is important in the late dry season (boreal springtime). The biomass burning pollution is uplifted by deep convection to the middle and upper troposphere, caught by the sub-tropical westerly flow, and transported to the western and central Pacific. Transoceanic transport of air pollutants from Southeast Asian biomass burning is thought to be one of the most significant sources of ozone and CO over the western and central Pacific in the springtime.

Acknowledgement We are grateful to M. Naja of FRSGC for his contribution of Fig. 1.

6 References

1. Kato N, Akimoto H (1992) Atmos Environ 26A:2997
2. OECD (1993) OECD environmental data compendium 1993. Paris, France
3. EPA (1999) National air pollutant emission trends, 1900–1998, Report EPA-454/R-00-002. EPA, Research Triangle Park, NC
4. EMEP (2001) Emission data reported to UNECP/EMEP: Evaluation of the spatial distribution of emissions. MSC-W Status Report, Norwegian Meteorological Institute, Oslo, Norway
5. Streets DG, Tsai NY, Akimoto H, Oka K (2002) Water Air Soil Pollut 130:187
6. Naja M, Akimoto H, Staehelin J (2003) J Geophys Res 108:doi:10.1029/2002JD002477

7. Hoell JM, Davis DD, Liu SC, Newell R, Shipham M, Akimoto H, McNeal RJ, Bendura RJ, Drewry JW (1996) J Geophys Res 101:1641
8. Hoell JM, Davis DD, Liu SC, Newell RE, Akimoto H, McNeal RJ, Bendura RJ (1997) J Geophys Res 102:28223
9. Jacob DJ, Crawford JH, Kleb MM, Connors VE, Bendura RJ, Raper JL, Sachse GW, Gille JC, Emmons L (2003) J Geophys Res 108:doi:10.1029.2002JD003276
10. Huebert BJ, Bates T, Russell PB, Shi G, Kim YJ, Kawamura K, Carmichael G, Nakajima T (2003) J Geophys Res 108:doi:10.1029/2003JD003550
11. Takigawa M, Sudo K, Takahashi M, Takegawa N, Kondo Y (2002) Agu Fall Meeting. San Francisco, USA
12. Takigawa M (2003) The PEACE international workshop 2003. Tokyo, Japan
13. Akimoto H, Mukai H, Nishikawa M, Murano K, Hatakeyama S, Liu CM, Buhr M, Hsu KJ, Merrill JT, Newell RE (1996) J Geophys Res 101:1999
14. Pochanart P, Hirokawa J, Kajii Y, Akimoto H, Nakao M (1999) J Geophys Res 104:3621
15. Pochanart P, Akimoto H, Kinjo Y, Tanimoto H (2002) Atmos Environ 36:4235–4250
16. Akimoto H, Nakane H, Matsumoto Y (1994) The chemistry of oxidant generation: tropospheric ozone increase in Japan. In: Calvert JG (ed) Chemistry of the atmosphere: its impact on global change. Blackwell Scientific Publications, Oxford, p 261
17. Lee SH, Akimoto H, Nakane H, Kurnosenko S, Kinjo Y (1998) Geophys Res Lett 25:1637
18. Berntsen TK, Isaksen ISA (1997) J Geophys Res 102:21239
19. Berntsen TK, Karlsdóttir S, Jaffe DA (1999) Geophys Res Lett 26:2171
20. Jacob DJ, Logan JA, Murti PP (1999) Geophys Res Lett 26:2175
21. Jaffe D, Anderson T, Convert D, Kotchenruther K, Trost B, Daneilson J, Simpson W, Berntsen T, Karlsdóttir S, Blake D, Harris J, Carmichael G, Uno I (1999) Geophys Res Lett 26:711
22. Jaffe D, Anderson T, Convert D, Trost B, Daneilson J, Simpson W, Blake D, Harris J, Streets D (2001) J Geophys Res 106:7449–7461
23. Jaffe D, McKendry I, Anderson T, Price H (2003) Atmos Environ 37:391
24. Yienger JJ, Galanter M, Holloway TA, Phadnis MJ, Guttikunda SK, Carmichael GR, Moxim WJ, Levy H II (2000) J Geophys Res 105:26931–26945
25. Newell RE, Evans MJ (2000) Geophys Res Lett 27:2509
26. Wild O, Akimoto H (2001) J Geophys Res 106:27729
27. Tanimoto H, Kajii Y, Hirokawa J, Akimoto H (2000) Geophys Res Lett 24:4073
28. Yurganov L, Akimoto H, Grechko E, Maksyutov S, Kondo Y (2003) IUGG2003, Sapporo, Japan
29. Pochanart P, Kreasuwun J, Sukasem P, Geeratithadaniyom W, Tabulanon MS, Hirokawa J, Kajii Y, Akimoto H (2001) Atmos Environ 35:2657
30. Pochanart P, Akimoto H, Kajii Y, Sukasem P (2003) J Geophys Res 108: doi: 10.1029/2002JD003360
31. Ogawa T, Miyata A (1985) J Meteorol Soc Jpn 63:937–946
32. Tsuruta H, Shinya K, Mizoguchi T, Ogawa T (1989) Seasonal behavior of the tropospheric ozone in rural Japan. In: Bojkov RD, Fabian P (eds) Ozone in the atmosphere. Deepak, Hampton, p 433
33. Zhang Y, Sperber KR, Boyle JS (1996) Climatology of East Asian winter monsoon and cold surges: Results from the 1979–1995 NCEP/NACR reanalysis. PCMDI Report No. 38. Lawrence Livermore National Laboratory, Livermore, CA
34. Pochanart P, Akimoto H, Kajii Y, Potemkin VM, Khodzer TV (2003) J Geophys Res 108:doi:10.1029/2001JD001412
35. Pochanart P, Kato S, Katusno T, Akimoto H (2004) Atmos Environ 38:1325
36. Austin JF, Midgley RP (1994) Atmos Environ 28:39
37. Parrish DD, Trainer M, Holloway JS, Yee JE, Warshawsky MS, Fehsenfeld FC, Forbes GL, Moody JL (1998) J Geophys Res 103:13357
38. Sunwoo Y, Carmichael GR, Ueda H (1994) Atmos Environ 28:25
39. Narita D et al. (1999) Chemosphere Global Chang Sci 1:137
40. Chameides WL et al. (1999) Geophys Res Lett 26:867

41. Luo C, St John JC, Zhou X, Lam KS, Wang T, Chameides WL (2000) J Geophys Res 105: 1889
42. Cheung VTF, Wang T (2001) Atmos Environ 35:4957
43. Wang T, Cheung VTF, Anson M, Li YS (2001) Geophys Res Lett 28:2373–2376
44. Raatz WE (1989) Atmos Environ 23:2501
45. Oltmans SJ, Levy H II (1992) Nature 358:392
46. Oltmans SJ, Levy H II (1994) Atmos Environ 28:9–24
47. Derwent RG, Simmonds PG, Seuring S, Dimmer C (1998) Atmos Environ 32:145
48. Pochanart P, Akimoto H, Maksyutov S, Staehelin J (2001) Atmos Environ 35:5553
49. Solberg S, Stordal F, Hov Ø (1997) J Atmos Chem 28:111
50. Laurila T (1999) J Geophys Res 104:26235
51. Novelli PC, Steele LP, Tans PP (1992) J Geophys Res 97:20731
52. Novelli PC, Masarie KA, Lang PM (1998) J Geophys Res 103:19015
53. Wang Y, Jacob DJ, Logan JA (1998) J Geophys Res 103:10757
54. Stohl A (2001) J Geophys Res 106:7263
55. Stohl A, Eckhardt S, Forster C, James P, Spichtinger N (2002) J Geophys Res 106: doi:1029/2001JD001396
56. Scheel HE et al. (1997) J Atmos Chem 28:11
57. Lawrence MG, Crutzen PJ (1999) Nature 402:167
58. Kajii Y, Someno K, Tanimoto H, Hirokawa J, Akimoto H, Katsuno T, Kawara J (1998) Geophys Res Lett 25:3305
59. Nagao I, Matsumoto K, Tanaka H (1999) J Geophys Res 104:11675
60. Chan LY, Liu HY, Lam KS (1998) Atmos Environ 32:159–168
61. Lam KS, Wang TJ, Chan LY, Wang T, Harris J (2001) Atmos Environ 35:3121
62. Chan CY, Chan LY, Lam KS, Li YS, Harris JM, Oltmans SJ (2002) J Geophys Res 107: doi: 10.1029/2002JD002131
63. Carmichael GR (2003) http://www.cgrer.uiowa.edu/people/carmichael/Carmichael.html
64. Bergamaschi P, Brenninkmeijer CAM, Hahn M, Röckmann T, Scharffle DH, Crutzen PJ, Elansky NF, Belikov IB, Trivett NBA, Worthy DEJ (1998) J Geophys Res 103:8227
65. Crutzen PJ et al. (1998) J Atmos Chem 29:179
66. Röckmann T, Brenninkmeijer CAM, Hahn M, Elansky NF (1999) Chemosphere Global Change Sci 1:219
67. Brunner D, Staehelin J, Jeker D (1998) Science 282:1305
68. Brunner D, Staehelin J, Jeker D, Wernli H, Schumann U (2001) J Geophys Res 106:27673
69. Marenco A et al. (1998) J Geophys Res 103:25631
70. Stohl A, James P, Forster C, Spichtinger N, Marenco A, Thouret V, Smit HGJ (2001) J Geophys Res 106:27757
71. Fehsenfeld FC, Trainer M, Parrish DD, Volz-Thomas A, Penkett S (1996) J Geophys Res 101:28869
72. Penkett PA, Law KS, Platt U, Volz-Thomas A (2003) European export of particulates and ozone by long-range transport: overview of subproject EXPORT-E2. In: Midgley PM, Reuther M, Verlag M (eds) Towards cleaner air for Europe – science, tools and applications, part 2. Weikersheim, Germany
73. Bey I, Jacob DJ, Logan JA, Yantosca RM (2001) J Geophys Res 106:23097
74. Staudt AC, Jacob DJ, Logan JA, Bachiochi D, Krishnamurti TN, Sachse GW (2001) J Geophys Res 106:32571
75. Jaegle L et al. (2003) J Geophys Res 108:doi:10.1029/2002JD003121
76. Liu H et al. (2003) J Geophys Res 108:doi:10.1029/2002JD003102
77. Takigawa M, Akimoto H, Sudo K, Takahashi M, Takegawa N, Kondo Y (2003) IUGG2003. Sapporo, Japan
78. Liu H, Jacob DJ, Chan LY, Oltmans SJ, Bey I, Yantosca RM, Harris JM, Duncan BN, Martin RV (2002) J Geophys Res 107:doi:10.1029/2001JD002005
79. Cahoon DR Jr, Stocks BJ, Levine JS, Cofer WR III, Pierson JM (1994) J Geophys Res 99:18627
80. Phadnis MJ, Carmichael GR (2000) Atmos Environ 34:483

81. Kato S, Pochanart P, Hirokawa J, Kajii Y, Akimoto H, Ozaki Y, Obi K, Katsuno T, Streets DG, Minko NP (2002) Atmos Environ 36:385
82. Wotawa G, Trainer M (2000) Science 288:324
83. Wotawa G, Novelli PC, Trainer M, Granier C (2001) Geophys Res Lett 28:4575
84. Röckmann T, Jöckel P, Gros V, Bräunlich M, Possnert G, Brenninkmeijer CAM (2002) Atmos Chem Phys 2:147
85. Shvidenko AZ, Nilsson S (2000) Extent, distribution, and ecological role of fire in Russian forests. In: Kasischke ES, Stocks BJ (eds) Fire, climate change, and carbon cycling in the boreal forests. Springer, Berlin Heidelberg New York, p 132
86. Shvidenko AZ, Nilsson S (2000) Fire and the carbon budget of Russian forests. In: Kasischke ES, Stocks BJ (eds) Fire, climate change, and carbon cycling in the boreal forests. Springer, Berlin Heidelberg New York, p 289
87. Kajii Y et al. (2002) J Geophys Res 107:doi:10.1029/2001JD001078
88. Kasischke ES, Bruhwiler LP (2003) J Geophys Res 107:doi:10.1029/2001JD000461
89. Yurganov LN, Grechko EI, Dzhola AV (1997) Geophys Res Lett 24:1231
90. Yurganov L, Grechko E, Dzhola A (2002) Recent Res Dev Geophys 4:249
91. Zhao Y, Kondo Y, Murcray FJ, Liu X, Koike M, Irie H, Strong K, Suzuki K, Sera M, Ikegami Y (2000) Geophys Res Lett 27:2085
92. Zhao Y et al. (2001) J Geophys Res 107:doi:10./1029/2001JD000748
93. Berntsen TK et al. (1996) Tellus B 48:13
94. Aardenne van JA, Carmichael GR, Levy H II, Streets D, Hordijk L (1999) Atmos Environ 33:633
95. Prather M et al. (2003) Geophys Res Lett 30:doi:10.1029/2002GL016285
96. Cooper OR, Moody JL, Parrish DD, Trainer M, Ryerson TB, Holloway JS, Hübler G, Fehsenfeld FC, Evans MJ (2001) J Geophys Res 106:5437
97. Grant WB et al. (2000) J Geophys Res 105:3757
98. Mari C, Evana M, Palmer P, Jacob DJ, Sachse GW, Escobar J, Gazen D (2004) J Geophys Res (submitted)
99. Hannan JR et al. (2003) J Geophys Res 108:doi:10.1029/2002JD003105
100. Kaneyasu N, Takeguchi K, Hayashi M, Fujita S, Uno I, Sasaki H (2000) J Geophys Res 105:17361
101. Kotamarti VR, Carmichael GR (1990) Atmos Environ 24a:1521
102. Liu CM, Buhr M, Merrill JT (1997) J Geophys Res 102:28613
103. Carmichael GR, Uno I, Phadnis MJ, Zhang Y, Sunwoo Y (1998) J Geophys Res 103:10649
104. Wild O, Sundet JK, Prather MJ, Isaksen ISA, Akimoto H, Browell EV, Oltmans SJ (2003) J Geophys Res 108:doi:10.1029/2002JD003283
105. Kawakami S et al. (1997) J Geophys Res 102:28367
106. Miyasaki Y et al. (2003) J Geophys Res 108:doi:10.1029/2002JD003248
107. Blake DR, Chen T-Y, Smith TW, Wang CJ-L, Wingenter OW, Blake NJ, Rowland FS, Mayer EW (1996) J Geophys Res 101:1763
108. Li Q et al. (2001) Geophys Res Lett 28:3235
109. Lelieveld J et al. (2002) Science 298:794
110. Zhang M, Uno I, Sugata S, Wang Z, Byun D, Akimoto H (2002) Geophys Res Lett 29: doi:10.1029/2001GL014368
111. Chan CY, Chan LY (2000) J Geophys Res 105:20707
112. Newell RE, Browell EV, Davis DD, Liu SC (1997) Geophys Res Lett 22:2733
113. Gregory GL, Merrill JT, Shipman MC, Blake DR, Sachse GW, Singh HB (1997) J Geophys Res 102:28275
114. Merrill JT (1989) Chem Oceanogr 10:15
115. Uematsu M, Duce RA, Prospero JM, Chen L, Merrill JT, McDonald RL (1983) J Geophys Res 88:5343
116. Uno I et al. (2001) J Geophys Res 106:18331
117. Gong SL, Zhang XY, Zhao TL, McKendry IG, Jaffe DA, Lu NM (2003) J Geophys Res 108:doi:10.1029/2002JD002633
118. Husar RB et al. (2001) J Geophys Res 106:18317

119. Tanimoto H, Furutani H, Kato S, Matsumoto J, Makide Y, Akimoto H (2002) J Geophys Res 107:doi:10.1029/2001JD0014196
120. Crawford J et al. (1997) J Geophys Res 102:28469
121. Russo R et al. (2003) J Geophys Res 108:doi:10.1029/2002JD003184
122. Heald HL et al. (2003) J Geophys Res 108:doi:1029/2003JD003507
123. Koike M et al. (2003) J Geophys Res 108:doi:10.1029.2002JD003284
124. Streets DG et al. (2003) J Geophys Res 108:doi:10.1029.2002JD003093
125. Jaffe D, Price H, Parrish D, Goldstein A, Harris J (2003) Geophys Res Lett 30:doi:10.1029/2003GL017024
126. Blake NJ et al. (2003) J Geophys Res 108:doi:10.1029/2002JD003367
127. Davis DD et al. (2003) J Geophys Res 108:doi:10.1029/2002JD003232
128. Browell EV et al. (2003) J Geophys Res 108:doi:10.1029/2002JD003290
129. Brasseur GP, Kiehl JT, Müller J-F, Schneider T, Granier C, Tie X, Hauglustaine D (1998) Geophys Res Lett 25:3807
130. Kirchhoff VWJH, Rasmussen RA (1990) J Geophys Res 95:7521
131. Crutzen PJ, Andreae MO (1990) Science 250:1669–1678
132. Roths J, Harris GW (1996) J Atmos Chem 24:157
133. Christopher SA, Chou J, Welch RM, Kliche DV, Connors VS (1998) J Geophys Res 103:19327
134. Goldammer JG (1999) Science 284:1782
135. Siegert F, Ruecker G, Hinrichs A, Hoffmann AA (2001) Nature 414:437
136. Thompson AM, Witte JC, Hudson RD, Guo H, Herman JR, Fujiwara M (2001) Science 291:2128
137. Crutzen PJ, Delany AC, Greenberg J, Haagenson P, Heidt L, Lueb R, Pollock W, Seiler W, Wartburg A, Zimmerman P (1985) J Atmos Chem 2:233
138. Sanhueza E, Crutzen PJ, Fernández E (1999) Atmos Environ 33:4969
139. Hurst DF, Griffith DWT, Cook GD (1994) J Geophys Res 99:16441
140. Nguyen BC, Mihalopoulos N, Putaud J-P (1994) J Geophys Res 99:16435
141. Jones SH (1997) Fire Ser Rep EUR 17282 EN. European Commission, Luxembourg
142. Lobert JM Harris JM (2002) J Geophys Res 107:doi:10.1029/2001JD000731
143. Naja M, Lal S (1996) Geophys Res Lett 23:81
144. Naja M, Lal S, Chand D (2003) Atmos Environ 37:4205
145. Holloway T, Levy H II, Kasibhatla P (2000) J Geophys Res 105:12123
146. Gros V, Bonsang B, Martin D, Novelli PC, Kazan V (1999) Chemosphere Global Change Sci 1:163
147. Lal S, Naja M, Jayaraman A (1998) J Geophys Res 103:18907
148. Naja M, Lal S (2002) J Geophys Res 107:doi:10.1029/2001JD000357
149. Chand D, Lal S, Naja M (2003) J Geophys Res 108:doi:10.1029/2001JD001589
150. Stehr JW, Ball WP, Dickerson RR, Doddridge BG, Piety CA, Johnson JE (2002) J Geophys Res 107:doi:10.1029/2001JD000446
151. Tsusumi Y, Matsueda H (2000) Atmos Environ 34:553
152. Buongiorno A, Arino O, Zehner C, Colagrande P, Goryl P (1997) Earth Observation Q 56:1
153. Goloub P, Arino O (2000) Geophys Res Lett 27:899
154. Blake NJ et al. (2001) J Geophys Res 106:32627
155. Liu H, Chang WL, Oltmans SJ, Chan LY, Harris JM (1999) Atmos Environ 33:2403
156. Chan LY, Chan CY, Liu H, Christopher S, Oltmans SJ, Harris JM (2000) Geophys Res Lett 27:1479
157. Chan CY, Chan LY, Zheng YG, Harris JM, Oltmans SJ, Christopher S (2001) J Geophys Res 106:14875
158. Chan CY, Chan LY, Harris JM, Oltmans SJ, Blake DR, Qin Y, Zheng YG, Zheng XD (2003) J Geophys Res 108:doi:10.1029/2001JD001555
159. Chan CY, Chan LY, Change WL, Zheng YG, Zheng XD, Qin Y, Li YS (2003) J Geophys Res 108:doi:10.1029/2003JD003427

The Handbook of Environmental Chemistry Vol. 4, Part G (2004): 131–172
DOI 10.1007/978-3-540-40037-0

Export of Air Pollution from Southern Asia and its Large-Scale Effects

Mark G. Lawrence

Max-Planck-Institute for Chemistry, Airchemistry Department, Junior Research Group, PO Box 30 60, 55020 Mainz, Germany
E-mail: lawrence@mpch-mainz.mpg.de

Abstract Southern Asia is one of the most heavily populated regions of the world. In contrast to much of the rest of the world, biofuel and biomass burning account for a large fraction of the total energy production in southern Asia compared to fossil fuel burning. This results in air masses which are relatively rich in carbon-containing aerosols, carbon monoxide, and hydrocarbons. The export of these polluted air masses is characterized by two main seasonal circulation patterns: the summer and winter monsoon. During summer, the heavy rains in this region effectively remove soluble gases and aerosols. Less soluble species, on the other hand, are lifted to the upper troposphere in deep convective clouds, and are then transported rapidly away from the region by the strong upper tropospheric winds, particularly towards northern Africa and the Mediterranean in the tropical easterly jet. During winter, the near-surface flow is mostly northeasterly, and the regional pollution forms a thick haze layer in the lower troposphere which spreads out over thousands of km between southern Asia and the Intertropical Convergence Zone (ITCZ). This chapter reviews the many observational and modeling studies which have examined the southern Asian outflow and its large scale effects.

Keywords Southern Asia · Intercontinental Pollution Transport · Monsoon Circulation · INDOEX · MINOS

List of Abbreviations

ABC	"Asian Brown Cloud" or "Atmospheric Brown Cloud"
AOD	Aerosol optical depth
BC	Black carbon (or soot)
BL	Boundary layer
CTM	Chemistry transport model
ECHAM, ECMWF	Forecast models, modified and extended in Hamburg
GCM	General circulation model
INDOEX	Indian Ocean Experiment
ITCZ	Intertropical Convergence Zone
KCO	Kaashidhoo Climate Observatory
LT	Lower troposphere
MATCH-MPIC	Model of Atmospheric Transport and Chemistry Max-Planck-Institute for Chemistry Version
MBL	Marine boundary layer
MINOS	Mediterranean Intensive Oxidants Study
MT	Middle troposphere
N	North
NE	Northeast
NH	Northern hemisphere
NMHC	Non-methane hydrocarbon
ppbv	Parts per billion volume, equivalent to nmol/mol
R/V	Research vessel
S	South
SE	Southeast
SH	Southern hemisphere
SSA	Single scatter albedo
SW	Southwest
UNEP	United Nations Environment Programme
UT	Upper troposphere
VOC	Volatile organic carbon

1 Introduction

Southern Asia is a region with a vibrant mixture of cultures which comprises over a quarter of the world's population. Stretching from Pakistan to Indonesia, the landscape includes seashores, tropical rainforests, vast plains, and the foothills of the Himalayas. People dominate this landscape, gathered in enormous megacities like Delhi and Bangkok, but also dot the land in small villages, some without electricity and running water. One activity which is common to all people amidst this immense diversity is energy consumption, from biomass burning in cookstoves to fossil fuel usage in trucks and rickshaws. Although the amount of energy consumption per capita still pales in comparison to industrial giants like the USA, the sheer size and rapid growth of the southern Asian population and the often poorly-controlled burning conditions lead to large emissions of pollutants such as CO, organic gases, and particles comprised of soot, sulfate, and organic compounds. These emissions are a menace for the health of the local populations and for local agriculture. A particularly poignant example is the "Delhi fog", which occurs intermittently every winter in the Indian capital. The combination of particle and gas emissions from biomass, biofuel, and fossil fuel burning in a stable and shallow planetary boundary layer leads to a smog so thick that it often shuts down the international airport and hinders many other facets of life for up to days at a time, in many ways reminiscent of the London smog events in the middle of the last century. Local pollution is not this extreme everywhere in southern Asia, and is not even omnipresent, but it is clearly a major problem which will need to be dealt with by local governments in the coming decades, a difficult task given the rapid increases in populations and energy consumption.

Recently southern Asian emissions have come into the international spotlight, since observations show that they can be transported over large distances away from their source regions. Furthermore, since these emissions are being released into the most photochemically active region of the world, their rapid growth may lead to unprecedented changes in the state of the tropical atmosphere. This chapter gives a review of some of the recent research findings regarding the export of pollutants from southern Asia and their large-scale effects. In the next section, some of the estimates of current emissions and future increases are discussed. Following that, the meteorological setting of the region is considered, focusing on the summer and winter monsoons, which dominate the seasonal evolution of southern Asian weather. The export of southern Asian emissions is then broken down into two main pathways, active at different times of the year: the wintertime Indo-Asian haze, and the Asian summer monsoon plume. Finally, the larger perspective and future research needs are discussed in the conclusions.

2
Estimates of Current Emissions and Future Increases

A summary of the current anthropogenic emissions from Asia compared with those from the rest of the world is given in Table 1. There is a considerable degree of uncertainty in many of these values; some are based on upscaling of only one or a few measurements of emissions factors appropriate to various combustion processes in different regions, and in many cases a serious, formal uncertainty estimate is not even possible at present. Often there are large discrepancies between bottom-up and top-down emissions estimates for the Asian region [1]. Nevertheless, a few basic features in the emissions distribution are relatively certain.

Most importantly, as will be discussed throughout this section, the emissions from southern Asia ("India" in the table) are characteristic of inefficient combustion processes, especially from biomass burning in cookstoves plus the use of two-stroke engines and mixing of kerosene with gasoline. Table 1 indicates that the fraction of the emissions from biofuel combustion and agricultural practices relative to the emissions from fossil fuel combustion is much greater in Asia, especially in India, than in North America and Europe (especially for CO). Compared to more efficient combustion (e.g., modern automobiles), these inefficient

Table 1 Global anthropogenic CO_2, CO, NO, SO_2, and NMHC emissions (the "India" region includes Bangladesh, Maldives, Sri Lanka, Myanmar, Nepal, and Pakistan, the "China" region includes Cambodia, Vietnam, Laos, Mongolia, and North Korea, and "East Asia" includes Japan, South Korea, Indonesia, Malaysia, Philippines, and Thailand); note that the units for nitrogen oxides are different to those in Fig. 1 (1 Tg(NO_2)/year = 3.3 Tg(N)/year). (Reproduced with permission from [21], copyright 2001, American Association for the Advancement of Science)

Source category	Global	North America	Europe	India	China	East Asia
Carbon dioxide (Pg of CO_2 per year)						
Total	29.8	6.2 (21%)	4.9 (16%)	2.2 (7%)	4.0 (13%)	2.5 (8%)
Fossil fuel use	21.9	5.6	4.5	0.7	2.6	1.7
Industrial processes	0.6	0.1	0.2	–	0.1	0.1
Biofuel use	5.5	0.5	0.2	1.4	1.2	0.5
Agriculture	1.8	–	–	0.1	0.1	0.2
Carbon monoxide (Tg of CO per year)						
Total	975	107 (11%)	85 (9%)	110 (11%)	111 (11%)	69 (7%)
Fossil fuel use	263	74	53	4	34	16
Industrial processes	35	2	8	1	5	6
Biofuel use	181	9	2	47	40	19
Agriculture	496	22	22	58	32	28
Nitrogen oxides (Tg of NO_2 per year)						
Total	102	26 (25%)	16 (16%)	6 (6%)	11 (10%)	6 (6%)
Fossil fuel use	72	24.3	13.6	2.6	7.2	4.3
Industrial processes	5	0.4	1.1	0.2	0.9	0.7
Biofuel use	5	0.5	0.2	1.1	1.5	0.4
Agriculture	20	0.8	0.7	2.0	1.1	1.0
Sulfur dioxide (Tg of SO_2 per year)						
Total	148	24.5 (17%)	333 (23%)	5 (3%)	28 (19%)	7 (5%)
Fossil fuel use	120	22.8	26.4	4.0	25.0	5.0
Industrial processes	23	1.2	6.4	0.3	2.8	1.7
Biofuel use	2	0.4	0.4	0.2	0.3	0.1
Agriculture	4	0.1	0.1	0.4	0.2	0.2
Nonmethane hydrocarbons (Tg of NMHC per year)						
Total	178	22 (12%)	21 (12%)	19 (11%)	17 (10%)	16 (9%)
Fossil fuel use	69	12	12	1.5	3	6
Industrial processes	34	7	7	3	4	4
Biofuel use	31	1	0.2	8.5	6	3
Agriculture	44	2	2	6	4	3

combustion processes do not burn as hot, resulting in less emissions of NO_x (NO_x=NO+NO_2), nor do they burn as well to completion, resulting in a higher CO/CO_2 ratio and a greater unburned hydrocarbon content. As a result, the CO_2 and NO_x emissions from the Indian region are only 1/2 as large as from the Chinese region, and only 1/3 (CO_2) or 1/4 (NO_x) as large as those of North America, while in contrast the CO and NMHC emissions from all three regions are roughly equal.

Another important product of inefficient burning is black carbon (BC). It is not yet clear, however, whether fossil fuel use or biomass and biofuel burning is the dominant source for BC from southern Asia. Several studies [2–4] have made use of measured ratios of BC to total carbon (the sum of BC and organic carbon), as well as ratios with other aerosol components such as potassium (a biomass burning indicator), to conclude that the main source of BC in the continental outflow over the northern Indian Ocean must be fossil fuel burning. On the other hand, other studies have concluded that biomass and biofuel burning is likely to be the primary source of BC aerosols, based on similar techniques using the ratios of aerosol components [5] and of BC to the trace gas CO [6], as well as by considering correlations of aerosol optical thickness with fire count data based on satellite observations [7]. Based on emissions inventories [3, 6, 8, 9], one would also expect the largest source to be from biomass and biofuel burning. Interestingly, though, there is also a notable discrepancy between these "bottom-up" inventories based on fuel consumption and emissions ratios, all of which indicate ~0.5 Tg(BC)/year from India (or ~0.7 Tg(BC)/year from southern Asia, with an upper limit of 1 Tg(BC)/year, based on [6]), compared with "top-down" estimates based on in situ measurements of BC [6], which suggest a much larger source of 2–3 Tg(BC)/year. These discrepancies are indicative of the need for better emissions inventories, which will only be resolved through a larger set of measurements of emissions factors for the major combustion processes along with measurements of BC and other compounds directly in the major source regions in southern Asia. Such improvements in emissions inventories will be crucial for enabling more accurate source attribution studies for the present as well as for future scenarios.

In the near future, the growing populations in Asia, accompanied by rapid industrialization and agricultural developments, are expected to result in emissions

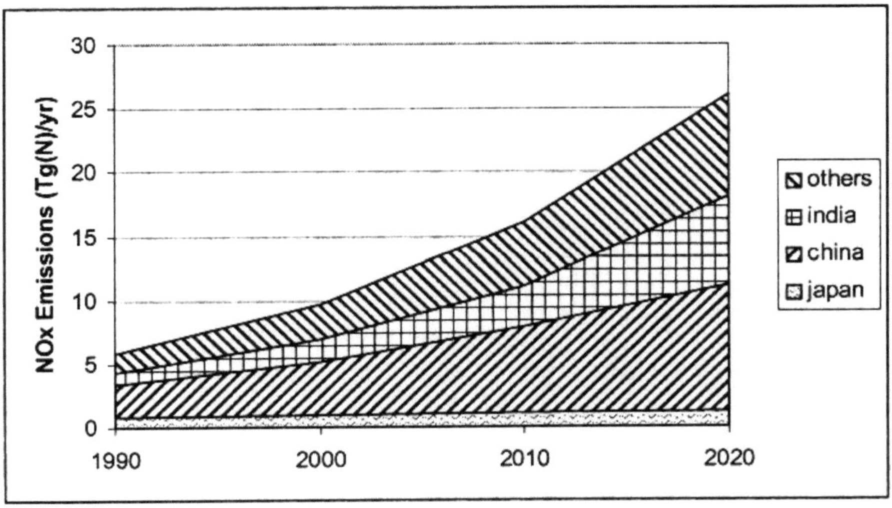

Fig. 1 NO_x emissions from Asia during the period 1990–2020. (Based on data from [10])

which will surpass those from the northern mid-latitudes for most gases, with the major exception of the important greenhouse gas CO_2. For instance, in Asia the emissions of the O_3 precursor NO_x due to fossil fuel burning could increase from about 6 Tg(N)/year in 1990 to over 25 Tg(N)/year in 2020, as depicted in Fig. 1; China and India are the largest contributors to this, followed by Indonesia, Thailand, and other southeast Asian nations [10]. For comparison, in the early 1990s the total amount of NO_x emitted due to fossil fuel burning worldwide was about 22 Tg(N)/year (Table 1). There is, however, even greater uncertainty in future scenario estimates than in current emissions inventories. For instance, Fig. 2 shows

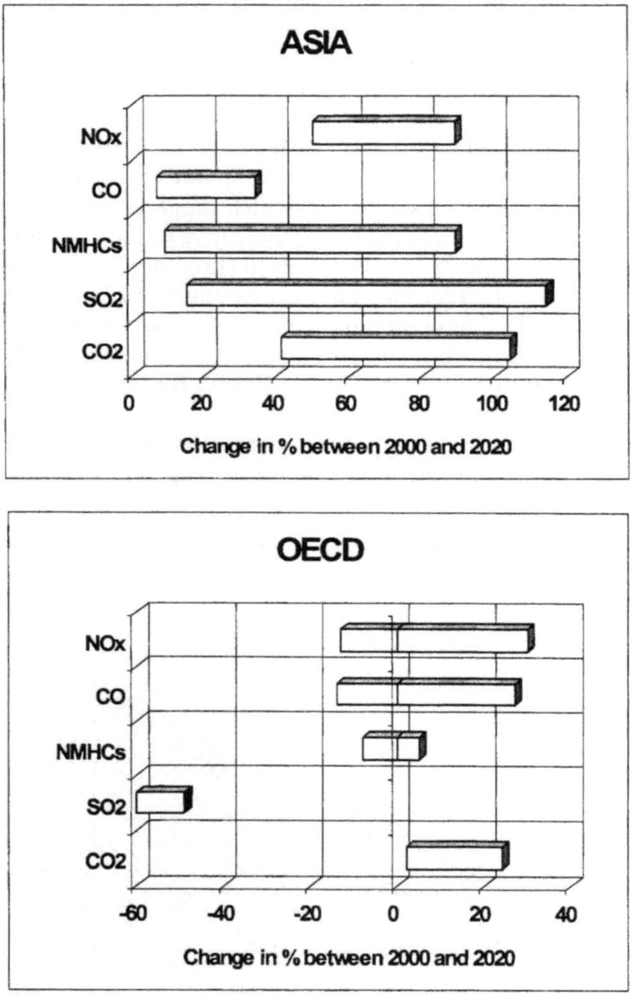

Fig. 2 Estimated range of likely relative changes in emissions of several trace gases between the year 2000 and 2020 in Asia and in the OECD countries. (Based on data published in IPCC Emission Scenarios 2000)

the anticipated changes in emissions of several gases from Asia compared with the OECD (Organisation of Economic Co-operation and Developments) nations. While there is a considerable range in the likely changes in each gas, it is clear that the emissions of nearly all gases are expected to increase in Asia, while emissions of most gases from the OECD nations could either increase or decrease (with the exception of SO_2, which will almost certainly decrease, and CO_2, which will almost certainly increase). On careful examination of the figures, one notices that the top of the range of anticipated increase of NO_x emissions given in Fig. 2 for Asia (up to about 85%) does not even include the estimate shown in Fig. 1 of about 150% (from 10 to 25 Tg(N)/year). This is indicative of the generally wide range in estimates for future emissions. In the end, the actual increase in anthropogenic emissions of NO_x and other gases will depend on which new regulations are implemented and enforced in the individual countries, which will in turn depend in part on the scientific information available to policy makers.

3 Meteorological Setting of Southern Asia

The torrential rains of the Asian summer monsoon are well known. Every summer, steady onshore winds bring air from the ocean over the continent (Fig. 3a), which has been heated by the tropical overhead sun. The moist, hot air in the surface layer rises, due to buoyancy as well as due to orographic forcing over features such as the Western Ghats mountains at the Indian coast. The rising air cools, causing water to condense, forming clouds and releasing latent heat, which reinforces the buoyant rising motion. In this phenomenon, known as moist, deep convection, the air parcels can rise several km, often reaching and sometimes even penetrating the tropopause. The uplifted air then spreads out, forming massive cirrus anvil clouds which can be clearly seen in satellite pictures. Deep convection can build up to virtually explosive energies, with updraft speeds often exceeding 10 m/s, and can transport largely intact air parcels from the surface to the upper troposphere (UT) in less than an hour.

Once the air reaches the UT, it typically encounters much faster winds than those which are found near the surface. Over southern Asia, the convergence of air masses from the outflow of the monsoon convection results in a high pressure region in the UT typically located over Tibet, around which air tends to flow in a clockwise (anti-cyclonic) direction. Another UT anticyclone located to the west of this over Arabia is also present most of the time. A snapshot of these two anticyclones is depicted in Fig. 4a. The two anticyclones meander back and forth, so that averaged over a month their central north-south branches tend to cancel each other out, and the mean circulation of the region is dominated by a large, elongated anticyclone, depicted in Fig. 4b. The combined east-to-west branches on the south side forms the tropical easterly jet, which is in contrast to the westerlies which usually prevail in the UT of the northern hemisphere tropics.

During the summer monsoon, the most frequent intense convection is found in the ITCZ, where air masses from the northern and southern meteorological hemispheres converge. Throughout most of the world the ITCZ is located within a few degrees of the equator. Over Asia in summer, in contrast, it is generally lo-

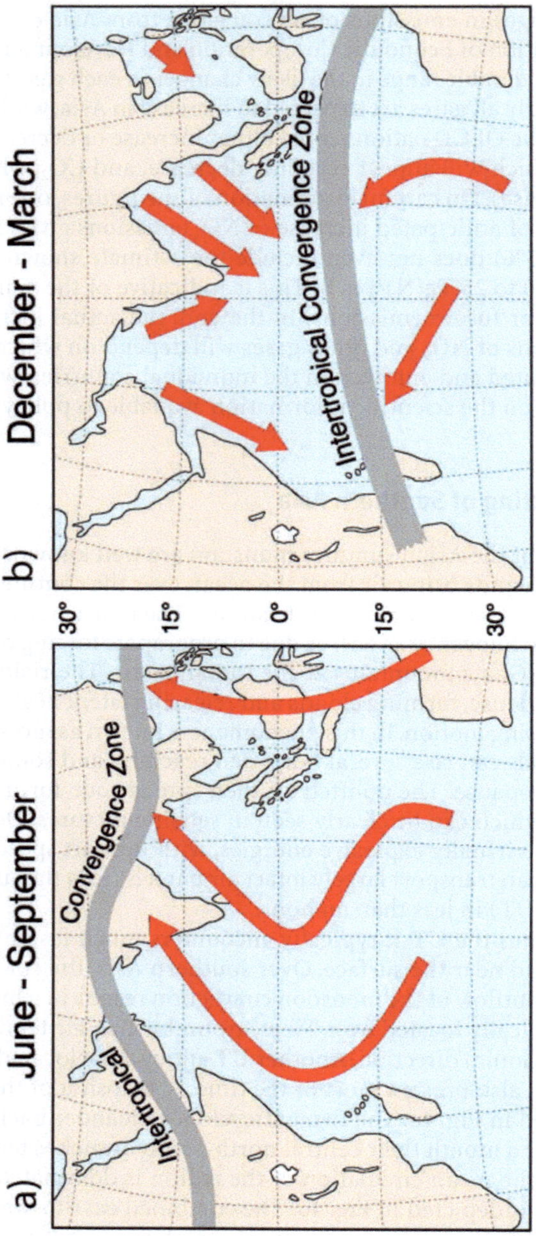

Fig. 3a, b Transport pathways for near-surface flow over the Indian Ocean during: **a** the summer monsoon period; **b** the winter monsoon period

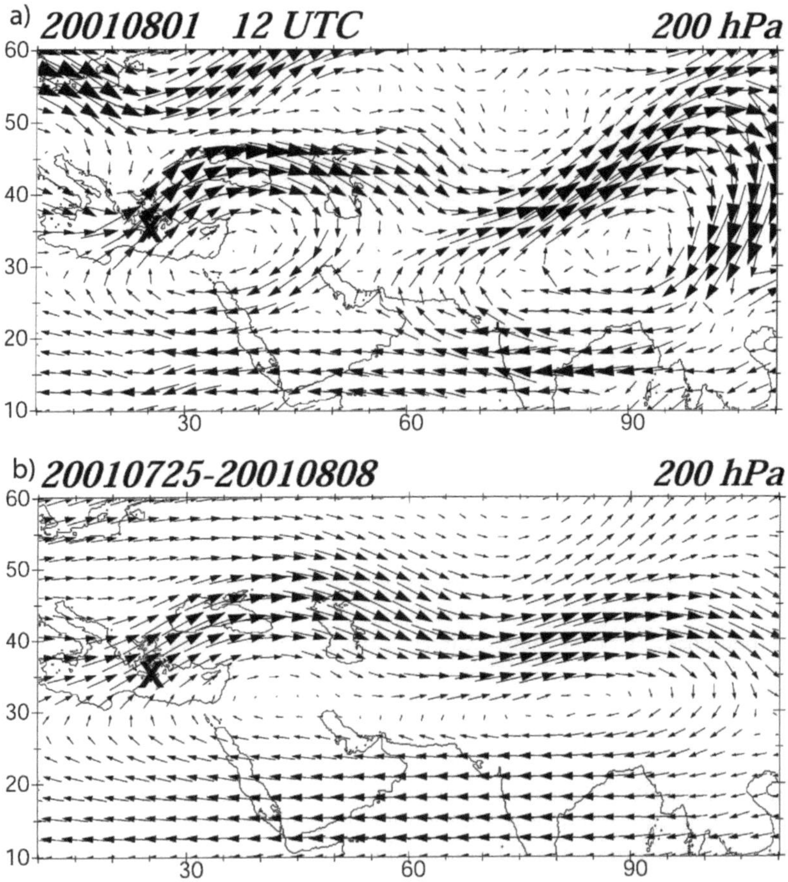

Fig. 4a, b NCEP AVN winds over southern Asia at 200 hPa: **a** average over the two-week period from 27 July to 8 August, 2001; **b** instantaneous winds for 12:00 UTC, 1 August. The "X" marks the location of Crete. (From [39], copyright 2003 by the European Geophysical Union, modified by permission of the EGU)

cated between 5°N and 30°N, directly over many of the highly populated regions of southern Asia.

During the winter, the ITCZ migrates south with the sun, and is typically found between 5°S and 15°S. The wintertime winds over much of southern Asia generally blow from the NE to the SW (see Fig. 3b), and the period is known as the NE monsoon (or winter monsoon). There is little rain from deep convection over southern Asia during this period, and the UT is largely disconnected from the surface layer [11]. Offshore flowing air encounters a largely cloud-free northern Indian Ocean down to the equator. Typical travel times are about seven to ten days to reach the ITCZ [12, 13], where much of the air is then carried upwards in deep convection. The typical flow patterns during this period are illustrated

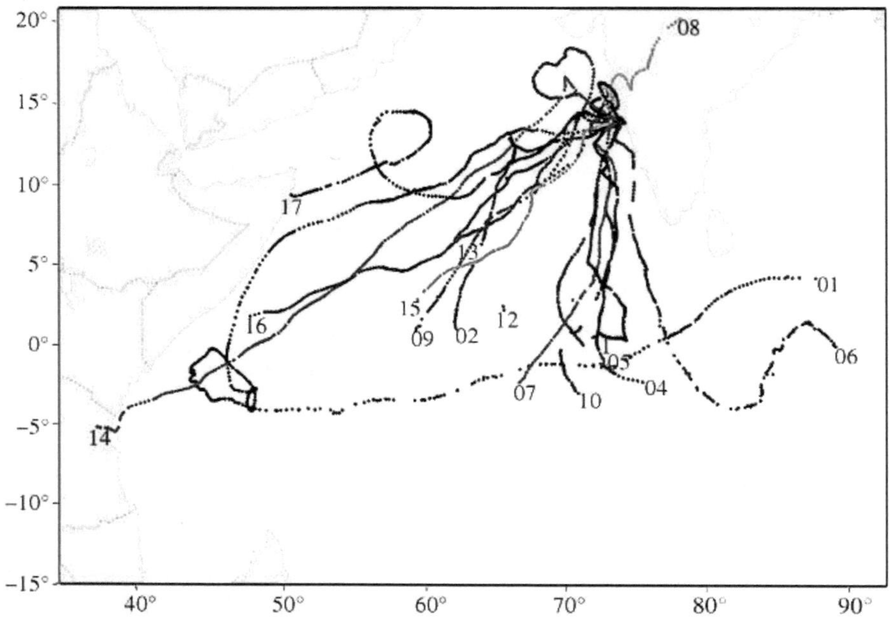

Fig. 5 Trajectories of the 17 constant-level balloons launched from Goa during INDOEX (January–March 1999) at pressure levels between about 900 and 800 hPa. (From [12], copyright 2002 by the American Geophysical Union, reproduced by permission of the AGU)

in Fig. 5, which shows the trajectories of 18 constant-level balloons which were released from the Indian coastal city of Goa (near Bombay).

The rest of this review will focus on the export of pollutants during the summer and winter monsoons. From a meteorological standpoint, the summer monsoon has been studied in far greater depth, particularly by Indian meteorologists over the last century, due to its essential role in the supply of fresh water and thus in human health and agriculture. There are various theories about the factors controlling the monsoon behavior, especially the characteristics of the summer rainy period, including the overall amount and distribution of precipitation (spatial and temporal, including the "breaks" in the monsoon rains), which are discussed in a number of reviews of the monsoons (e.g., [14, 15]). On the other hand, from the perspective of atmospheric chemistry and the export of pollution, the winter monsoon has thus far received the largest amount of attention. This is largely due to the INDOEX and pre-INDOEX field campaigns, which are discussed in the next section. The few measurements and modeling studies of long-range export of pollution during the summer monsoon are discussed after that.

4 Export in the Wintertime Indo-Asian Haze

It is not surprising that the immense populations of southern Asian nations lead to difficult urban air pollution problems, in some ways resembling those experienced in European and American population centers over the past century. Until recently, however, it was unclear whether the southern Asian air pollution was confined to only being a local problem, or whether it had significant effects on the regional or global scale. It has now become evident that the export of pollution from southern Asia during the winter monsoon is immense in magnitude and extent, with significant impacts on the climate and chemistry of the surrounding region. This message has come from a wide range of experimental and theoretical studies. The greatest amount of information about the outflow of pollution from southern Asia during the winter monsoon has been provided by the Indian Ocean Experiment (INDOEX), with an intense field phase in January to April 1999. Prior to that, annual pre-INDOEX ship cruises were conducted starting in 1995. A number of other smaller field campaigns and independent modeling studies have also provided information on the composition of the regional atmosphere and ocean during the winter months, and there are significant, ongoing efforts to further explore these issues.

The pollution outflow from southern Asia contains compounds that can be classified into several different types, of which three main ones are: 1) aerosol particles and their precursors, 2) moderate- and short-lived gases (with lifetimes less than about a year) related to tropospheric ozone chemistry, and 3) long-lived greenhouse gases. The amount of the latter is highly uncertain but is thought to pale in comparison to emissions from Europe and America, and, due to their long lifetimes (mostly greater than ten years), greenhouse gases are relatively well-mixed in the troposphere and thus are not thought of in terms of regional outflow or intercontinental transport. This section will only focus on the first two classes. One of the most widely discussed findings of INDOEX was the thick layer of sunlight-absorbing aerosols which covers much of the region, which can easily be seen in visible satellite images (Fig. 6) and which is responsible for a significant perturbation to the regional atmospheric energy balance. Following INDOEX this aerosol layer came to be known as the "Asian Brown Cloud," or "ABC" for short. Since then, it has been suggested that the term generally be replaced with "Atmospheric Brown Clouds" [16], in recognition of the fact that aerosol outflow layers like this are by no means confined to solely being a problem for the southern Asian region (see Chap. 8). Here the term "wintertime Indo-Asian haze" (based on [17]) is used as a more accurate classification of the overall outflow, which includes various types of aerosols and trace gases.

4.1 Outflow Channels

The wintertime Indo-Asian haze over the Indian Ocean is due to the outflow of pollutants along several different channels. Three primary flow channels have been suggested by [18, 19], and a fourth was added by [20]. The four channels, fol-

Fig. 6 Visible satellite image of the INDOEX outflow from the SeaWiFS satellite. (Provided courtesy of the SeaWiFS Project, the NASA/Goddard Space Flight Center and ORBIMAGE)

lowing [20], are depicted in Fig. 3b, originating from: 1) the Arabian desert with flow over the western Arabian Sea, 2) the Indo-Pakistan desert and the west coast of India, 3) central India and the Ganges Valley, flowing over the western Bay of Bengal, and 4) southeast (SE) Asia. The channels are not all active all the time, but rather intermittently, depending on the weather systems present in the region. The number of days during which each of the channels was present during February to March, 1999 was analyzed in [20] (Table 2), which shows that channel 1 was the most persistent, being active over 80% of the time, while channels 2 and 3 were active about half of the time, and channel 4 only about 1/3 of the time. Channels 2 and 3 were most relevant to the measurements during the INDOEX campaign, which were mainly in the region south of the Indian Peninsula, as discussed in [21], based on the back trajectory analysis in [13].

Table 2 Frequency distribution of the presence of the four flow channels over the north Indian Ocean area during February and March 1999. (From [20], copyright 2001 by the American Geophysical Union, reproduced by permission of the AGU)

Period	Channel 1 (West Arabian Sea)	Channel 2 (Indo-Pakistan Desert of West Coast of India)	Channel 3 (West Bay of Bengal)	Channel 4 (SE Asia)
February 1999	26	12	26	15
March 1999	24	21	6	5
Total	50	33	32	20

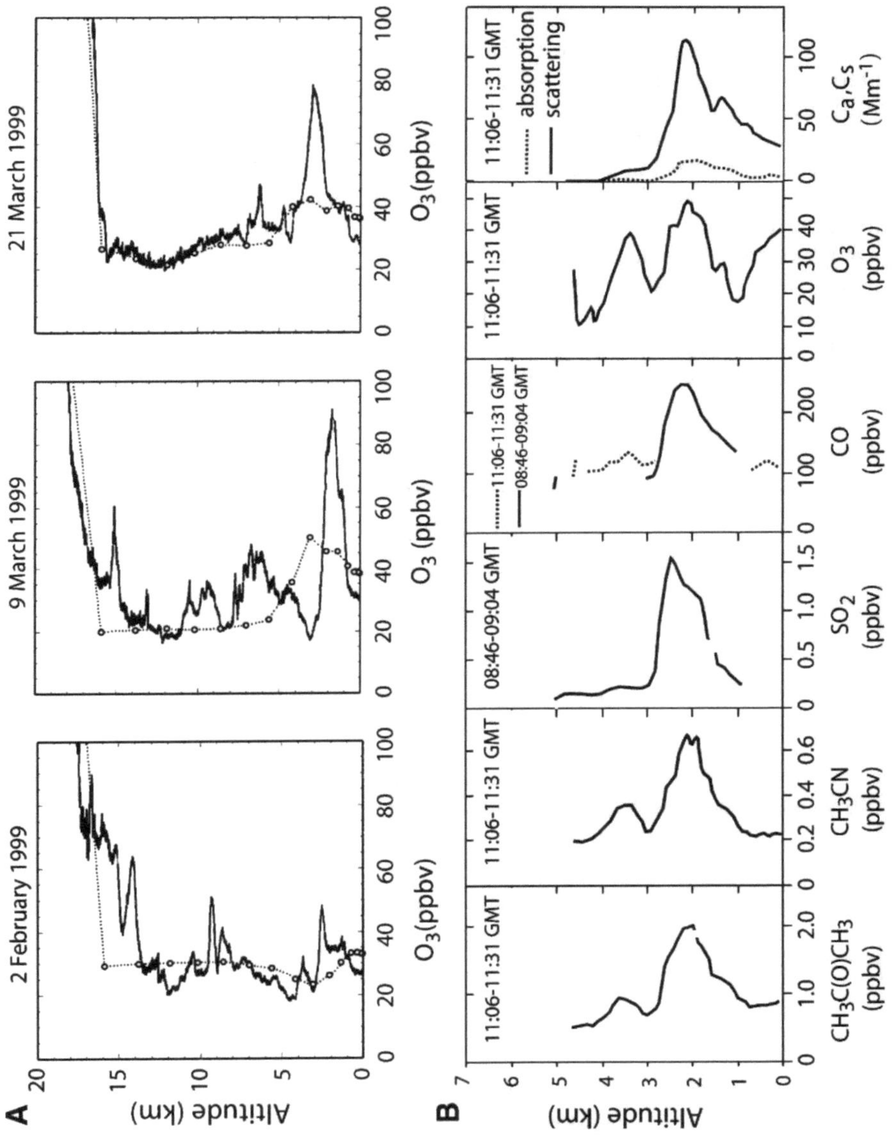

Fig. 7a, b Examples of the elevated layers observed during INDOEX: **a** O_3 profiles over KCO from balloon sondes (*solid line*) and computed with the ECHAM model (*dashed lines*); **b** profiles of various gases and aerosol quantities at 7.5°N, 72°E, observed on a flight of the C-130 aircraft on 13 March 1999. (Reproduced with permission from [21], copyright 2001, American Association for the Advancement of Science)

While the monsoon winds provide for a significant outflow of continental air along these channels near the surface, often (though not always) the most polluted outflow is in an elevated layer between ~1–3 km height; the flow at this altitude is along similar channels to the flow at the surface, though generally with a more easterly origin [17,22]. These elevated layers have been observed in measurements of numerous trace gases and aerosols, from aircraft, balloon soundings, and lidar; examples are shown in Fig. 7. Various explanations have been proposed for this phenomenon. One of the main possibilities (e.g., [7,21,23,24]) is that the outflow is influenced by the strong land-sea breeze circulation which is present over much of the west coast of India (Fig. 8). During the day, the flow from the ocean onto the land causes the low-level outflow to be blocked, and pollution from the plateau near the coast (the Western Ghats) is lifted into the residual layer above the marine boundary layer (MBL) cap. This circulation occurs on a scale of ~100 km, and thus is not resolved in most global model simulations

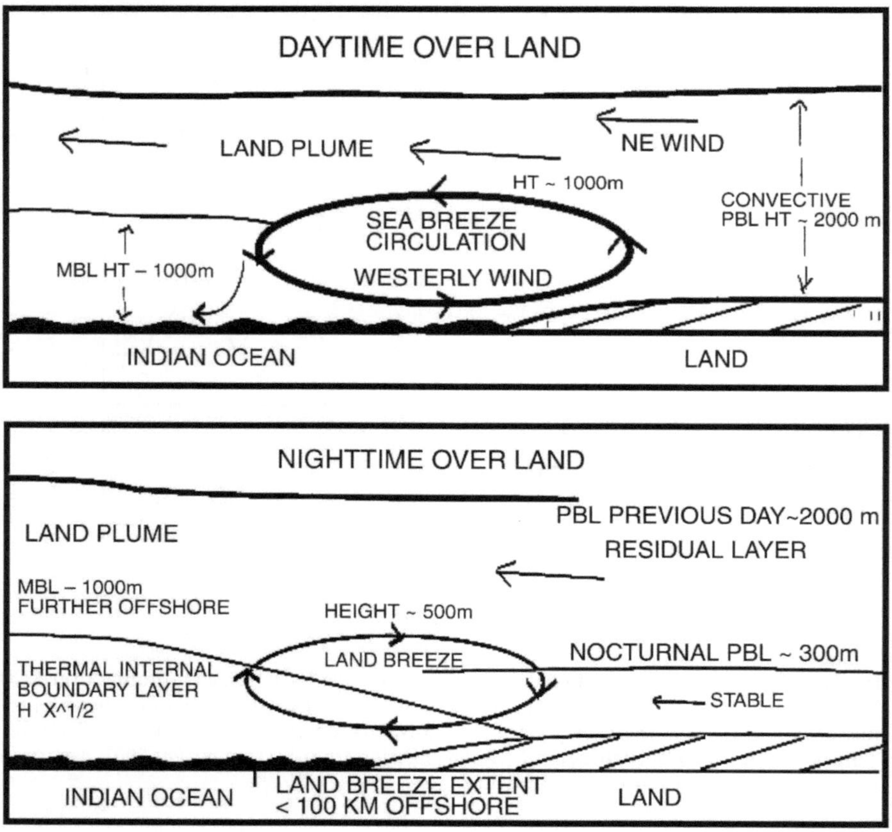

Fig. 8 Conceptual depiction of the interaction between the coastal circulations and the land plume over the Indian Ocean for day and night. (From [24], copyright 2002 by the American Geophysical Union, reproduced by permission of the AGU)

(e.g., the ECHAM model fails to reproduce the layers in O_3 shown in Fig. 7a). This feature of the wintertime Indo-Asian haze requires more study. It will have important implications for the large scale effects, due in particular to differences in the impact of warming by absorbing aerosols located near the surface (which would tend to substantially increase atmospheric instability) compared to warming in the residual layer (which can strengthen the MBL inversion cap). Furthermore, it makes long-term monitoring of the pollutant outflow more complicated, since island stations with only ground-based instruments may not be observing the most important part of the outflow (thus arguing for the need for balloon sondes and lidar measurements where possible).

4.2 Air Mass and Pollution Origins Assessments

Global chemistry-transport model (CTM) simulations of tagged CO tracers [25, 26] have been used to examine the regional origins of the pollution in the wintertime Indo-Asian haze (Fig. 9). The model results show that the main source of the CO in the MBL over the northern Indian Ocean, especially the Arabian Sea, is from India, predominantly from biomass and biofuel burning. Arabian and SE Asian sources also make notable contributions, particularly over the northern Arabian Sea and the eastern Bay of Bengal, respectively. These results depend on the regional transport characteristics as well as the spatial distribution of CO emissions. A more generic analysis of the continental air mass origins using tagged radon tracers [27] also indicates that the Indian contribution dominates over much of the region covered by the wintertime Indo-Asian haze, though the other sources, particularly Arabia and Africa, also contribute significantly to the total radon levels, especially at the altitude (~2 km) of the elevated pollutant layers discussed earlier.

More detailed classifications of the origins of individual air masses have been made based on their chemical characteristics combined with back trajectories [28–30]. NH tropical continental emissions were shown to strongly affect the atmospheric chemistry and radiation budgets in the region up to 1500 km south of India. Tracer ratios could be used [30] to distinguish between fossil fuel and biomass burning sources for the trace gases, indicating that air from western India was mainly influenced by biomass burning, while air from the Middle East was more characteristic of fossil fuel use. Interestingly, a significant amount of pollution imported from the NH extratropical region was found [29], mainly from Arabia/Europe along the westward edge of outflow channel 2 (Fig. 3b). In these air masses the highest levels of O_3 and several organic trace gases such as methane, acetylene, ethane, and propane during the entire 1999 R/V Ronald H. Brown cruise were measured. Similarly, some of the highest O_3 mixing ratios (60–70 nmol/mol) which were measured during the Indian R/V Sagar Kanya cruise off the west coast of India have been attributed to the same flow channel [31]. In contrast, the highest CO levels measured in 1999 [29] were instead in air masses from India and SE Asia. CO measurements at the Kaashidhoo Climate Observatory (KCO) on the Maldives (to the southwest of India) also indicate that the highest mixing ratios (150–200 nmol/mol) are associated with air from India and

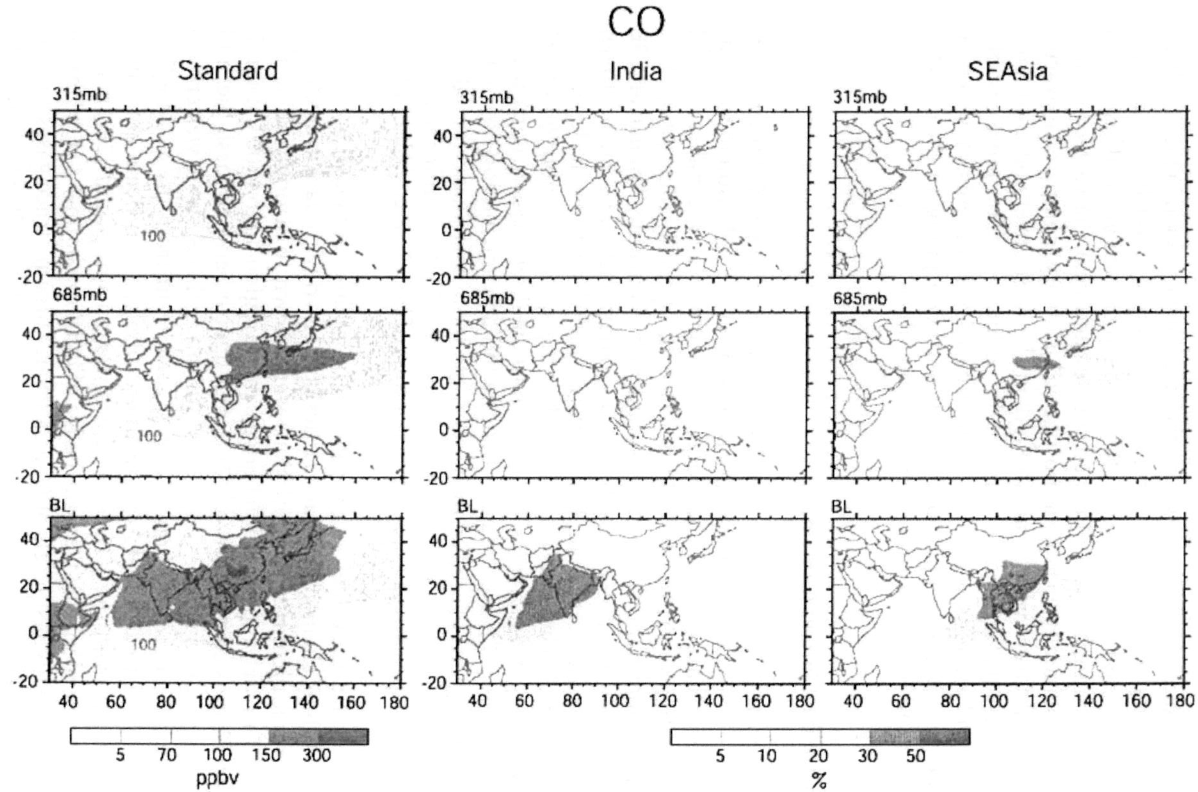

Fig. 9 Simulated mixing ratios of total CO (*left*) during January, February, and March. Contributions from South Asian (*middle*) and Southeast Asian (*right*) fossil fuel and biomass burning are given as a percent of the total CO. BL-boundary layer mixing ratios are calculated as weighted average of the three lowermost model layers (990 mb, 940 mb, and 850 mb). (From [26], copyright 2002 by the American Geophysical Union, reproduced by permission of the AGU)

SE Asia, while somewhat lower levels (100–150 nmol/mol) came from Arabia/Europe [22], consistent with the differences in emissions characteristics in India and other regions discussed above.

4.3
Chemical Processing

Significant chemical processing takes place in the southern Asian outflow, especially due to the high solar radiation levels and the generally cloud-free conditions. Directly off the coast of India this includes photochemical production of O_3 in the outflow air masses, which can lead to considerably higher mixing ratios (50–70 nmol/mol) than generally observed in major coastal cities (~10–35 nmol/mol on average, up to ~50 nmol/mol maximum daytime peak values) [31]. Based on this observation it was hypothesized that this also might apply to rural regions of India downwind of major cities. Recent observations support this [32]. Measurements made in the west coast city of Ahmedabad and ~100 km downwind in the outflow show average daytime peak values of 52 nmol/mol in the city vs 80 nmol/mol downwind during one week in January, 2001, and somewhat higher daytime peak values (averaging 55 and 88 nmol/mol in the city and outflow, respectively) during the week of 30 December 2001 to 5 January 2002. Maximum values exceeding 100 nmol/mol at the downwind site indicate that rural O_3 levels may already be high enough to have negative impacts on human health and agriculture, and the situation is likely to become worse in the near future if India moves towards greater mobility and more high-temperature fossil fuel burning. Satellite observations [33] also support the suggestion that high O_3 concentrations are already prevalent in parts of India, indicating a band of enhanced tropospheric O_3 over a large region of northern India. Many more in situ observations throughout India are needed to investigate this further.

Further downwind in the outflow over the Indian Ocean, chemical loss of O_3 takes over, and the lifetime of O_3 in the MBL falls to about five days (estimated in [34] based on measurements of the N-S gradient in O_3 and the average wind speed). The daytime loss of O_3 and its replenishment by advection of polluted upstream air masses at night [35] results in a diurnal cycle with minimum concentrations during the day. Interestingly, the observed magnitude of the diurnal cycle of O_3 in the MBL during the 1995 R/V Malcolm Baldridge cruise was about 30% [28], which implies an O_3 loss rate which is faster than can be explained with traditional HO_x-NO_x photochemistry. Based on this, it has been suggested [36] that halogen chemistry might play an important role in the Indian Ocean MBL. This issue has been revisited in two studies which used box models to examine the data from the 1999 R/V Ronald H. Brown cruise, resulting in a mixed picture, with one study [37] finding that a small amount of Cl could help improve the comparison between their box model and the observations, while the other study [38] concluded that the diurnal cycle of CH_2O is apparently not strongly influenced by additional halogen reactions. As noted in [38], there are still considerable uncertainties in models of the photochemical processing of air in the Indian Ocean MBL.

Due to the high levels of incident solar irradiation and moisture, the loss of O_3, which largely occurs via photolysis followed by the reaction of excited oxygen atoms with water vapor, results in significant production of the hydroxyl radical (OH). OH in turn initiates the breakdown process of most hydrocarbons and CO. The OH concentration in the Indian Ocean MBL was estimated to be 3.7×10^6 molec/cm^3 (24-h average) based on the rate of decrease of various gases in a quasi-Lagrangian experiment performed with the Dutch Citation aircraft [13]. Model output from ECHAM [25] and from MATCH-MPIC [39] are in basic agreement with this high OH concentration, computing 2.6×10^6 molec/cm^3 and $(3–4)\times10^6$ molec/cm^3, respectively over the measurements region. These are several times higher than the global average OH concentration of about 1×10^6 molec/cm^3 [40], and are sufficient to provide significant breakdown rates of trace gases in the haze outflow, resulting for instance in a CO lifetime of only ~15 days near the surface.

The high oxidant levels in the haze are also important for chemical processing of aerosols. In particular, most of the sulfur in sulfate aerosols in this region originates as SO_2, which is then oxidized to sulfate by reaction with OH in the gas phase, or by reaction with H_2O_2, O_3, HOCl, and HOBr [41] in cloud droplets and aerosols. Due to the rarity of clouds, oxidation by OH plays a disproportionately large role here compared to other cloudier regions [27]. Furthermore, the lack of precipitation results in long aerosol lifetimes, ~7 days for sulfate and ~8 days for carbonaceous aerosol, compared to the global average of ~5 days [27]. This prolongs the time during which oxidation reactions can take place, and results in a greater "sulfate potential" for the SO_2 emitted from southern Asia during the wintertime, which partly compensates for the lower relative SO_2 emissions for this region compared to most other parts of the world (Table 1). This also provides a longer time during which aerosol-aerosol interactions can take place. Electron microscope analysis of the haze aerosol clearly shows that "internally mixed" aerosols containing sulfate, organic carbon, and soot are present all the way down to the ITCZ [17]. In contrast, models generally treat the aerosols as separate, "externally mixed" components (e.g., [27, 42]). Future work on this issue will be important, since the internally vs externally mixed composition of aerosols can influence their optical properties and thus the regional radiative energy budget.

4.4
North-South Flow Times, Gradients, and the "Chemical" ITCZ

The amount of chemical processing in the outflow depends not only on the tracer lifetimes but also on the amount of time the air masses are underway. Typical air mass ages, computed based on the ratios of pollutant concentrations measured onboard the Dutch Citation, increase from a couple days near the southern coast of India to around ten days near the ITCZ [13], consistent with the regional aerosol lifetime estimate [27] discussed above (some loss of aerosol occurs underway by dry deposition, explaining the somewhat shorter aerosol lifetime than the transport time prior to reaching the ITCZ, where loss due to washout is generally rapid [43]). This transport time, combined with the effects of chemical pro-

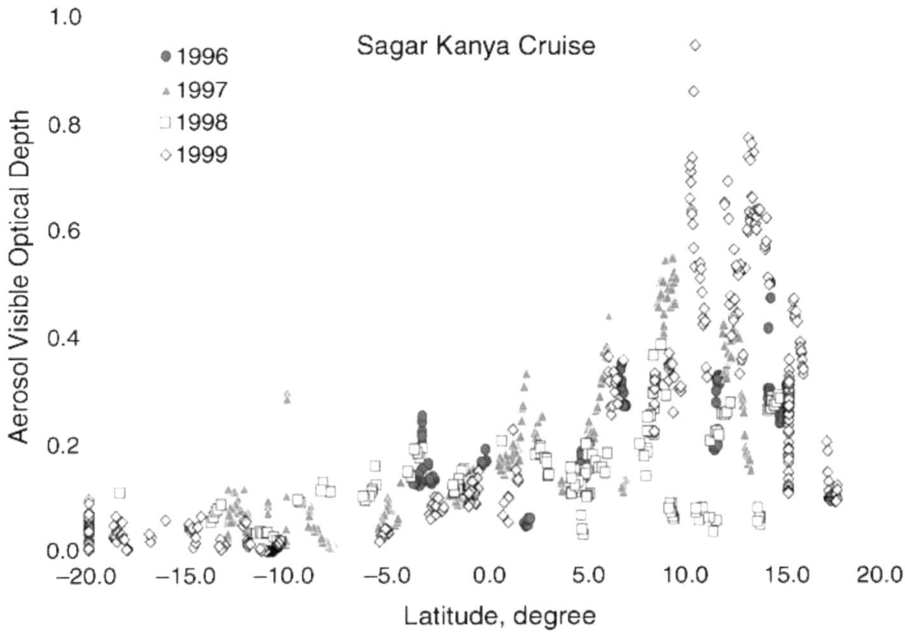

Fig. 10 Variation of aerosol optical depth (AOD) as a function of latitude. (From [17], copyright 2001 by the American Geophysical Union, reproduced by permission of the AGU)

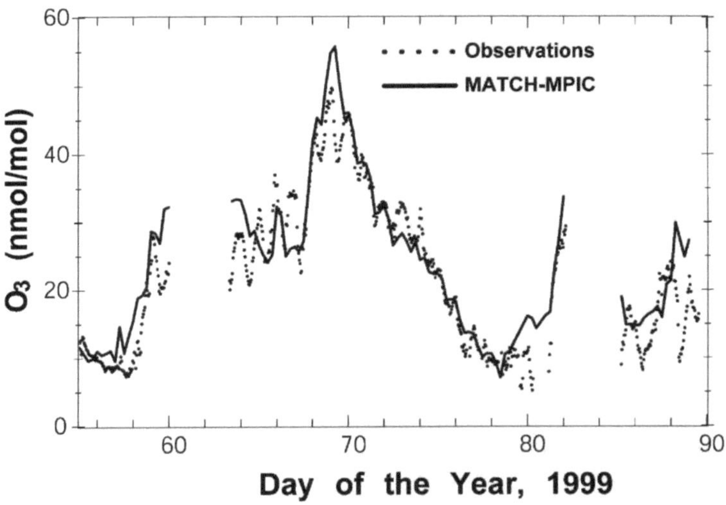

Fig. 11a–c Observations (dots) and modeled values (*solid lines*, from MATCH-MPIC) of: **a** O_3; **b** CO during the 1999 R/V Ronald H. Brown cruise; **c** the cruise route, with indicators for the two encounters with the ITCZ: the crossing on 27 February 1999 (Julian day 58, see Fig. 12a), and the brief encounter on 19 March 1999 (Julian day 78, see Fig. 12b). (From [39], copyright 2003 by the European Geophysical Union, modified by permission of the EGU)

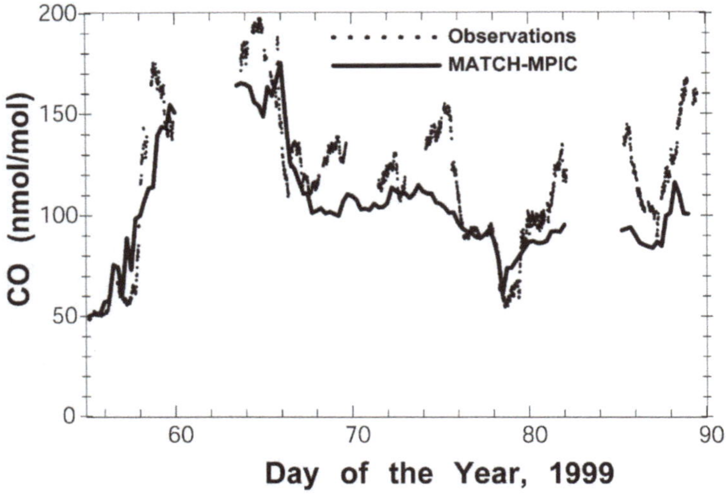

b

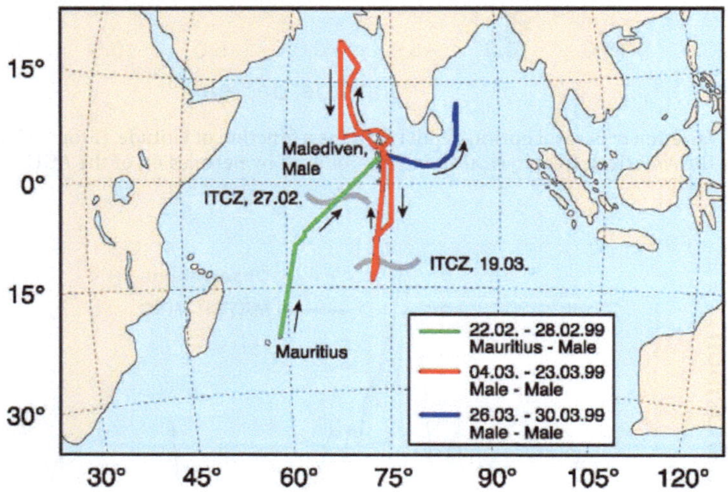

c

Fig. 11b, c

cessing, dilution, and surface deposition, results in substantial north-south gradients of various gases and aerosols, which have been measured on research ships repeatedly (e.g., [17, 34, 44, 45]). Examples of the gradients measured for the aerosol optical depth (AOD) are shown in Fig. 10; similar gradients have also been measured for O_3 and CO (see Fig. 11).

The ITCZ separates the meteorological and chemical northern and southern hemispheres. Near the surface, the ITCZ is characterized by an abrupt transition between the polluted NH air masses and the more pristine SH air. This "chemical ITCZ" was seen, for instance, in the O_3 and CO observations made on the R/V Ronald H. Brown, which are depicted in Fig. 11, along with output from a

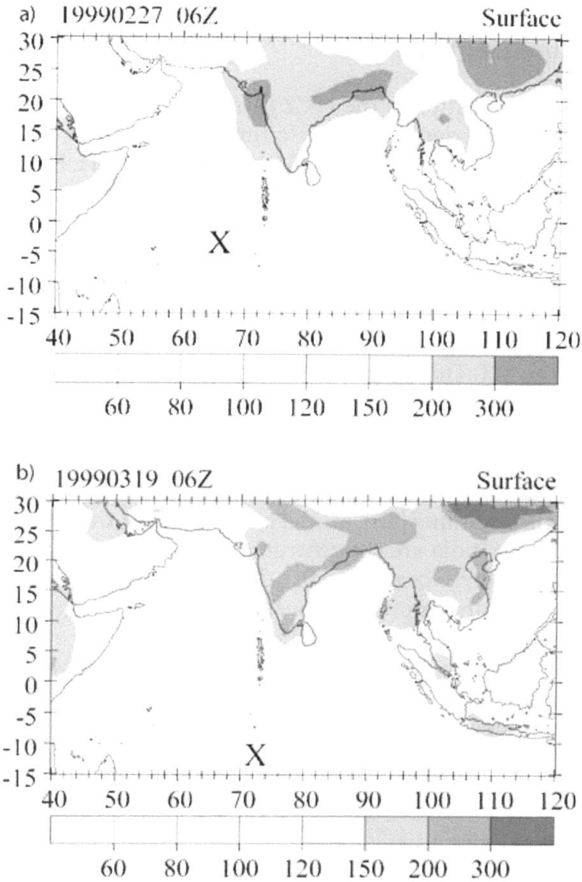

Fig. 12a, b Simulated CO surface level mixing ratios (nmol/mol) from the MATCH-MPIC model at the times of the two encounters of the R/V Ronald H. Brown with the ITCZ; the "X" marks the location of each encounter. (Based on [39], copyright 2003 by the European Geophysical Union, modified by permission of the EGU)

global chemistry-transport model [39]. The model was able to capture the observed variations quite well; it was also able to forecast these variations, which was used in support of flight and cruise route planning [39]. Abrupt jumps in CO and to an extent in O_3 are seen on Julian days 58 and 78, when the ITCZ was encountered. During the three weeks between these two encounters, the ITCZ had migrated nearly 10° southwards, from 2°S to 11°S (Figs. 11c and 12), at an average rate of about 50 km/day. The tendency of the ITCZ to shift positions regularly during the winter monsoon has been studied extensively by meteorologists in the past, and this particular southward migration during INDOEX can be seen in the meteorological analyses of the wind fields discussed in [20].

4.5
Transport in and Across the ITCZ

The deep cumulus convection of the ITCZ lifts trace gases and aerosols out of the boundary layer (BL) and deposits them in the MT and UT. At the same time, soluble gases and aerosols are taken up in the condensate, and a fraction is removed with the precipitation [43]. This fraction depends largely on the solubility of the tracers [46–49]; only a small fraction (<10%) of highly soluble species like HNO_3 and H_2SO_4 is normally able to escape precipitation scavenging in cumulonimbus clouds, although there are important exceptions, particularly for gases that are not retained well in ice when supercooled water freezes. Deep convection characteristically transports air rapidly from the BL to the UT, compensated by slower subsidence into the MT. This is seen in the average age profile of air masses sampled by the Dutch Citation (Fig. 13, based on pollutant ratios), which shows a pronounced maximum of about ten days at 4–6 km, dropping to six to eight days above 8 km. CO and several other gases also frequently had UT mixing ratios which exceeded their MT values, in some cases (e.g., propane) even averaged over the entire set of observations [13]. Model simulations show that the main region in which tracers from India are concentrated in the UT is over the equator in the central to western Indian Ocean (Fig. 14), due to convective uplifting in the ITCZ [25, 26]. Finally, one of the important consequences of the convective lifting, supported by observations [50], is that the convective outflow region is very conducive to the formation (nucleation) of new particles, since washout in the convective cores significantly reduces the outflow concentration of large aerosol particles (on which new particles would otherwise rapidly condense).

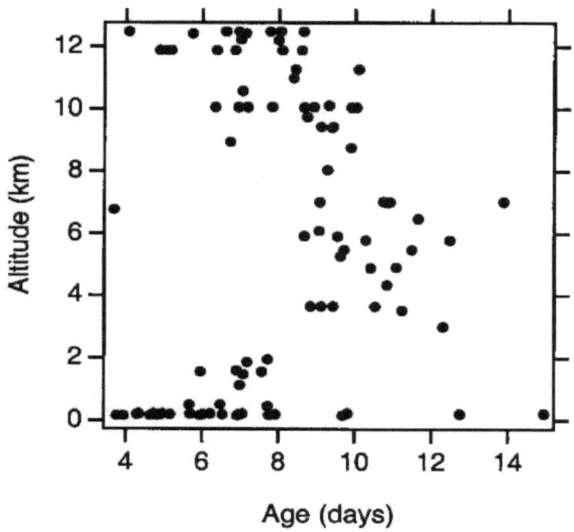

Fig. 13 Altitude profile of the estimated age of sampled pollution during INDOEX. (From [13], copyright 2001 by the American Geophysical Union, reproduced by permission of the AGU)

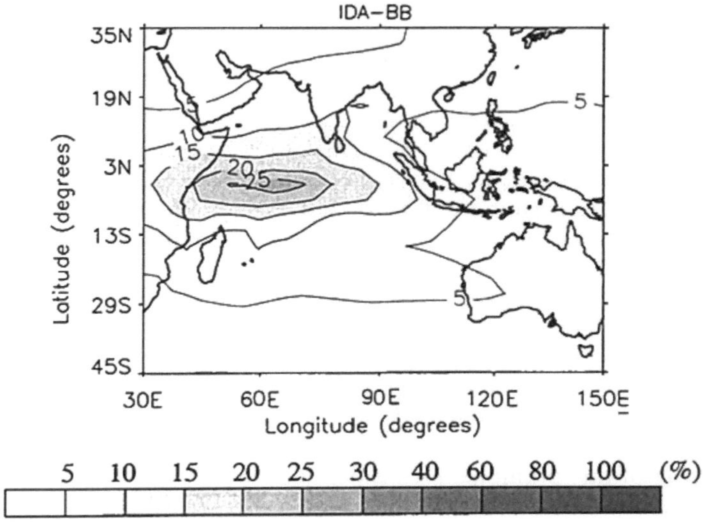

Fig. 14 Indian biomass- and biofuel-burning CO tracer at 10 km altitude simulated with the ECHAM chemistry-GCM. (From [25], copyright 2001 by the American Geophysical Union, reproduced by permission of the AGU)

The ITCZ is the gateway for interhemispheric transport. The occurrence of interhemispheric transport was inferred [51] based on the observation that in the UT south of the ITCZ, the mixing ratios of CO (>100 nmol/mol) can exceed the typical SH MBL mixing ratios (about 50 nmol/mol) by more than a factor of two, but are very consistent with the typical NH MBL mixing ratios in the inflow region of the ITCZ deep convection. The physical mechanisms controlling interhemispheric exchange in this region are only roughly understood [52]. Most of the exchange appears to take place in the MT and UT, since the strong Hadley Cell convergence near the surface makes the transport of air masses from one hemisphere deep into the other hemisphere difficult. A preferred region for interhemispheric mixing in the UT has been identified using back trajectories [52] as the converging region between two anticyclones, one in the SH over northern Australia and the other in the NH east of the Philippines, from which air then enters the Indian Ocean region, passing over the highly convective Indonesian region. A more general mechanism which has been discussed as a possibility for interhemispheric exchange is the mixing of air masses from the NH and SH in the deep convective updrafts of the ITCZ, which then can enter the opposite hemisphere in the UT outflow. This mechanism, however, is very difficult to evaluate quantitatively based on currently available models, and will be one of the major future challenges in understanding the more global impacts of southern Asian pollution export.

4.6 Interannual Variability

Superimposed on the short-term variability in the ITCZ position and the large seasonal cycle discussed above, there is also a considerable interannual variability in the wintertime Indo-Asian haze. This can be seen, for instance, in the AOD measurements in Fig. 10, as well as in several other measurements, e.g., the ground-based CO observations at KCO [22]. Interestingly, taking the region as a whole ([53], based on [54] and [55]), the variability is much smaller (Fig. 15); the wintertime AOD only varies by about ±15% from year to year, with the AOD for 1999 being only slightly above the average from 1995 to 2001. Nevertheless, there is a substantial interannual variability in measurements at specific locations (e.g., a factor of 2 difference between 1998 and 1999 in the AOD averaged over the 1000×1000 km region around KCO [53]. Based on these observations, it appeared that 1999 was a particularly polluted year, especially during February, with higher aerosol and trace gas levels than during the pre-INDOEX years 1996–1998. Recognition of this prompted a considerable amount of discussion, including a special session at the INDOEX workshop in Utrecht in September, 1999, with 15 presentations on various aspects of the contrast between 1998 and 1999.

The high pollutant levels observed during February, 1999, were a result of two main factors acting together [56]. First, early 1999 was a La Niña period, during which the high pressure system over southern Asia was stronger than normal, thus enhancing the pressure gradient between the Indian subcontinent and the northern Indian Ocean and strengthening the continental outflow. This helps explain the difference with 1998, an El Niño period with a weakened high pressure

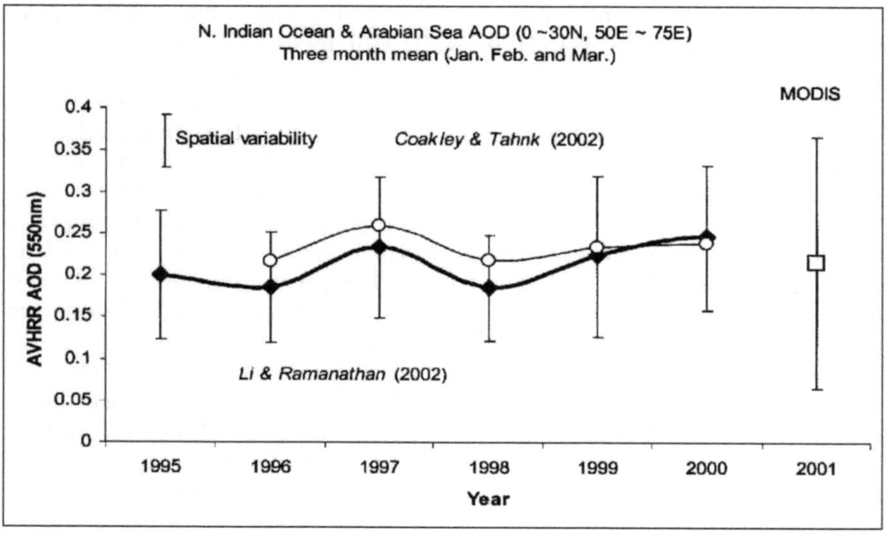

Fig. 15 Interannual variability of the mean AOD averaged over the northern Indian Ocean and the Arabian Sea; *closed symbols* are from [54], *open symbols* from [55]. (From [53], copyright 2002 by the Indian Academy of Sciences, reproduced by permission of the IAS)

system over southern Asia, but does not explain the large difference with 1996 and 1997. The second factor which combined with this was that during this period, the convection in the ITCZ was considerably weaker than normal, which could be seen for instance in observed and modeled precipitation fields for the region [56]. This resulted in ineffective ventilation of the pollutants out of the continental outflow, and thus in a buildup of trace gas and aerosol levels in the lower troposphere (LT) in the INDOEX measurements region. The weakened ITCZ convection was associated with a passive phase of the Madden-Julian Oscillation (MJO), a westerly planetary wave which represents one of the major modes of intraseasonal variability in this region, occurring on a 30–50 day timescale (the other major modes are on a 3–6 day timescale due to baroclinic disturbances, along with diurnal variations; [19]). Although the MJO can be simulated successfully by global models [27, 56], the physical mechanisms which govern it are still a major topic of research (e.g., [57]). These same two factors can also be used to explain most of the variations on a monthly-average basis which were observed throughout the INDOEX and pre-INDOEX campaigns [56].

4.7 Large-Scale Effects

The large scale effects of the wintertime Indo-Asian haze are mostly due to the radiative and cloud-modifying properties of aerosols and the resulting perturbations to the regional atmospheric and oceanic energy balance and climate. The aerosol effects depend on both the amount of aerosol present and the aerosol composition, and thus in turn on the emissions and chemical processing, with the uncertainties noted above. Despite these uncertainties, one message has become clear: the aerosol forcing is large, and it is of a highly absorbing nature. Aerosol composition measurements onboard the C-130 and at KCO (Table 3) indicate that

Table 3 Mean fine and coarse mass fractions of aerosols collected on filters on board the C-130 aircraft in the boundary layer (34 samples) and at KCO (24 samples). *D* is diameter. MSA is methane sulfonic acid. "Rest" includes magnesium, calcium, oxalate, formate, and unidentified material. (Reproduced with permission from [21], copyright 2001, American Association for the Advancement of Science)

Compound	D<1 μm(%)	D>1 μm(%)
Sufate	32	25
Organics	26	19
Black carbon	14	10
Mineral dust	10	11
Ammonium	8	11
Fly ash	5	6
Potassium	2	1
Nitrate	<1	4
Sea salt, MSA	<1	12
Rest	2	1
Total mass (μg/m^3)	22	17

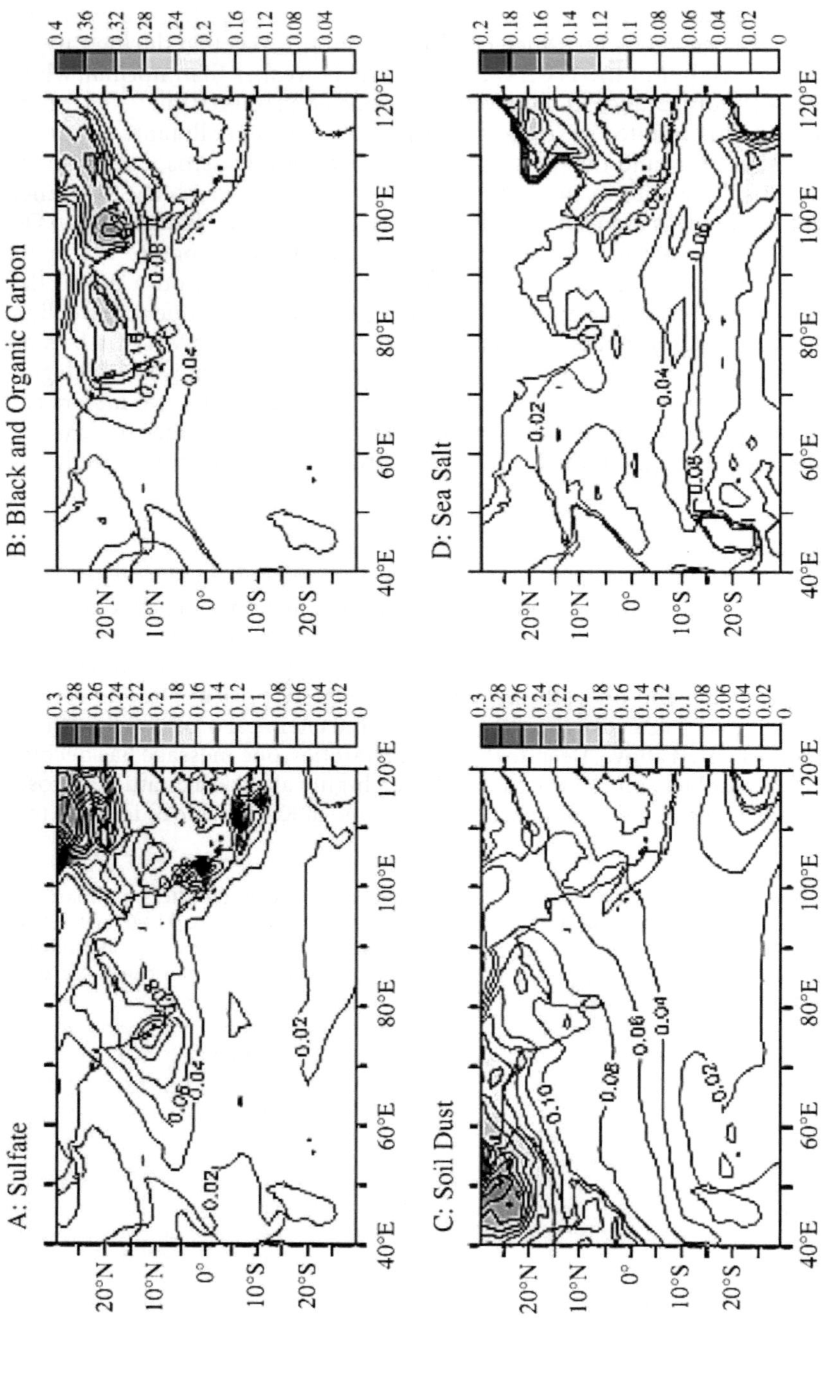

Fig. 16A–D Contribution of individual aerosol components to the total AOD simulated with the MATCH model, averaged over January-March, 1999. (From [58], copyright 2002 by the American Geophysical Union, reproduced by permission of the AGU)

sulfate accounts for 1/3 of the submicron particle mass and 1/4 of the supermicron particles. Carbonaceous aerosols account for 30–40%, while soil dust is about 10% of the total aerosol mass for both size classes, as is sea salt for the large particles. Simulations with the MATCH aerosol model [27, 42, 58] are in general agreement with these observations for the region SW of India (Fig. 16), and also indicate the spatial variability in the aerosol composition: all four modeled aerosol types contribute significantly to the total wintertime AOD over the Indian Ocean, each dominating in a different region.

The high black carbon fractions observed at KCO have been noted to be comparable to what is found in US cities [59], with the important difference that KCO is over 500 km downwind of the nearest major urban centers. These high carbon contents result in aerosols which are highly absorbing (in contrast to sulfate aerosols, which are nearly perfectly reflecting), with an average SSA of ~0.9 for the aerosol mixture in the region (both continental and maritime), computed from several independent methods based on observations [17]. Model simulations [58] indicate particularly low values (SSA<0.8) over continental northern India, SE Asia, and the northern Bay of Bengal.

The AOD over land during INDOEX was very high, about 0.5, and remained high in the outflow, with values of up to 0.2 near the equator. The AOD regional distribution based on the MATCH simulation [58] is depicted in Fig. 17. In this case, the degree of uncertainty in the model results is lower than usual for global

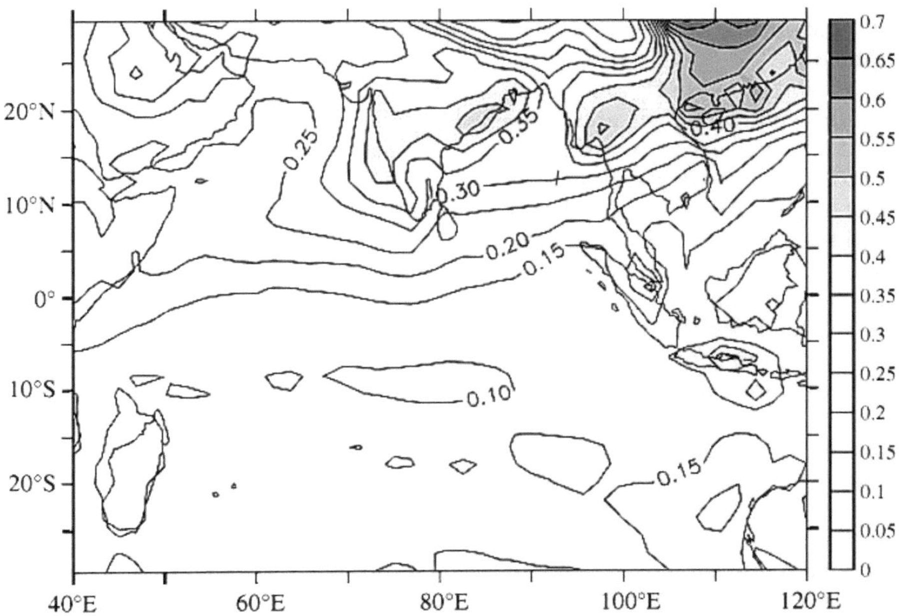

Fig. 17 Total AOD simulated with the MATCH model, averaged over January–March, 1999. (From [58], copyright 2002 by the American Geophysical Union, reproduced by permission of the AGU)

Table 4 Aerosol Radiative Forcing (W m^{-2}): North Indian Ocean, January–March 1999; 0–20°N, $\tau_a = 0.3$. (Basedon plate 13 of [17], copyright 2001 by the American Geophysical Union, reproduced by permission of the AGU)

	Direct (clear sky)	Direct (cloudy sky)	First indirect
TOA	–7.0±1	–2.0±2	–5±2.5
Atmosphere	+16.0±2	+18.0±3	+1±0.5
Surface	–23±2	–20±3	–6±3

models, due to the assimilation of observed AOD over the ocean from the AVHRR (Advanced Very High Resolution Radiometer) satellite instrument into the model simulation. Combined with the low SSA values, this yields a direct aerosol forcing of –23(±2) W/m^2 at the surface averaged over the northern Indian Ocean and the southern tip of India (Table 4), which is about 15% of the solar heating of the northern Indian Ocean, and is over half as large as the net downward heat flux into the ocean, which is estimated at 30–50 W/m^2 [17]. Only about 1/3 of this surface forcing is due to the reflection of sunlight and is thus evident to satellites at the top of the atmosphere; the other 2/3 is due to the absorption of solar radiation by aerosols in the troposphere. This immense energetic input to the LT amounts to a heating rate of 0.4 K/day for the layer from the surface to 3 km, which accounts for about half of the climatological solar heating of the LT in this region [17]. The presence of clouds modifies the direct aerosol forcing slightly, but the qualitative distribution and magnitude of the forcing remains the same (Table 4).

Finally, the aerosols can modify cloud properties via the so-called "Twomey" or "first indirect" effect: greater aerosol particle number densities cause water to be distributed over a greater number of droplets, thus increasing the surface area of the cloud droplets and thereby the reflectivity of the clouds. Though this effect has been discussed for three decades [60, 61], some of the first clear observational evidence of it was gathered during INDOEX. Two studies [62, 63] have shown that the NH clouds that are affected by the haze had both aerosol number densities and cloud droplet number densities which were much higher than in SH clouds with "background" aerosol levels, while the cloud water content was nearly the same for clouds in both regions. This basic effect has been reproduced in simulations with a cloud resolving model for the INDOEX conditions [43]. This results in effective droplet radii that were ~20% smaller in the polluted clouds. These data were used in [17] along with a radiative transfer model to estimate the impact on solar forcing to be –6±3 W/m^2 at the surface (–5±3 W/m^2 at the top of the atmosphere). The combined forcing from the direct aerosol effect and the first indirect effect varies as a function of latitude (Fig. 18), from about 5 W/m^2 south of the ITCZ to over 30 W/m^2 north of about 10°N. Two other indirect effects, namely the reduction of drizzle due to smaller droplets ("second indirect") and the reduction of trade cumuli due to warming in the residual layer ("semi-direct") have also been postulated. These effects will have opposite signs (the second indirect effect cools while the semi-direct effect warms), and the ef-

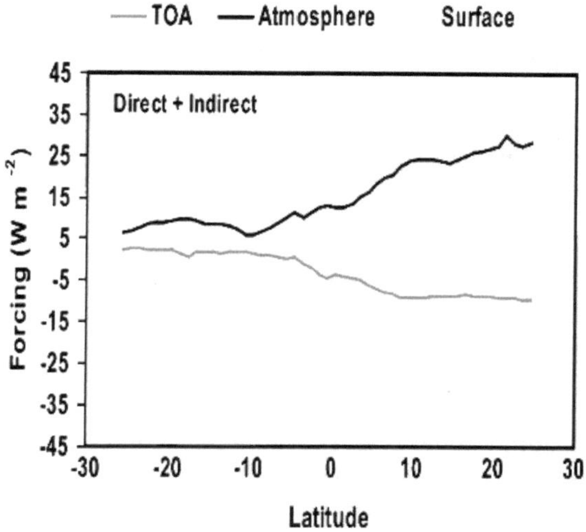

Fig. 18 Latitudinal variation of the aerosol radiative forcing (including natural and anthropogenic emissions), averaged over the ocean in the region from 40°E to 100°E. The *top line* is the atmospheric forcing, the *middle line* is the TOA (top of the atmosphere) forcing, and the *bottom line* is the forcing at the surface. (From [17], copyright 2001 by the American Geophysical Union, reproduced by permission of the AGU)

fects have been estimated to be much smaller, of the order of 2–5 W/m^2 with an uncertainty exceeding 50% [17].

Recently, first attempts have been made to estimate the large scale climate effects of the wintertime Indo-Asian haze by including the aerosol radiative forcing terms in global general circulation models [64–66]. The absorbing nature of the haze causes it to cool the land surface and warm the atmosphere, resulting in a greater stabilization of the BL and a reduction of energy fluxes (latent and sensible heat) from the surface. Dynamically, the models respond with a northward shift in the ITCZ and an increase in precipitation (by up to 20%) over the haze regions, accompanied by a decrease in precipitation over the Indonesian and Pacific warm pool regions [64]. The simulated changes in precipitation and temperature are similar to observations, in particular of a small cooling trend in China and India, contrasted with the warming trend throughout most of the rest of the world [66]. A preliminary link has been made to changes in the Indian summer monsoon rainfall patterns [64] and to the El Niño Southern Oscillation [65], though further studies with improved models (e.g., with interactive aerosols or with a fully coupled ocean-atmosphere system) are needed to reduce the uncertainties, and to eventually allow a viable assessment of the possible socio-economic consequences.

5 Export in the Asian Summer Monsoon Plume

The southern Asia summer monsoon deep convection discussed above carries along with it pollutants from the surface layer (except those which are rained out). An example of the simulated 3D structure of this outflow plume is given in Fig. 19. An analysis of the general transport characteristics in the plume [39] using the global CTM MATCH-MPIC has shown that it exhibits a considerable short-term and interannual variability (Figs. 20 and 21). The plume is most prevalent over northeast Africa. The degree to which it extends over the Mediterranean varies significantly from day to day and from year to year. The variability of the plume is influenced by several factors, including the strength and distribution of convection over Asia, the intensity and location of the Tibetan and northern Arabian anticyclones (Fig. 4), the chemical processing of the plume during transport, and dilution due to mixing with other UT air masses [39]. The frequency of occurrence of the plume over the eastern Mediterranean can be classified by considering how often the simulated mixing ratio of Asian CO exceeds some threshold [39]. Taking Crete as a reference point, during 2001 a strong plume was present above the island about a quarter of the time (between 12% and 40%, depending on the threshold value chosen for the Asian CO); it is about 1.5 times as prevalent to the south and east of Crete, while it is relatively rare (<10%) that the plume extends much past Crete (due to the curvature on the western side of the Arabian anticyclone).

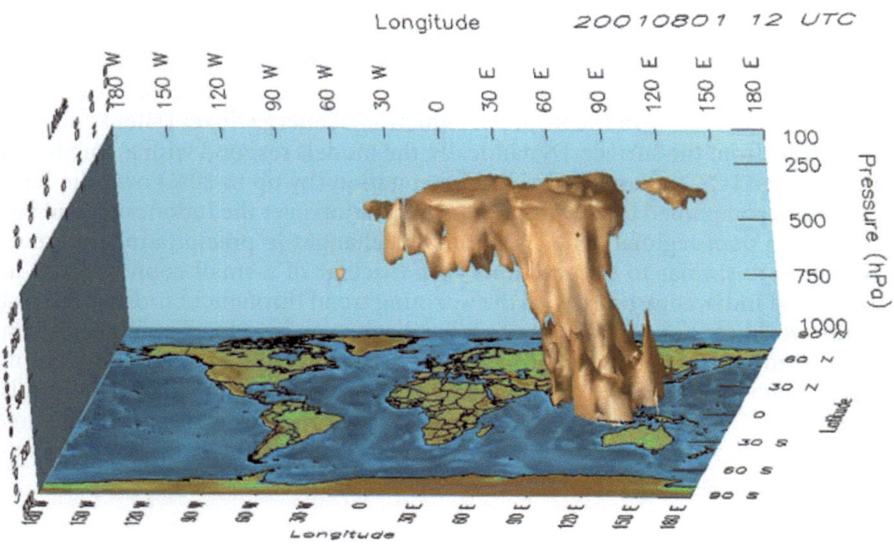

Fig. 19 Simulated 3D structure of the Asian monsoon plume on 1 August 2001, based on simulation of a southern Asian CO tracer with the global CTM MATCH-MPIC. (From [39], copyright 2003 by the European Geophysical Union, reproduced by permission of the EGU)

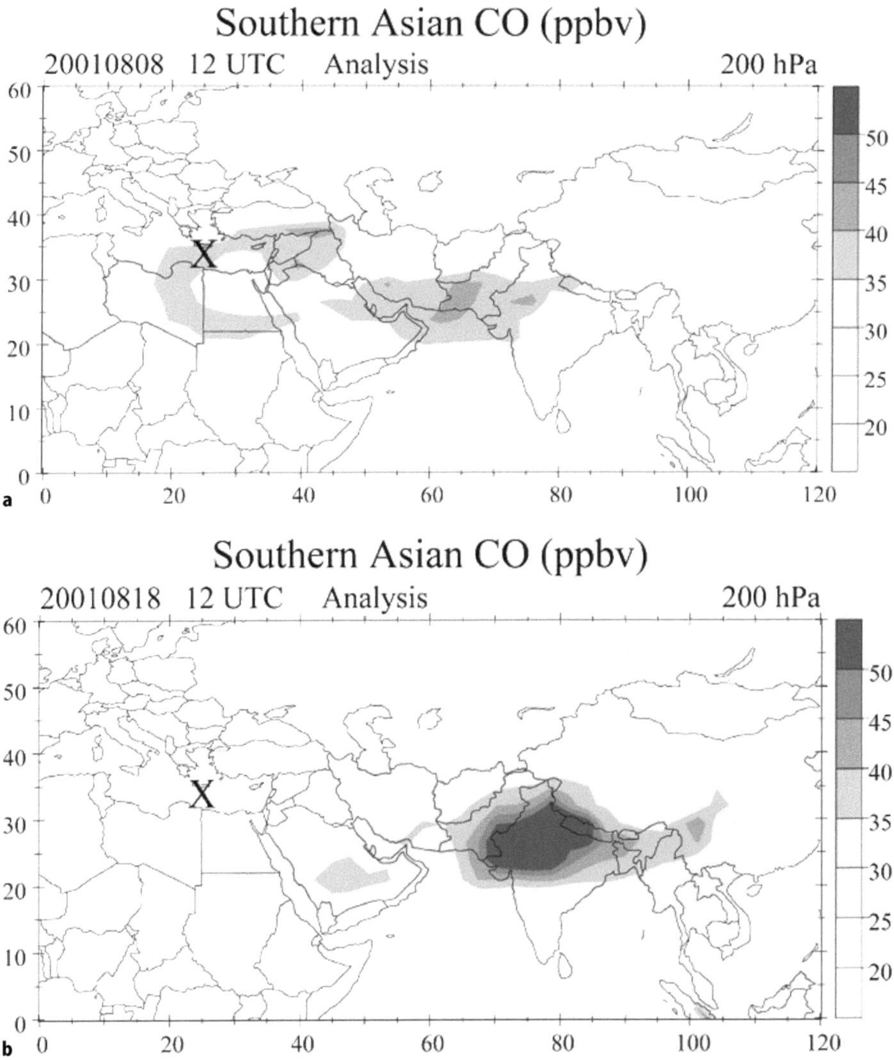

Fig. 20a, b Short-term variability of the Asian summer monsoon plume at 200 hPa based on the simulation of a southern Asian CO tracer at two points in time (8 and 18 August 2001) with the global CTM MATCH-MPIC. (Based on [39], copyright 2003 by the European Geophysical Union, modified by permission of the EGU); an animation of the outflow is available in the supplementary material of [39] (http://www.copernicus.org/EGU/acp/acp/3/267/acp-3-267_supp.zip)

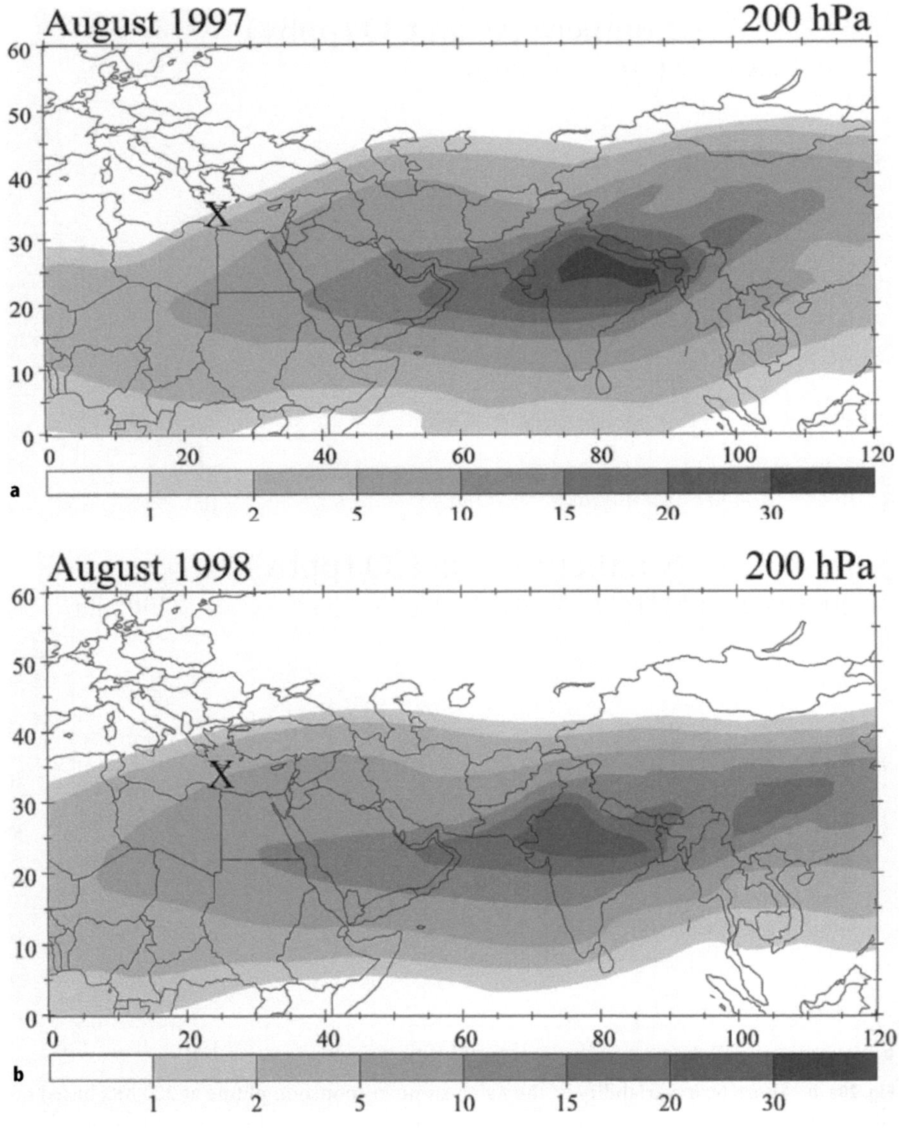

Fig. 21a, b Interannual variability of the Asian monsoon plume, illustrated with the averages for August 1997 and 1998 of a southern Asian CO tracer simulated with the global CTM MATCH-MPIC. (From [39], copyright 2003 by the European Geophysical Union, modified by permission of the EGU)

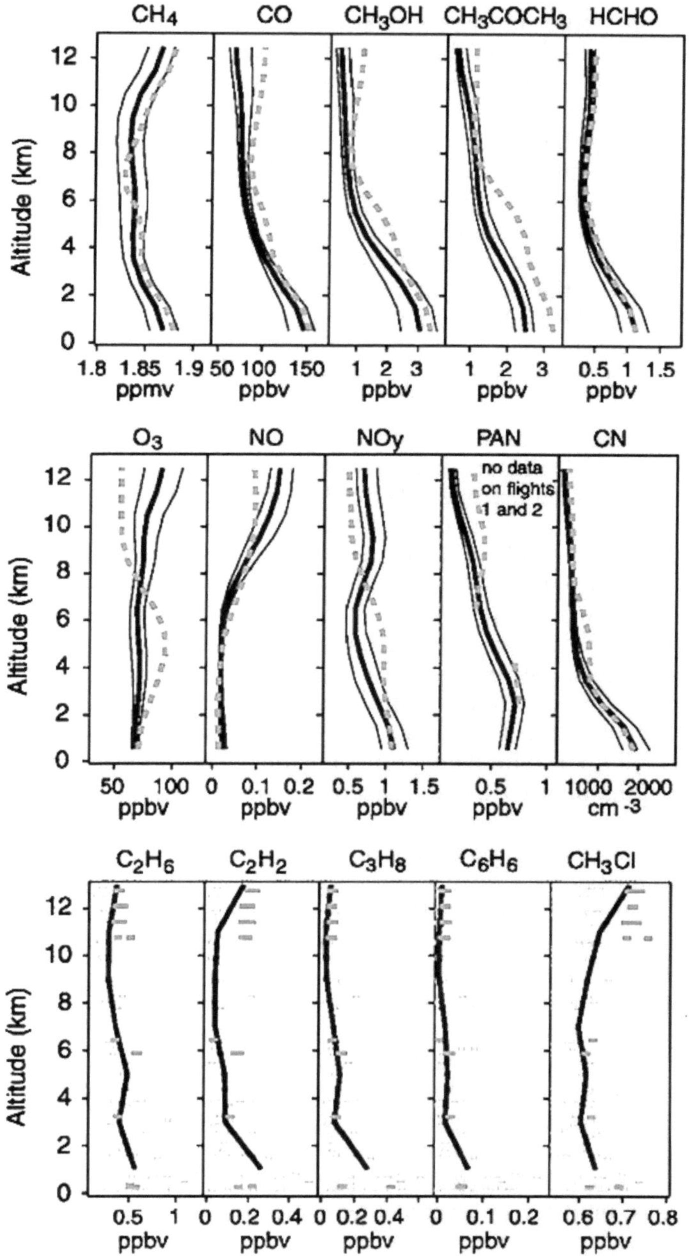

Fig. 22 Vertical profiles of trace gases measured during the 14 MINOS flights in August 2001. *Thick solid lines* are median values, *thin solid lines* are $\pm 1\sigma$, *dashed lines* are the median values for only the first three flights (which sampled the Asian summer monsoon plume), and *small boxes* (*bottom row*) are individual canister samples. (Reproduced with permission from [67], copyright 2002, American Association for the Advancement of Science)

This section focuses on the westward branch of the plume. Little is known about the potential effects of the eastward branch on cirrus clouds in the western Pacific, though it is likely that any effects are small in comparison to the more direct input of pollution to that region from eastern Asia (in particular from China). Observing the plume over the Mediterranean was one of the objectives of the field campaign MINOS (Mediterranean Intensive Oxidants Study), which had as a primary objective the investigation of the European pollutant outflow into the Mediterranean region, as well as investigating the import of pollution from North America into the region [67].

During MINOS, the first three research flights targeted the Asian summer monsoon plume. The median of the observed profiles for several gases measured on these three flights compared to the profiles averaged over the entire campaign are plotted in Fig. 22. In the UT numerous gases such as CO, CH_4, CH_3OH, HCHO, C_2H_6, and CH_3Cl were substantially enhanced above the average levels, in some cases by considerably more than one standard deviation. When ten-day backward trajectories are used to determine the longitude of origin of each of the air samples (Fig. 23), it is seen that the enhanced mixing ratios of these tracers are associated with air coming from the east. The high levels of indicator gases for biomass burning, in particular CH_3Cl, C_2H_2, and CH_3CN [68], low levels of the new air conditioning agent HFC-134a in comparison to air masses at lower altitudes of North American origin [68], and high levels of CH_4, likely from rice fields [67], give experimental evidence for the Asian origin of this pollution. High levels of CH_4 have also been observed over the eastern Mediterranean and the Arabian regions (Fig. 24) during commercial aircraft flights between Germany and the northern Indian Ocean (Sri Lanka and the Maldives) as part of the project CARIBIC (Civil Aircraft for Regular Investigation of the atmosphere Based on an Instrument Container, [69]). The elevated CH_4 mixing ratios support the findings of MINOS, and show that the plume is visible from late June through mid October, corresponding with the timing of the summer monsoon deep convection over southern Asia.

On the other hand, during MINOS it was found that O_3, NO, and NO_y in the Asian summer monsoon plume were significantly below the average UT levels, consistent with the inefficient burning and NO_x limitation of O_3 production seen in the wintertime Indo-Asian haze discussed in the previous section. Interestingly, this is in contrast with model results [70] which show a maximum in the tropospheric O_3 column during the summer over the Middle East. Model sensitivity runs [70] indicate that this maximum is largely associated with Asian anthropogenic emissions and lightning in the monsoon convection, although farther to the northwest (over the southwestern Mediterranean) anthropogenic emissions from North America and Europe become the dominant contributor. This feature is simulated by several other contemporary CTMs, and there is some evidence for it in O_3 profiles measured onboard commercial aircraft flights to Middle Eastern airports (as part of the project MOZAIC, Measurements of Ozone and Water Vapor by Airbus In-service Aircraft, [71]). However, this Middle Eastern O_3 maximum is not present in a recently-developed tropospheric O_3 column climatology based on satellite retrievals (see Fig. 7 in [33]), which instead shows maximum O_3 columns in the regions adjacent to the simulated Middle Eastern

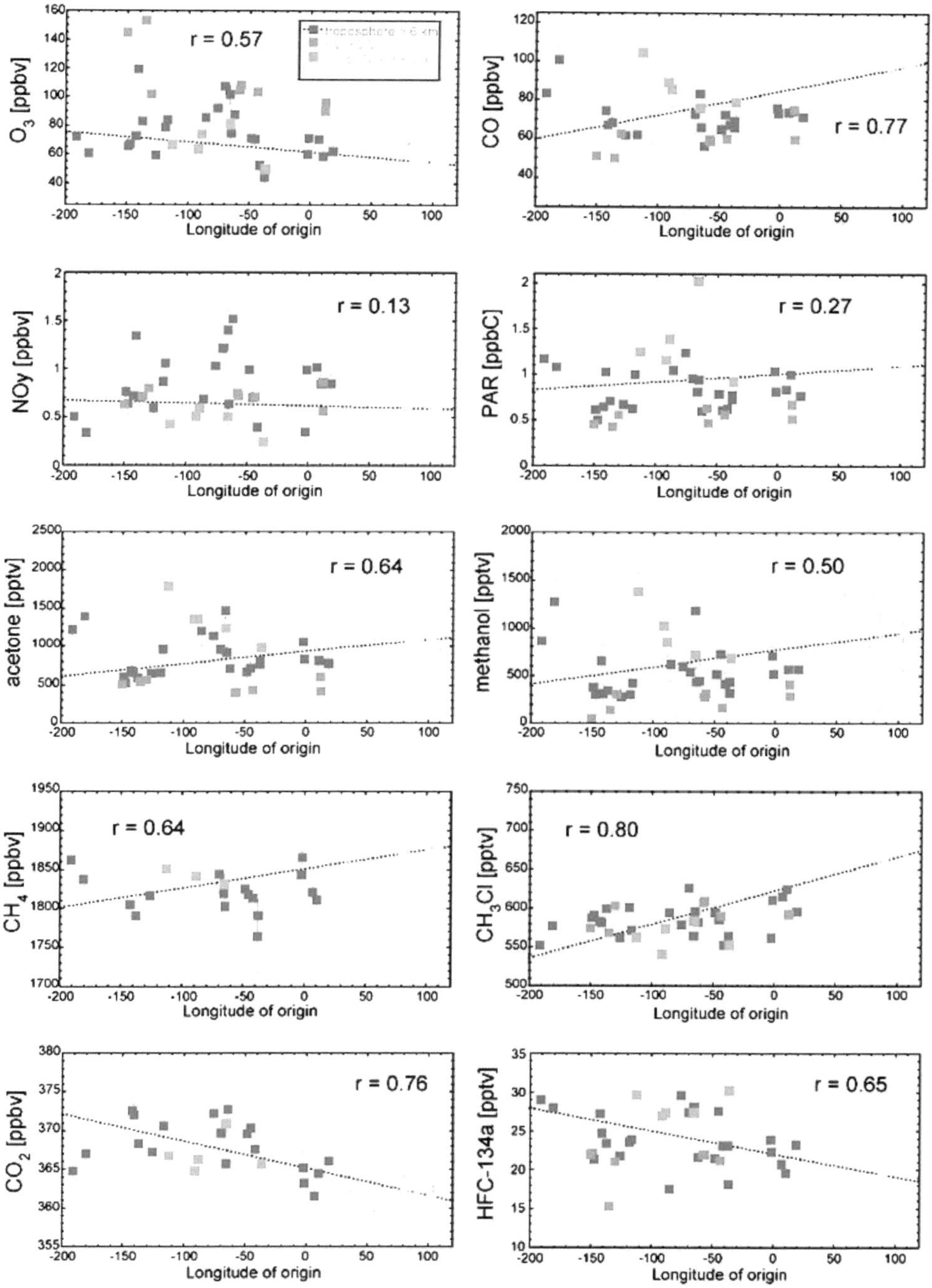

Fig. 23 Trace gas measurements in the UT (6–13 km altitude) made during MINOS plotted as a function of longitude of origin, determined from ten-day backward trajectories. (From [68], copyright 2003 by the European Geophysical Union, reproduced by permission of the EGU)

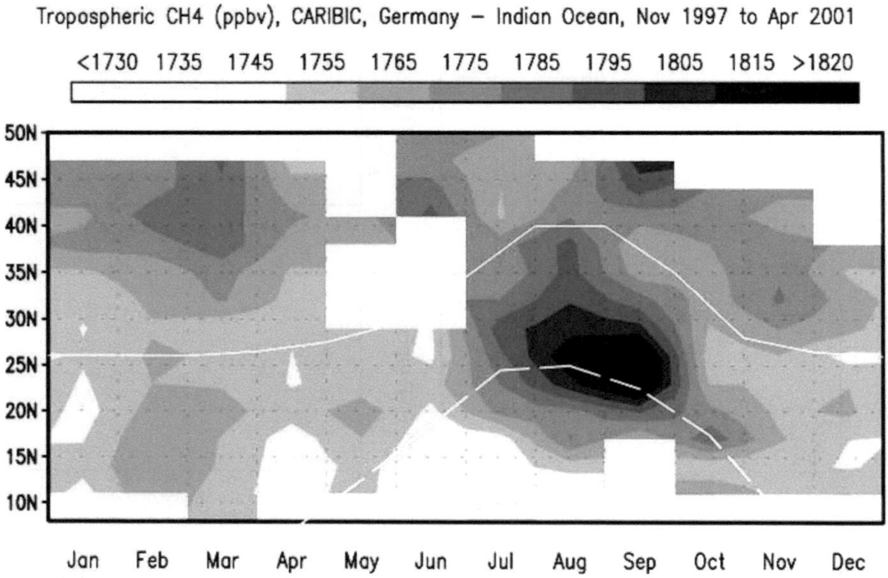

Fig. 24 Time-latitude distribution of CH_4 (in nmol/mol) observed during the CARIBIC flights from Germany to the Indian subcontinent averaged over 1997–2001. Only tropospheric data are considered. Flight level is typically 10–11 km. The data are binned by the month of the year and by 2° of latitude intervals. The *dashed line* indicates the mean location of the ITCZ, and the *solid line* marks the location of the subtropical jet stream. Figure provided courtesy of C. Brenninkmeijer, Max Planck Institute for Chemistry, Mainz, and Andreas Zahn, Forschungszentrum Karlsruhe

maximum (i.e., over northern India and the central to eastern Mediterranean). Thus, the behavior of O_3 in the Asian summer monsoon outflow – whether it should be enhanced or whether it is at reduced levels compared to other subtropical regions (or how this varies spatially) – is at present a mystery that will require further observations and detailed analysis of O_3 and other key trace gases in this region to sort out.

The age of the Asian air masses by the time they reach the Mediterranean can be estimated based on the ratios of certain gases along with knowledge about their ratios at the source and their chemical lifetimes. This method was applied in [68], based on the measured ratios of CH_3CN, CH_3Cl, C_2H_2, and C_6H_6 to CO in the Asian outflow during MINOS, and the assumption that the initial ratios could be taken from the INDOEX outflow measurements near the Indian coast [13, 72], with the exception of C_2H_2, for which the emission ratio to CO measured for fresh biomass burning was used. A chemical age of about ~13 days is inferred from the measured ratios of both CH_3Cl and C_6H_6 to CO, and of ~15 and ~20 days based on C_2H_2 and CH_3CN, respectively. The travel time from the Asian UT to the Mediterranean region is about one week (based on both trajectories [73] and analysis of the tracer animations in [39]). Thus, on average the air masses reside in the BL over India for about one week before they are rapidly transported to the

UT, which fits with the episodic nature of deep convection during the summer monsoon (e.g., [14]).

Global models simulate the presence of the Asian summer monsoon plume over the Mediterranean and its differences with air masses from other sources. The CO from MATCH-MPIC was found to correlate well (r=0.59) with the MINOS measurements in the UT during the three flights into the plume, and even better for the overall set of measurements during MINOS (r=0.73), supporting the use of the modeled CO tracers for source region characterization. CO tracer forecasts from the ECHAM GCM, used along with the MATCH-MPIC forecasts in the flight planning for MINOS (discussed in [39]), showed a similar degree of agreement with the observations. Furthermore, the ECHAM model with a full tropospheric chemistry scheme [68, 74] was found to reproduce the variation in CO and O_3 levels as a function of source longitude (like Fig. 23). On the other hand, the models have difficulties with the absolute levels of several gases. For instance, ECHAM tends to underestimate NO_y levels by 30–50% [68]; a similar discrepancy has been found for MATCH-MPIC (unpublished data, available via the MINOS data server). Both models also have difficulties with VOCs. ECHAM underestimates the C2–C5 alkanes by up to a factor of 5 during some of the MINOS UT flights, though it simulates the levels in the Asian plume fairly well [68]. MATCH-MPIC has been found to have particular difficulties with CH_2O, which is substantially underestimated in the Asian plume [75, 76], for as yet unknown reasons; CH_2O has frequently proven difficult for models to simulate accurately, since it is a near-end-product for the oxidation of many gases (before conversion to CO), and can be significantly influenced by precipitation scavenging as well as by the washout of intermediate oxidation products of gases like isoprene [77]. Several other VOCs are also generally underestimated by MATCH-MPIC during MINOS by factors of 3 or more [78, 79]. On the other hand, these modeled VOCs were generally well correlated with the observations, which implies that the discrepancies are likely due to errors in the emissions factors (the amount of each gas produced per amount of fuel burned), which largely control the absolute levels of the gases, while the geographical source distributions and meteorology (which mainly determine the correlations) are relatively well simulated.

There are several important potential effects of the Asian summer monsoon outflow. The most discussed thus far is the transport of the pollutants into the stratosphere. Since the tropopause slopes from about 15 km altitude over northern Africa down to about 11 km over Europe, isentropic transport across the tropopause may be possible; the likelihood of this is enhanced by the weak nature of the subtropical jet stream in this region [67]. During MINOS, a few measurements could be made in the lower stratosphere. During one of these flights, enhanced levels of CO, VOCs, and CH_3Cl were encountered, similar to the characteristics of the air masses measured directly within the plume in the UT, which is suggestive of troposphere to stratosphere transport of the polluted air masses [68]. There may be various chemical consequences of this, including increased O_3 loss due to the release of halogens from CH_3Cl and other short-lived halocarbons [67].

Another possible consequence of the Asian summer monsoon outflow is an enhancement of the tropospheric O_3 column (and thus the greenhouse warming) over the Mediterranean in the future, when NO_x emissions from Asia are likely to increase [68]. The ECHAM model was used to estimate that the southern Asian emissions contributed about 20–25% of the O_3 in the UT during the first week of MINOS when the Asian plume was present over the western Mediterranean [74]. An ECHAM simulation for the year 2025 emissions shows an increase in O_3 in the plume of 14%, and also indicates that the O_3 formation in the plume is NO_x-limited. Depending on the future developments of fossil fuel burning in southern Asia, particularly from the transport sector, and the emissions controls which are implemented, a substantial increase in NO_x emissions and consequently O_3 production in the outflow plume could result over the next several decades.

Finally, as discussed above, INDOEX and other studies of the wintertime Indo-Asian haze have pointed out the remarkable role of BC aerosols in the LT. If a sufficient fraction of the BC is transported through the convective clouds in the Asian summer monsoon plume, it may have radiative consequences in the UT and perhaps the lower stratosphere as well. Assessments of this will depend on measurements of BC in the UT, as well as knowledge of the scavenging efficiencies of deep convective clouds for these aerosols, which are presently poorly understood.

6
Conclusions and Outlook

This chapter has given an overview of the key aspects of the export of pollutants from southern Asia during the summer and winter monsoons. While an immense amount about these issues has been learned over the last decade, there are still many unanswered questions. Foremost in many readers' minds will not be the impact of southern Asian emissions on the rest of the world, which has been the focus of this chapter, but instead the impact of southern Asian emissions plus the emissions from the rest of the world on southern Asia itself. It is clear that long-lived greenhouse gases from Europe and North America have global effects, and certainly other gases with intermediate lifetimes such as CO and O_3 may also be strongly affected by trans-boundary transport, particularly between various parts of Asia. Considerable work is needed to deepen our scientific understanding of atmospheric interconnections and to develop effective strategies for dealing with the air pollution problems facing the local populations, which will almost inevitably help alleviate many of the superregional pollution problems discussed here and in other chapters of this book.

In addition to the emissions from biomass, biofuel and fossil fuel burning in southern Asia, which have been focused on in this chapter, another major problem facing the region is the intermittent severe biomass burning (forest fires) on the southeast Asian islands (especially Indonesia), particularly in connection with El Niño. These fires can lead to very large amounts of trace gas and aerosol emissions on short time scales in a region which is one of the most convectively active worldwide. For further discussions of biomass burning in this region, the reader is referred to Chap. 9.

The next decade of research on this subject will be driven by a number of key questions, many of which have been brought up already in the discussions in this chapter:

Open Issues – wintertime Indo-Asian haze:

- How much of the pollution over various parts of the Indian Ocean is from India vs other regions (especially Africa, Indonesia, and Arabia/Europe)?
- What are the effects of the haze on deep convection in the ITCZ, on the transition period, and on the summer monsoon onset and intensity? What are the key factors that need to be focused on to reduce the uncertainty in the initial GCM studies of this done so far?
- How much pollution in general (e.g., as a function of lifetime) makes it into the ITCZ? What are the details of the convective and large-scale transport in the ITCZ, such as the amount of interhemispheric exchange which occurs, and the fraction of the soluble gases and aerosols that are washed out during the transport?

Open Issues – Asian summer monsoon outflow:

- What quantities of various trace gases in the Asian monsoon plume are actually transported into the lower stratosphere?
- How much of the immense amount of carbonaceous aerosols which are produced in southeast Asia (evidenced by the wintertime haze) is able to survive transport to the UT through deep convective clouds? What happens to the aerosols once they are in the UT in terms of aging, coating, and primary loss mechanisms, and how do they affect solar radiation and possibly cirrus cloud properties?
- Is O_3 enhanced or depleted in the Asian summer monsoon plume, and how does this vary spatially within the outflow?
- What is the role of breaks in the monsoon in allowing pollution buildup and transformations (especially photochemical O_3 formation and aerosol aging) prior to the convection (i.e., how different do things look in years when the convection is more even vs years when there are lots of breaks between intense rainy spells)?
- How significant is the eastward component of the monsoon outflow? Can it also be observed (in situ or by satellites)? Does it affect clouds in the western Pacific? Does it join up with the outflow from northern China and head across the Pacific towards North America, and if so, does it represent a non-negligible additional source for any particular gases or aerosols?

Open Issues – general:

- How do the emissions from southern Asia influence global mean OH levels and the tropospheric oxidizing efficiency [40]? What are the effects on global tropospheric O_3, particularly on surface O_3 over the US and Europe, which is predicted to increase notably in the future due to the impact of Asian emissions on the hemispheric background O_3 levels [80]?
- What does the region look like during the transition periods? Where does the outflow go then? Is this of global significance compared to the longer summer and winter monsoon periods?
- How do we go about improving emissions scenarios? Can combinations of bottom-up (e.g., [6, 8–10, 81] and top-down (e.g., [1, 11, 6, 82]) approaches lead to major advances in the near future?

Finally, it is important to consider how future research will be organized, and who will lead it, so that it will most effectively address the issues that are most important for the populations of southern Asia, as well as those that are critical to the rest of the world. A major step towards developing such a focused international and interdisciplinary research initiative is the UNEP-sponsored ABC (Atmospheric Brown Clouds) project, in which it is currently planned to investigate the wintertime Indo-Asian haze as the first case study [16, 53].

Acknowledgments My appreciation goes to the many colleagues, especially Paul Crutzen, V. Ramanathan, Jos Lelieveld, Phil Rasch, Rolf von Kuhlmann, Horst Fischer, Jonathan Williams, and Russ Dickerson, who have shared their curiosity and wisdom with me during the course of INDOEX and MINOS.

7 References

1. Kasibhatla P, Arellano A, Logan JA, Palmer PI, Novelli P (2002) Geophys Res Lett 29: 10.1029/2002GL015561
2. Novakov T, Andreae MO, Gabriel R, Kirchstetter TW, Mayol-Bracero OL, Ramanathan V (2000) Geophys Res Lett 27:4061
3. Mayol-Bracero OL, Gabriel R, Andreae MO, Kirchstetter TW, Novakov T, Ogren J, Sheridan P, Streets DG (2002) J Geophys Res 107:10.1029/2001JD000039
4. Salam A, Bauer H, Kassin K, Ullah SM, Puxbaum H (2003) J Environ Monit 5:483
5. Guazzotti SA et al. (2003) J Geophys Res 108:10.1029/2002JD003277
6. Dickerson RR, Andreae MO, Campos T, Mayol-Bracero OL, Neusuess C, Streets DG (2002) J Geophys Res 107:10.1029/2001JD000446
7. Leon J-F et al. (2001) J Geophys Res 106:28427
8. Reddy MS, Venkataraman C (2001) Atmos Env 36:677
9. Reddy MS, Venkataraman C (2001) Atmos Env 36:699
10. van Aardenne JA, Carmichael GR, Levy HL II, Streets D, Hordijk L (1999) Atmos Environ 33:633
11. Kunhikrishnan T, Lawrence MG, von Kuhlmann R, Richter A, Ladstätter A, Burrows JP (2003) Atmos Environ (in press)
12. Ethé C, Basdevant C, Sadourny R, Appu KS, Harenduprakash L, Sarode PR, Viswanathan G (2002) J Geophys Res 107:10.1029/2001JD001120
13. de Gouw JA et al. (2001) J Geophys Res 106:28453
14. Gadgil S (2003) Ann Rev Earth Planet Sci 31:429
15. Webster PJ, Magana VO, Palmer TN, Shukla J, Tomas RA (1998) J Geophys Res 103:14451
16. Ramanathan V, Crutzen PJ (2003) Atmos Environ 37:4033
17. Ramanathan V et al. (2001) J Geophys Res 106:28371
18. Krishnamurti T, Jha B, Rasch P, Ramanathan V (1997) Meteorol Atmos Phys 64:123
19. Krishnamurti T, Jha B, Rasch P, Ramanathan V (1997) Meteorol Atmos Phys 64:151
20. Verver GHL, Sikka DR, Lobert JM, Stossmeister G, Zachariasse M (2001) J Geophys Res 106:28399
21. Lelieveld J et al. (2001) Science 291:1031
22. Lobert JM, Harris JM (2002) J Geophys Res 107:10.1029/2001JD000731
23. Mohanty UC, Niyogi DS, Raman S, Sarkar A (2001) Curr Sci 80:60
24. Raman S, Niyogi DS, Simpson M, Pelon J (2002) Geophys Res Lett 29:10.1029/2001GL014193
25. de Laat ATJ, Lelieveld J, Roelofs GJ, Dickerson RR, Lobert JM (2001) J Geophys Res 106: 28481
26. Phadnis MJ, Levy HL II, Moxim WJ (2002) J Geophys Res 107:10.1029/2002JD002190
27. Rasch PJ, Collins WD, Eaton BE (2001) J Geophys Res 106:7337

28. Rhoads KP, Kelley P, Dickerson RR, Carsey TP, Armer M, Savoie DL, Prospero JM (1997) J Geophys Res 102:18981
29. Mühle J, Zahn A, Brenninkmeijer CAM, Gros V, Crutzen PJ (2002) J Geophys Res 107: 10.1029/2001JD000730
30. Wisthaler A, Hansel A, Dickerson RR, Crutzen PJ (2002) J Geophys Res 107: 10.1029/ 2001JD000576
31. Lal S, Lawrence MG (2001) Geophys Res Lett 28:1487
32. Chand D (2002) PhD thesis Mohan Lal Sukhadia University, Udaipur, India
33. Fishman J, Wozniak AE, Creilson JK (2003) Atmos Chem Phys 3:893
34. Chand D, Lal S, Naja M (2003) J Geophys Res 108:10.1029/2001JD001589
35. de Laat ATJ, Lelieveld J (2000) J Geophys Res 105:11547
36. Dickerson RR, Rhoads KP, Carsey TP, Oltmans SJ, Burrows JP, Crutzen PJ (1999) J Geophys Res 104:21385
37. Burkert J et al. (2003) J Geophys Res 108:10.1029/2002JD002790
38. Wagner V, von Glasow R, Fischer H, Crutzen PJ (2002) J Geophys Res 107:10.1029/ 2001JD000722
39. Lawrence MG et al. (2003) Atmos Chem Phys 3:267
40. Lawrence MG, Jöckel PJ, von Kuhlmann R (2001) Atmos Chem Phys 1:37
41. von Glasow R, Sander R, Bott A, Crutzen PJ (2002) J Geophys Res 107:10.1029/ 2001JD000943
42. Collins WD, Rasch PJ, Eaton BE, Khattatov BV, Lamarque J, Zender CS (2001) J Geophys Res 106:7313
43. Andronache C, Donner LJ, Seman CJ, Hemler RS (2002) J Geophys Res 107:10.1029/ 2001JD900248
44. Naja M, Lal S, Venkataramani S, Modh KS, Chand D (1999) Curr Sci 76:931
45. Stehr JW, Ball WP, Dickerson RR (2002) J Geophys Res 107:10.1029/2001JD000446
46. Crutzen PJ, Lawrence MG (2000) J Atmos Chem 37:81
47. Barth MC, Stuart AL, Skamarock WC (2001) J Geophys Res 106:2381
48. Yin Y, Parker DJ, Carslaw KA (2001) Atmos Chem Phys 1:19
49. Yin Y, Carslaw KS, Parker DJ (2002) Atmos Chem Phys 2:293
50. de Reus M, Krejci R, Williams J, Fischer H, Scheele R, Ström J (2001) J Geophys Res 106: 28629
51. Williams J, Fischer H, Wong S, Crutzen PJ, Scheele MP, Lelieveld J (2002) J Geophys Res 107:10.1029/2001JD001126
52. Zachariasse M, Smit HGJ, van Velthoven PFJ, Kelder H (2001) J Geophys Res 106:28441
53. Ramanathan V, Crutzen PJ, Mitra AP, Sikka D (2002) Curr Sci 83:947
54. Li F, Ramanathan V (2002) J Geophys Res 107:10.1029/2001JD000949
55. Tahnk WR, Coakley JA Jr (2002) J Geophys Res 107:10.1029/2001JD000183
56. deLaat ATJ, Lelieveld J (2002) J Geophys Res 107:10.1029/2001JD001483
57. Krishnamurti TN, Chakraborty DR, Cubukcu N, Stefanova L, Vijaya Kumar TSV (2003) Q J R Met Soc 129:2559
58. Collins WD, Rasch PJ, Eaton BE, Fillmore DW, Kiehl JT, Beck CT, Zender CS (2002) J Geophys Res 107:10.1029/2000JD000032
59. Chowdhury Z, Hughes LS, Salmon LG, Cass GR (2001) J Geophys Res 106:28592
60. Twomey S (1974) Atmos Environ 8:1251
61. Twomey S (1977) J Atmos Sci 34:1149
62. Heymsfield AJ, McFarquhar GM (2001) J Geophys Res 106:28653
63. Liu G, Shao H, Coakley JA Jr, Curry JA, Haggerty JA, Tschudi MA (2003) J Geophys Res 108:10.1029/2001JD001395
64. Chung CE, Ramanathan V, Kiehl JT (2002) J Climate 15:2462
65. Chung CE, Ramanathan V (2003) J Climate 16:1791
66. Menon S, Hansen J, Nazarenko L, Luo Y (2002) Science 297:2250
67. Lelieveld J et al. (2002) Science 298:794
68. Scheeren HA et al. (2003) Atmos Chem Phys Discuss 3:2285
69. Brenninkmeijer CAM et al. (1999) J Atmos Oceanic Technol 16:1373

70. Li Q et al. (2001) Geophys Res Lett 28:3235
71. Marenco A et al. (1998) J Geophys Res 103:25631
72. Scheeren HA, Lelieveld J, de Gouw JA, van der Veen C, Fischer H (2002) J Geophys Res 107:10.1029/2001JD001121
73. Traub M et al. (2003) Atmos Chem Phys Discuss 3:459
74. Roelofs G-J, Scheeren B, Heland J, Ziereis H, Lelieveld J (2003) Atmos Chem Phys Discuss 3:1247
75. Kormann R et al. (2003) Atmos Chem Phys Discuss 3:1303
76. Ladstätter-Weissenmayer A et al. (2003) Atmos Chem Phys Discuss 3:2051
77. von Kuhlmann R (2001) PhD thesis, Johannes Gutenberg-Universität Mainz, Mainz, Germany
78. Salisbury G et al. (2003) Atmos Chem Phys 3:925
79. Gros V et al. (2003) Atmos Chem Phys Discuss 3:1893
80. Lelieveld J, Dentener F (2000) J Geophys Res 105:3531
81. Streets DG, Waldhoff ST (1998) Energy 23:1029
82. Leue C, Wenig M, Wagner T, Klimm O, Platt U, Jähne B (2001) J Geophys Res 106:5493

The Handbook of Environmental Chemistry Vol. 4, Part G (2004): 173–195
DOI 10.1007/978-3-540-40037-0

Characteristics of Atmospheric Transport of Air Pollution for Africa

Stuart J. Piketh · Nicola M. Walton

Climatology Research Group, University of the Witwatersrand, Gauteng South Africa,
Private Bag 3x, Wits, 2050 Johannesburg, South Africa
E-mail: stuart@crg.bpb.wits.ac.za

Abstract The African continent is an important source of atmospheric constituents including both trace gases and aerosols. Emissions from Africa impact on at least five other continents. These emissions are derived from wind blown dust, biomass burning, biogenic emissions and anthropogenic industrial activities. Atmospheric transport to and from the continent are affected by both the easterly and westerly circulation patterns of the northern and southern hemispheres. Wind blown dust is the most important emission from the Sahel region of Africa. At different times of the year, dust from the Sahara may impact on South America, the Caribbean, the Mediterranean as well as other parts of Europe. Transport over southern Africa is complex and includes recirculation of air over the subcontinent. Transport off the subcontinent may be to the west to the Atlantic Ocean or to the east. Eastward movement of air has been shown to either impact directly on the central Indian Ocean through deposition or reach as far east as Tasmania. Atmospheric transport towards the African continent is extensive and includes emissions that have originated in Asia, North America, Europe and South America.

Keywords Africa · Air pollution · Long-range transport · Air quality · Ocean fertilization · Dust transport

1 Introduction

Africa is the only continent that has a significant amount of land mass in both the southern and northern hemispheres. Trace gas and aerosol emissions from this vast continent have important climatic implications. Significant advances have been made over the past two decades in understanding the sources of emissions as well as the nature of the atmosphere and its constituents. In a simplistic model it is possible to divide Africa into emissions regions in which one particular source dominates the emissions to the atmosphere. This is true of both trace gas and aerosol emissions. Three emission region types can be clearly defined in Africa. The first region is associated with wind blown dust. Due to the large arid regions on the continent wind blown dust is suspended from the surface into the atmosphere. The Sahara desert and to a lesser extent the Namib desert are the

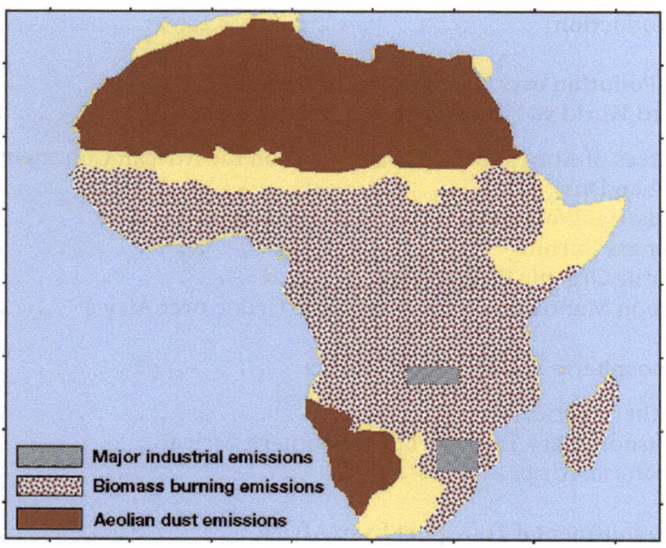

Fig. 1 Emissions from the African continent defined in regions by major source type. Note that many areas do have multiple emission sources

most important sources of wind-blown dust. The second region is the densely vegetated forests of central Africa as well as the Savanna regions in the southern parts of Africa. The dominant source of atmospheric constituents in these areas is biomass burning, which makes up for approximately 30% of global biomass burning. The regions extend from approximately 10°N to about 25°S. The third emission type is anthropogenic sources which are clustered on the continent. The largest of these are found in South Africa, which is the most industrialised region of the continent. Figure 1 summarises the source regions for the three aforementioned source categories. Aeolian dust and biomass burning emissions change strength and shift geographically with the season. This will be discussed in more detail below.

2 Air Pollution over the African Continent: Third World vs First World

2.1 Sources of Atmospheric Constituents on the African Continent

Africa is largely an underdeveloped continent with concentrations of intensive industrial activity occurring in isolated pockets. In general the main sources emitting gases and aerosols to the atmosphere include aeolian crustal material consisting of mineral soil dust, marine aerosols from the two adjacent oceans, biomass burning particles occurring mainly north of 20°S, aerosols from industrial emissions and finally emissions from the biosphere.

2.1.1 *Aeolian Dust*

Emissions of crustal aeolian dust material, often termed aeolian or mineral dust, is caused primarily by the surface winds acting on dry soils where vegetation cover is or has become sparse [1]. The term mineral dust refers to a large range of species that are highly variable in their chemical composition and include such diverse compounds as quartz, clay, calcite, gypsum, haematite and others. The African continent has vast tracks of desert that are significant sources of this type of atmospheric aerosol. Emissions from the Sahara desert as well as the south Western Coast of Namibia are some of the few sources that are easily detectable using satellite remote sensing data in the visible range. Dust is a regional scale climatic forcing agent. This is particularly true of regions of the globe from which vast plumes of dust are suspended to the atmosphere and transported over hundreds to thousands of kilometres (Fig. 2), as discussed in Chap. 11 of this book. The presence of dust in the atmosphere alters surface radiation budgets which, in turn, affects surface temperatures, surface-air exchange processes, thermodynamic structure and hence atmospheric dynamics [2, 3]. The complexity of the radiative effects of mineral dust on climate is determined by the uneven distribution of sources and sinks and the relatively short atmospheric life time of mineral dust which lead to a complex spatial and temporal pattern [4, 5]. Radiative properties of dust need to be investigated in detail in the future at many sites

Fig. 2 The African continent is a significant contributor to global atmospheric Aeolian crustal dust. The dust is typically blown off the west coast of Africa from the Sahara (*top left*) or the deserts of Namibia (*bottom right*) as well as of the Mediterranean (*bottom left*) and Red Sea (*top right*). The images are SEAWIFS images (http://visibleearth.nasa.gov/)

around the world to reduce uncertainties of the climatic effects of mineral dust aerosols in the atmosphere [2, 5].

2.1.2
Industrial Emissions

The most significant industrial trace gas emitted into the atmosphere over southern Africa is SO_2. Particulate sulfate (SO_4^{2-}) is also emitted. Sulfur emissions are concentrated in two major source regions, the Mpumalanga Highveld of South Africa and the Zambian copperbelt. In addition, there are several large, isolated point sources.

Siversten et al. [6] have compiled the most comprehensive inventory of sulfur emissions to date, covering southern Africa south of 17°S. The spatial distribution of emissions over the subcontinent is shown in Fig. 3. Of the total of 1.1 million tons of sulfur emitted into the regional atmosphere annually (excluding Zambia), 66% originates in South Africa and of this about 90% is from the

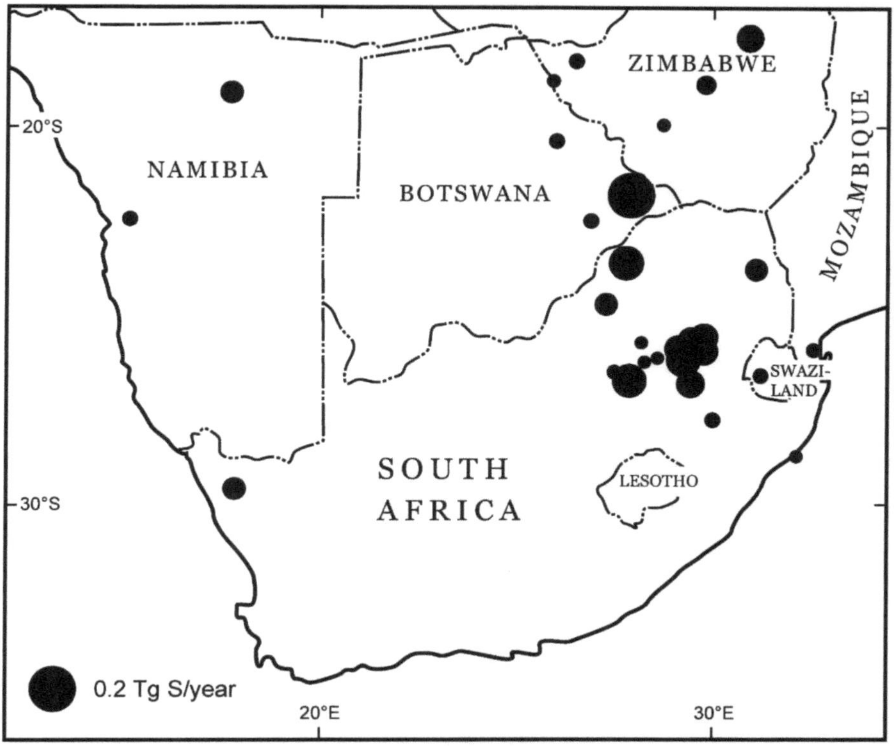

Fig. 3 Distribution of major sulphur emissions in southern Africa south of 17°S (after Sivertsen et al. 1995)

Mpumalanga Highveld [6, 7] (globally approximately 74 million tons are emitted into the atmosphere per annum). This is the region of the country where most of the major coal-fired power plants are situated. The limited distribution of other sulfur sources over the subcontinent makes emissions from the Mpumalanga Highveld an ideal tracer for transport studies. This is also true of other important pollutants such as the oxides of nitrogen. Power production in South Africa accounts for 55% of the total emissions of sulfur in southern Africa. In Botswana only one significant source of sulfur emissions exists and it accounts for 0.19 million tons per annum.

The Zambian copperbelt is also an important emitter of SO_2. Currently it is thought that the emissions from these industrial emissions are up to 2,24 million tons per annum [8–11].

2.1.3
Biomass Burning

Savanna fires are the single largest source of biomass burning emissions worldwide [12] and account for about 75% of all fire related emissions [13]. Savanna

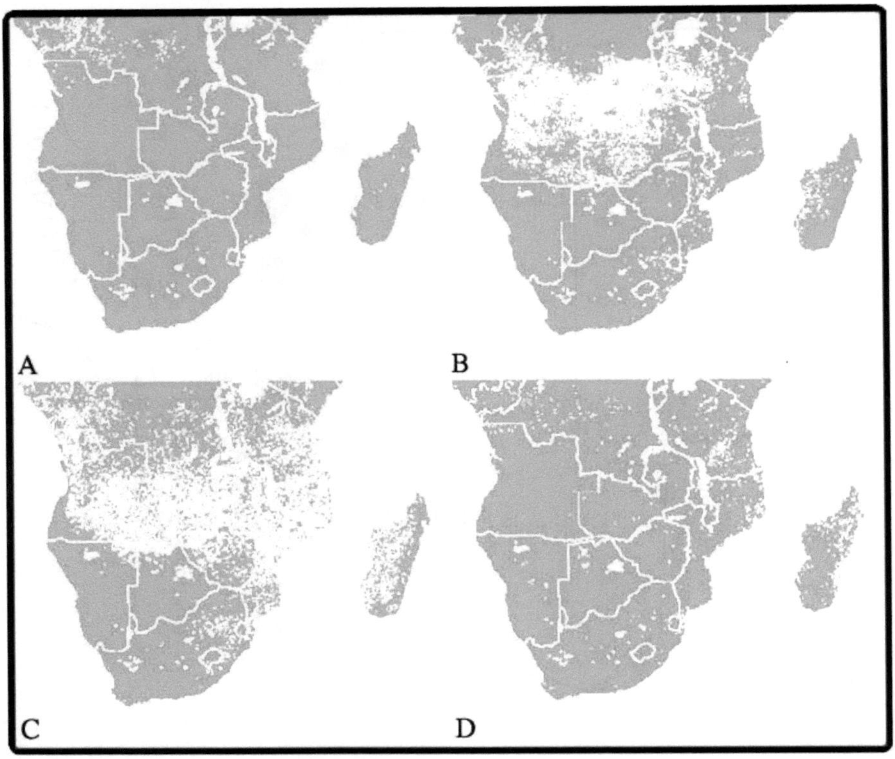

Fig. 4A–D Seasonal variation of fire activity in southern Africa during 1993: **A** March; **B** June; **C** September; **D** December 1993. The *white spots* on the continent represent pixels in which fire was detected during the given month (European Space Agency, 1999)

fires account for about 57% of the total biomass burnt on the African continent [14]. Biomass burning has received much attention from the international scientific community over the past decade. In particular, SAFARI-92 was designed to study the emissions and transport of trace species from fires. The distribution of fires in southern Africa has been well documented [15] (Fig. 4).

Biomass burning is a seasonal source with its highest intensity between June and October. In the non-burning season (November to May) few fires are detected south of 20°S (central Zimbabwe). The fire intensity increases from May onwards in the West Coast region south of the equator. A high frequency of burning events is shown in northern and central Angola, the Democratic Republic of the Congo and north-western Zambia. During the southern African burning months (June to October), large numbers of fires are detected north of 20°S with the highest intensity being in September [16, 17]. For latitudes south of 20°S pyrogenic emissions increase slightly from May onwards and then more rapidly during July and August, and then slowly decrease to the end of October.

During the burning season it has been shown that the industrial and biomass burning emissions become well mixed over the subcontinent. They are transported over the entire region with a clear regional source signature [14]. The burning of wood, coal, dung and residual crops for domestic cooking and space heating is an important source of emissions to the atmosphere over Africa. These emissions are difficult to distinguish from biomass burning and are less seasonal. Marufu et al. (2000) [26] have estimated that these emissions account for at least as much as biomass burning and industrial emissions.

2.1.4 *Volatile Organic Compounds*

Volatile Organic Compounds (VOCs) are emitted into the atmosphere from both anthropogenic sources (fossil fuel burning, industrial processes, waste treatment and agricultural activities) as well as biogenic sources [18, 19]. Biogenic volatile organic compound (BVOC) emissions are very important, with vegetation producing over 90% of the total global VOC budget [20]. Tropical regions with their high temperatures and strong solar radiation are the largest contributors to the global VOC budget. The biogenic VOC emissions for the area south of the equator were estimated to total 80 Tg C per annum with isoprene and monoterpenes contributing 56 and 7 Tg C per annum, respectively. In the tropical regions, the seasonal variations in VOC emissions are small while in subtropical regions, variations are large due to the highly seasonal savannas and grasslands.

2.2 Carbon Monoxide and Tropospheric Ozone over Africa

Regional CO mixing ratios over southern Africa are not well known. High background CO values (>135 ppbv) are known to exist over sub-equatorial Africa [21], but Novelli et al. [22] found a global decline over time in CO. Values range from 45 ppbv in the austral summer to 230 ppbv in the boreal winter [22]. CO concentrations in the lower troposphere are mostly below 70 ppbv, while moderate values are considered to lie between 70 and 135 ppbv [21].

Hardly any long term measurements of ozone are available on the African continent. Airborne ozone measurements have shown that vertical concentrations over southern Africa exhibit strong layering with height [23, 24]. This has been confirmed by ozone sondes launched from various sites in southern Africa during SAFARI-92 [25, 26]. Swap and Tyson [21] have shown that the development of elevated concentrations of trace gases, including ozone, are facilitated by the presence of inversion layers in the atmosphere, some of which remain intact for weeks, preventing vertical mixing [27]. Ozone concentrations in these layers range from 55 ppbv between 700 and 500 hPa, to 80 ppbv between 300 and 200 hPa [28]. Surface ozone concentrations over southern Africa are typically between 20 and 40 ppbv [29]. More details on ozone and its relation to biomass burning in the tropics are given in Chap. 9 of this book.

3 Atmospheric Transport over Africa

3.1 Southern Africa

Southern Africa is situated geographically beneath the planetary-scale, southern hemisphere descending limb of the Hadley circulation throughout much of the year. The general circulation of the atmosphere over southern Africa determines the transport of aerosols and trace gases. Four major synoptic circulation types are responsible for most of the atmospheric transport of aerosols and gaseous pollutants over southern Africa: the semi-permanent subtropical continental anticyclones, transient mid-latitude ridging anticyclones, westerly baroclinic disturbances and barotropic quasi-stationary tropical easterly disturbances [30–34]. The frequency of occurrence of each of these circulation patterns varies seasonally. The most frequent synoptic-scale circulation type over southern Africa is the continental anticyclonic circulation which occurs up to 65% of the time in winter. Anticyclonic circulation and westerly disturbances result in air transport from Africa to the southeast over the Indian Ocean in the Natal plume. This anticyclonic flow results in considerable recirculation, which may either be confined to the continent or extend over and return from the Indian Ocean. Easterly wave conditions favour transport to the west over the tropical Atlantic Ocean in the Angolan plume. Transient mid-latitude ridging anticyclones result in maximum transport originating in the mid-latitudes and flowing towards the southern Atlantic Ocean [31, 32]. The general flow over southern Africa is summarised in Fig. 5.

At the 800 hPa geopotential height level, the mean atmospheric circulation over southern Africa is anticyclonic throughout the year [33], resulting in adiabatic warming and drying of the air as it descends. Extensive and persistent inversions and stable layers can therefore form over southern Africa. Stable conditions associated with anticyclones are responsible for the development of the haze layer over southern Africa for most of the year on non-rainy days [27, 29, 34]. Within the haze layer are frequently-occurring and spatially ubiquitous stable layers observed most frequently at the 300 hPa, 500 hPa, 700 hPa and 850 hPa level. These stable layers significantly inhibit vertical mixing and dispersion, leading to potentially severe pollution problems [35].

3.2 Transboundary Transport over Southern Africa

A large percentage of transport from the Highveld affects remote parts of South and southern Africa. This has been observed in chemical source apportionment analyses [33, 36] and as far north as Kenya [37]. Air masses from the interior not exiting the subcontinent into the Indian and Atlantic Oceans are advected over South Africa's neighbouring countries. For political reasons and to assure the integrity of regional ecosystems, knowledge of the transboundary transport of pollutants between South Africa and its neighbours is required. The implications for

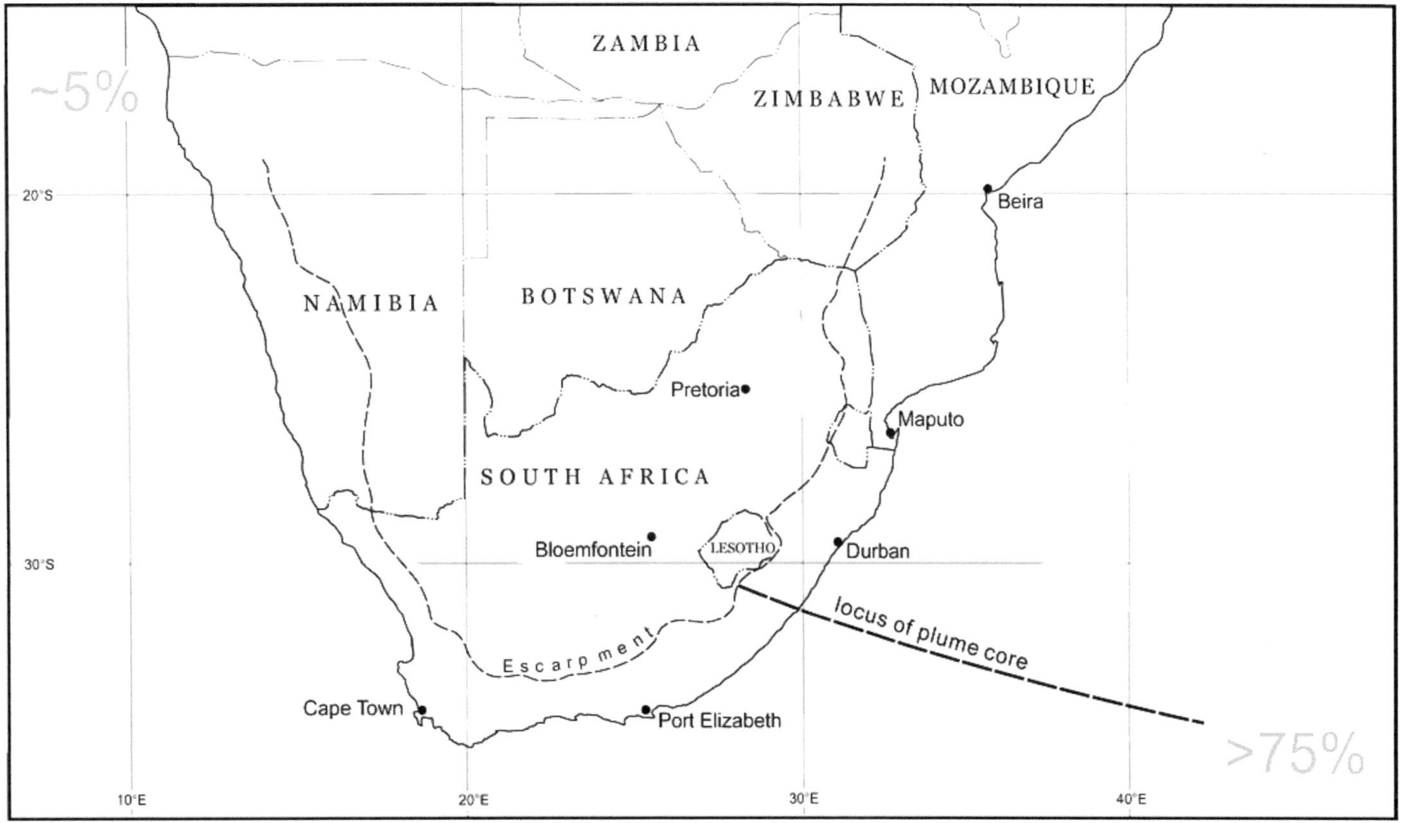

Fig. 5 Average transport of air over southern Africa. The flow patterns have been summarised from all transport trajectory climatologies undertaken for this part of the continent

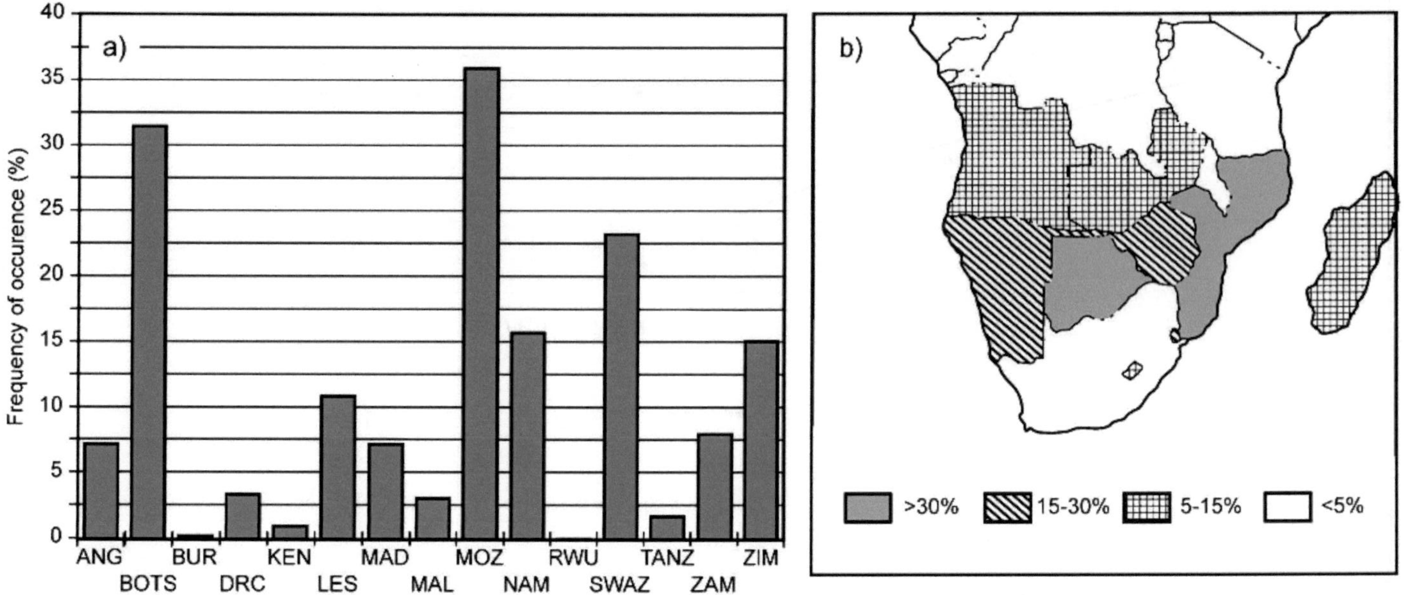

Fig. 6a, b Percentage frequency with which the borders of each country in southern Africa are crossed by trajectories originating over the Highveld for 1990–1994: **a** countries: Angola (ANG); Botswana (BOTS); Burundi (BUR); Democratic Republic of the Congo (DRC); Kenya (KEN); Lesotho (LES); Madagascar (MAD); Malawi (MAL); Mozambique (MOZ); Namibia (NAM); Rwanda (RWU); Swaziland (SWAZ); Tanzania (TANZ); Zambia (ZAM) and Zimbabwe (ZIM); **b** a graphic representation thereof

anthropogenic emission transport from the Highveld are very important and have previously not been quantified but rather dealt with in an integrated fashion. Remote sites of southern Africa are affected by industrial aerosols and trace gases from the Highveld to a large degree [38, 39]. Countries closest to South Africa are most frequently affected by the transport, direct and especially recirculated, of air having acquired aerosols and trace gases over the Highveld (Fig. 6). Transport to Mozambique occurs more than one third of the time. Roughly one out of three trajectories exiting the Highveld impacts Mozambique or Botswana, whilst Swaziland is affected, on average, by one out of four trajectories. Direct northward transport from the Highveld to Zimbabwe occurs during about 15% of the time. The fact that so little air from the Highveld impacts Lesotho (11%) is indicative of the large anticyclonic transport vortex over the region, where Lesotho is located near the centre of the vortex. Further north, the effect of transport from the Highveld diminishes. The spatial pattern of transboundary transport from the South African Highveld is given in Fig. 6b.

3.3 Equatorial Africa

An air transport climatology for Kenya was developed by Gatebe [37] at the synoptic scale. Air reaching Kenya originates mainly from the Saharan region and north-western Indian Ocean during January and March while transport is mainly from the southwest Indian Ocean during May and November. Atlantic westerly transport occurs in the upper troposphere during January, March and November. Airflow away from Kenya is dominated by the equatorial easterly to the Atlantic Ocean and the recurved easterly to the Arabian Peninsula during January, March and November. As a result, airflow off the continent is via the Atlantic Ocean and Saudi Arabia. During May, air transport is mainly from the south-western Indian Ocean, and air exits to the Atlantic Ocean. However, the transport of air from Kenya is both spatially and temporally invariant in the tropical easterlies to the Congo Basin and Atlantic Ocean when compared to transport to the country [37].

4 Intercontinental Transport from Africa

4.1 Transport from South Africa to Australia

Prior to SAFARI-92, the transport of aerosols and trace gases out of southern Africa was not fully understood. Previous studies recognized the fact that aerosols and trace gases may be transported from southern Africa into the tropical South Atlantic Ocean as well as into the Indian Ocean [14, 40].

The plume to the Indian Ocean is well defined and has been observed to pass south of Australia in both the summer (January) and winter (July) seasons. In winter, about 30% of the air from southern Africa passes south of Australia at 140°E while in summer less transport from southern Africa to Australia occurs

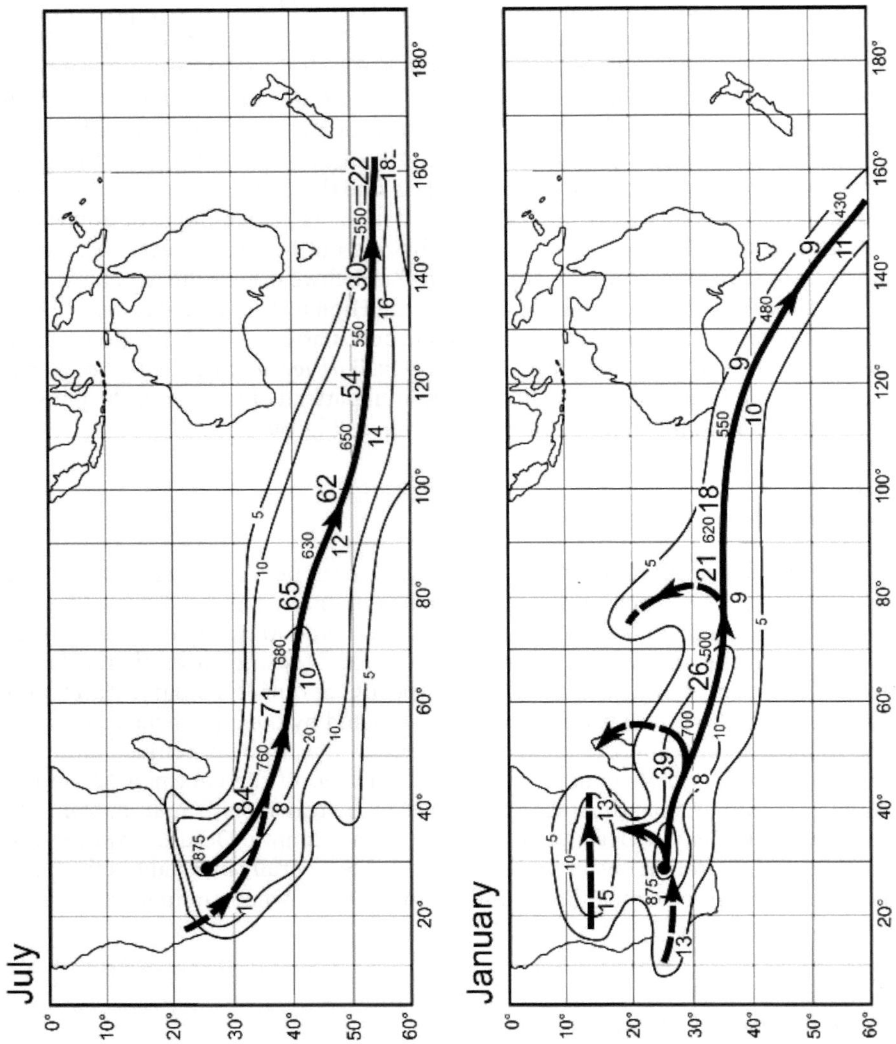

Fig. 7 Mean 850–800 hPa trajectory transport fields for winter and summer from the South African Highveld. The *dark line* indicates the centre of most frequent transport. Mean percentage transports within the plume to given longitudes are indicated in heavy numbers. *Light numbers* along the most frequent pathway indicate days of travel to the point, geopotential height of which is given in italic (after [35])

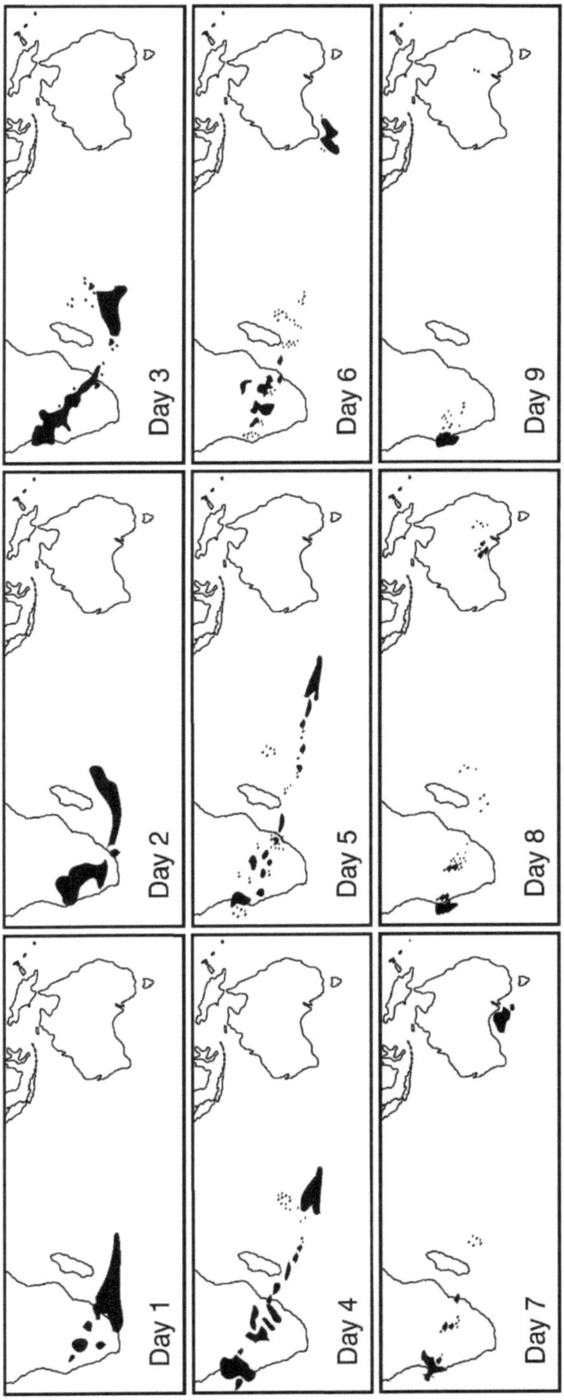

Fig. 8 Dust transport from southern Africa to Australia as observed over a period of nine days in September using TOMS data (after [35])

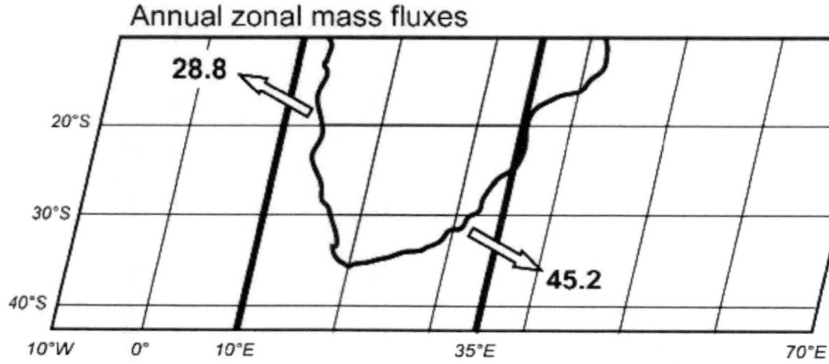

Fig. 9 Annual estimated flux of aerosol over the Atlantic Ocean (10°E) and Indian Ocean (35°E) from southern Africa (after [34])

as more air recirculates out of the plume prior to reaching Australia [7] (Fig. 7). Herman et al. [41] noted from TOMS satellite data that smoke plumes from southern Africa may be transported across the Indian Ocean to Australia in eight to nine days (Fig. 8). Zencker et al. [23] provided a first estimate of mean annual aerosol mass fluxes over and off the southern Africa subcontinent. A total mass flux of aerosols of 134 million tons is generated annually over the subcontinent, of which 60 million tons per annum is deposited locally, 29 million tons per annum is transported westward over the Atlantic Ocean and 45 million tons per annum is transported eastward over the Indian Ocean (Fig. 9) [32].

A more recent study by Wenig et al. [55], using satellite data from the Global Ozone Monitoring Experiment (GOME), showed that nitrogen oxides originating from the South African Highveld can reach Australia. During May 1998 such an intercontinental transport event was observed when a low pressure system arrived from the west. Pollution accumulated in the subtropical high over southern Africa, before it was lifted to the upper troposphere and transported by strong winds towards Australia in about six to ten days (Fig. 10) [55].

4.2 Transport from North Africa

The large scale movement of dust aerosol plumes originating over northern Africa has been extensively studied by in situ and remote sensing observations [42–44] as well as model calculations [2]. Prospero [45] indicates that North Africa is the largest source of aeolian dust in the world. Deposition of dust from the Sahara desert is estimated to be 170 Tg per annum to the Atlantic Ocean, 25 Tg per annum to the Mediterranean region and 5 Tg per annum to the Caribbean. A major route of transport of aeolian dust from the Sahara is across the North Atlantic Ocean westward towards the American continent [42, 44]. The dust transport from the Sahara has an important seasonal signal. During the winter months of the northern hemisphere the dust plume travels south of 20°N towards the Amazon basin [46]. During the summer transport from Africa is distinctly toward central and North America (north of 20°N) or alternatively the

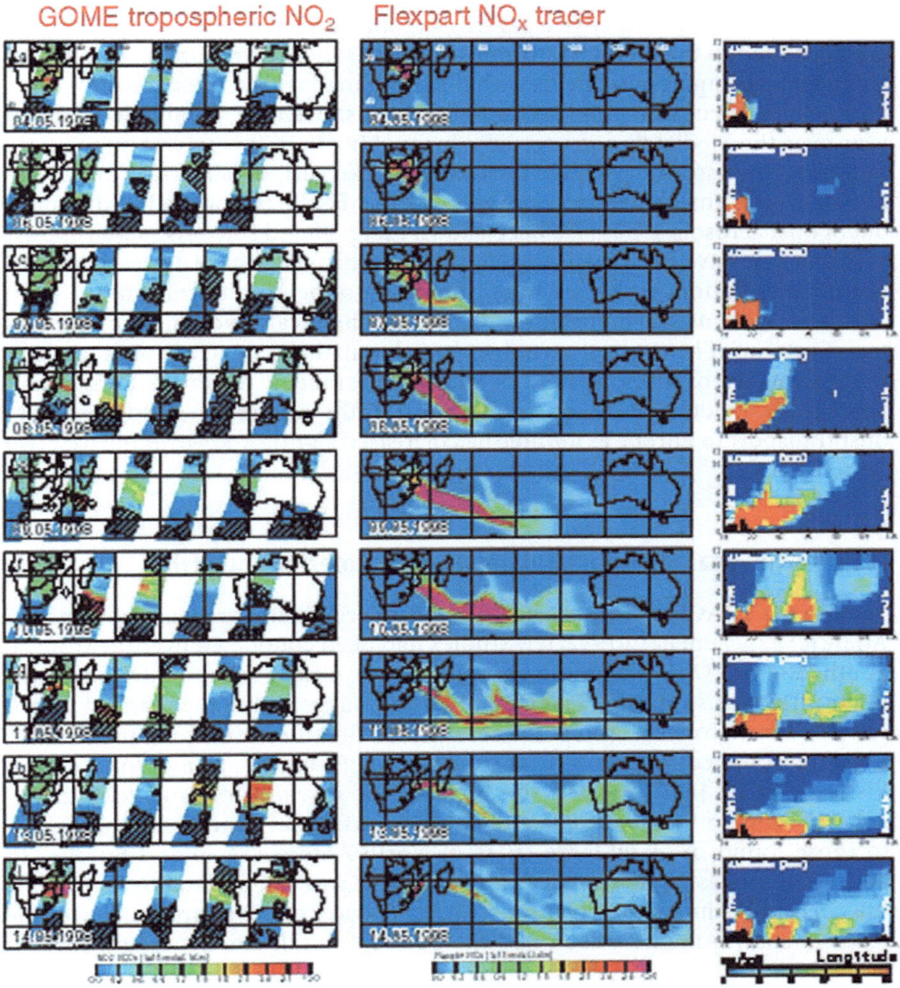

Fig. 10 Transport of tropospheric NO_2 from southern Africa towards Australia as shown from GOME on May 1998 (*left*), evolution of the NOx plume using FLEXPART (*centre*) and vertical profiles of the FLEXPART NOx tracer concentrations for emissions from the Highveld (*right*) (Wenig et al. 2003)

Caribbean [44]. It seems from the long-term records of atmospheric dust loadings measured at Barbados from 1965, that the amount of dust transported across the North Atlantic Ocean has steadily increased. A good relationship exists between the total amount of dust detected at Barbados and the previous season's rainfall amounts received as well as the sustained long-term drought in the Sahel region [45].

Another significant transport route from northern Africa is northwards towards Europe whereby large quantities of African dust are carried into the

Mediterranean basin every year [47]. An estimated 9 to 25 tons km^{-2} per annum are deposited to the Mediterranean surface waters. The African dust transport is driven by complex wind fields with the resultant dust layers exhibiting a complex vertical structure [48]. Strong Saharan dust events generally occur every year over the Mediterranean regions during May [47]. It is interesting to note that direct transport of Aeolian dust to the ocean surface is not necessarily associated with enrichment of elements such as Fe. Fe is not readily dissolved out of dust aerosols. It has also been shown that polluted aerosols as well as dust aerosols that have been transported over long distances have a greater potential for dissolution in the sea water. Inputs into the Mediterranean Sea from Africa are potentially important to the biogeochemical functioning of the system. Certainly it has been shown that the Saharan dust impacts in an important way to the atmospheric system of the Mediterranean region of Europe and beyond [4]. Evidence from dust deposition in the Alpine snows suggests that dust transport to Europe as a whole has increased significantly over the past two decades [45].

5 Intercontinental Transport Between South America and Africa

Prior to the TRACE A (Transport and Atmospheric Chemistry near the Equator-Atlantic) experiment in 1992, few studies had been conducted over the tropical South Atlantic Ocean and the adjacent South American and African continents. Previous studies by Fishman et al. [40] suggested that enhanced tropospheric ozone concentrations found at Natal, Brazil originated from biomass burning in the cerrado as well as from the long-range transport from Africa. Fishman et al. [40] reported the presence of a region of enhanced tropospheric ozone off the west coast of southern Africa at concentrations similar to those found in many large cities in North America, Asia and Europe. To address these issues, the TRACE A experiment was undertaken to investigate the chemical composition, transport and chemistry of the atmosphere over the tropical south Atlantic Ocean and the adjacent South American and African continents.

A major observation from this experiment was that biomass burning in South America and southern Africa is the principal source of the precursor gases (NOx, CO and hydrocarbons) that lead to the formation of large amounts of ozone over the South Atlantic Ocean. These high ozone levels coincide with the southern hemisphere's spring, a time when biomass burning is intense. Biomass burning products are transported from Brazil at high altitudes (8–10 km) via the westerlies near 25°S and around transient anticyclones to the tropical south Atlantic Ocean (Fig. 11). Deep convective activity from near the surface transports pollutants to the upper troposphere where ozone is photochemically produced [49]. Over Africa, in the absence of deep convection, biomass burning products are transported at low altitudes (below ~6 km) off the west coast via the easterlies towards the south Atlantic Ocean. However, South America's contribution of ozone and ozone precursor gases is reduced as most of the enhanced ozone equatorward of 10°S was from Africa [49].

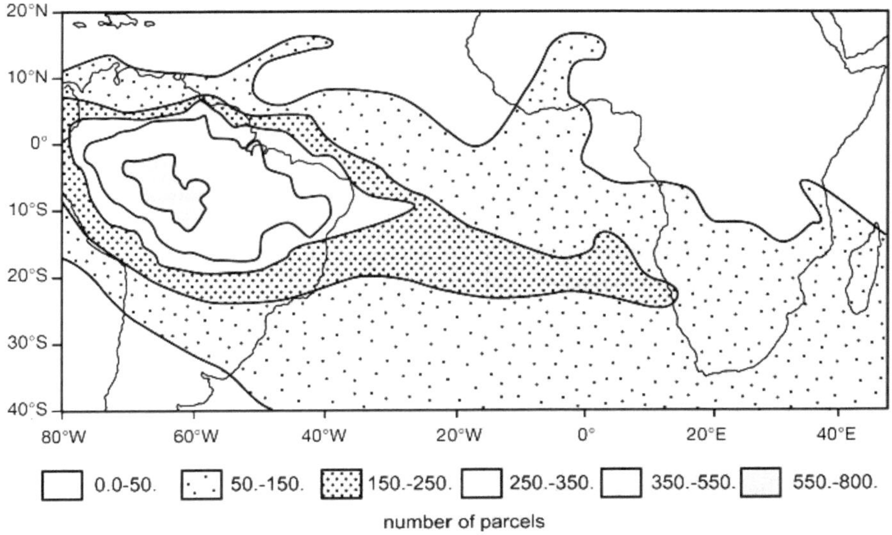

Fig. 11 Distribution of air parcels released over a region in Brazil where biomass fires and deep convection occurred, as derived from isentropic ECMWF trajectory analysis (after [50])

However, biomass burning alone cannot account for the high ozone concentrations but rather, a combination of meteorological processes and photochemistry that interact to produce a region of enhanced ozone. Intense periods of sunlight favour the production of ozone while large scale subsidence over the South Atlantic Ocean promotes the accumulation of ozone over the region. It has been [40, 50] similarly concluded that the ozone maximum over the Atlantic Ocean was caused by the buildup of ozone precursors from biomass burning over the Atlantic Ocean, stable layering of air over the eastern Atlantic and southern Africa together with deep convection of ozone precursor gases to the upper troposphere. See Chap. 9 for more details on the transport of tropical biomass burning emissions and their impact on ozone.

6 Impacts of Export of Atmospheric Constituents from Africa

6.1 Fertilisation of the Amazon Basin

In the north-eastern and central parts of Amazonia, where nutrient poor oxisols and utisols are present and nutrient supply by rivers is limited, deposition of airborne dust particles is needed to achieve a nutrient balance [51, 52]. The amounts of aeolian influx have been estimated to be in the order of millions to tens of millions of tons per year. Swap et al. [46] demonstrated that a considerable amount of dust found in the Central Amazon Basin originates from the Sahara desert. Synoptic-scale weather systems, with associated low-level horizontal convergence, gather the dust into the basin in intermittent pulses. Individual storm events deposit about 480,000 tons of dust onto the north-eastern Amazon Basin. Storm and dust climatologies suggest that the annual import of dust into the basin is on the order of 13 million tons. In the north-eastern basin, this may amount to as much as 190 kg ha^{-1} per annum. Many of the nutrient species associated with Saharan dust are believed critical to the physiology of tropical rain forests [52]. Deposition of trace elements, such as phosphate, associated with this dust ranges from 1 to 4 kg ha^{-1} per annum. Part of the productivity of the Amazon rain forest is dependant upon the critical trace elements contained in the soil dust originating in the Sahara/Sahel region.

6.2 Fertilisation of the Southern Indian Ocean

Sub-continental-scale recirculation is a characteristic of the transport climatology of the southern Africa region. Recirculation occurs on average 44% of the time [32] before air, and whatever is carried in it, exits in a minor plume to the Atlantic Ocean at about 20°S and in a major plume to the Indian Ocean centred at about 30°S [28]. It crosses, and completely covers, the area covered by the hitherto unexplained carbon sink (Fig. 12). As the plume leaves South Africa it tends

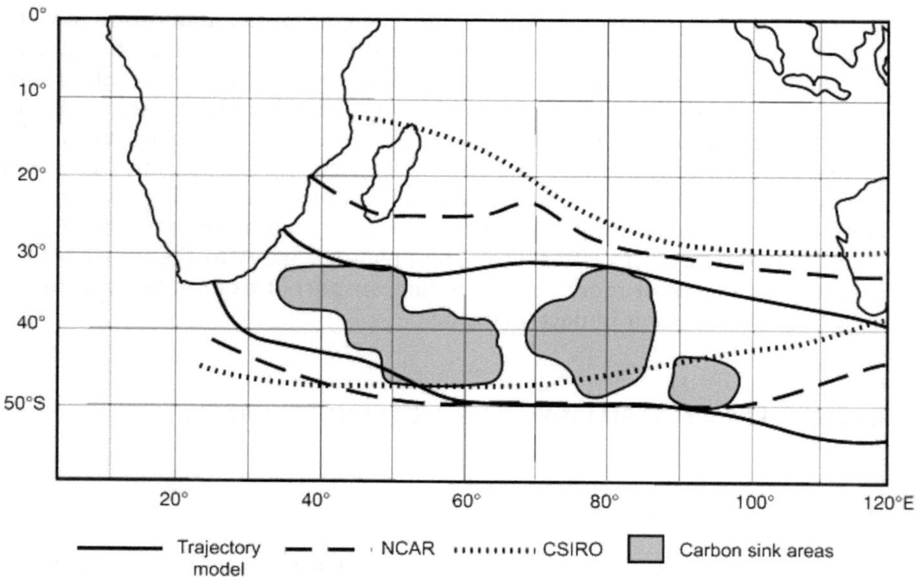

Fig. 12 Aerosol and trace gas transport plumes from southern Africa to the South Indian Ocean. The *shaded areas* represent areas of enhanced biological productivity and carbon sinks (Takahashi et al. 1997). Observed plume outlines are indicated by *solid lines* (after Garstang et al. 1996[31]; Sturman et al. 1997). *Broken lines* indicate modelled plumes for CO_2 transport derived from the NCAR transport model and dotted lines the CO_2 transport from the CSIRO transport model (Rayner and Law 1995)

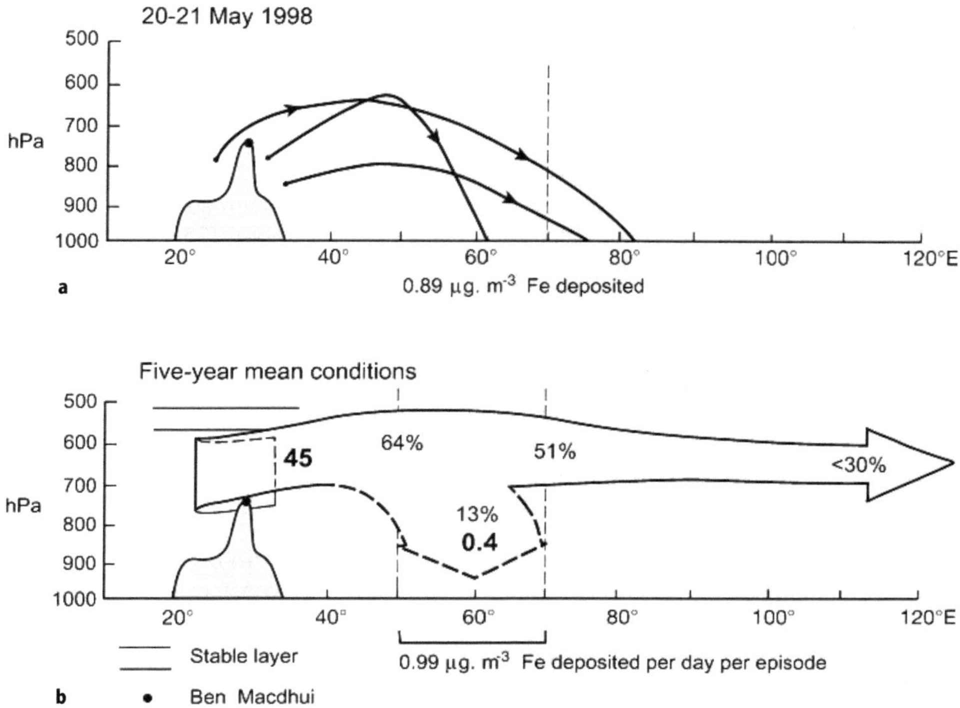

Fig. 13a, b Iron fertilisation: **a** on 20–21 May 1998 resulting in 0.89 μg m^{-3} iron deposition between 60°E and 85°E (c.f., 0.11 μg m^{-3} deposited over 64 km^2 in the eastern Pacific Ocean iron fertilisation experiment (Behrenfeld et al. 1996[5]; Cooper et al. 1996); **b** mean daily deposition of 0.99 μg m^{-3} from 50° to 70°E estimated from a five-year trajectory climatology and PIXE measurements of iron loadings in the plume passing over Ben Macdhui at 3000 m in South Africa. Approximate annual mass fluxes of total aerosol loading are given as million tons per annum in *bold*; numbers of trajectories passing given meridians or having subsided to the surface are given as percentages

to rise, because of low-level convergence and ascent of air with poleward geostrophic flow and a constant pressure gradient occurring in the mean wind field on the western margin of the South Indian Anticyclone [35]. As air is transported towards the centre of the South Indian Anticyclone in the vicinity of 70°E it tends to subside. On occasions, when subsidence is particularly strong, air parcels may subside to the surface. It is estimated that 13% of all transport trajectories reach the surface this way between 50° and 70°E.

A case study of transport and deposition to the central Indian Ocean has been undertaken using data from the Aerosol Recirculation and Rainfall Experiment (ARREX) [53]. Trajectories from peak atmospheric aerosol concentrations indicated that the plumes ascended as they exited the landmass and then subsided to the ocean surface between 60° and 85°E over the carbon sink area, i.e. at a distance of 3500–6000 km from South Africa (Fig. 13a). The iron fraction deposited nine days later into the ocean has been estimated as 0.89 μg m^{-3} between 60° and

85°E. In comparison, the application of ferrous sulfate in the iron enrichment experiment which produced significant phytoplankton blooming in the equatorial Pacific Ocean was 0.11 $\mu g\ m^{-3}$ [5]. Extrapolation from this case study yields an annual aerosol flux from South Africa to the Indian Ocean of around 40 million tons per annum [54].

Using trajectory analysis to determine mean air volumes transported, and observed background aerosol concentrations, climatological aerosol mass fluxes have been estimated over southern Africa and adjacent oceans [28]. A five-year regional transport climatology reveals that the aerosol flux to the Indian Ocean off the east coast of South Africa is about 45 million tons per annum. Given that most recirculation takes place near the coast, that 64% of all trajectories reach 50°E and 51% 70°E (Fig. 13b), it may be inferred that, on average, ~13% of trajectories subside to the surface between 50° and 70°E. The dissolvable fraction of the transported aerosols from southern Africa in ocean water has not yet been determined. Based on the observations of aerosol deposition to the Mediterranean Sea it is assumed that, due to the fact that the aerosols are characterised by anthropogenic emissions, a significant fraction of the deposited burden will be soluble.

Estimates have also been made for the transport of aeolian dust off the west coast of Namibia. A trajectory climatology from the west coast, centred around both Luderitz and Walvis Bay, show that during strong offshore flow, wind blown dust will be transported off the coast and deposited within approximately 100 km of the coast. Much work remains to be done in quantifying the deposition to the ocean system as well as establishing the possible contribution of these atmospheric constituents to the nutrient enrichment of the Benguela upwelling current.

7 Discussion

Although not all aspects of the atmospheric transport to and from the African continent have been determined, a fairly good understanding of the general flows exists. The continent stretches across the equator and is therefore subject to atmospheric transports influenced by both the easterly and westerly winds of the general circulation in both hemispheres (Fig. 14). The transport patterns are highly seasonal. Transport for example from Africa to the Caribbean or South America is determined by the seasonal position of the ITCZ. Transport over southern Africa is very distinctly divided in easterly and westerly transport. The westerly circulation dominates the flow over the southern part of the subcontinent. Easterly waves determine that movement of air over central southern Africa. The easterly transport of air is important to southern Africa. Transport to the subcontinent within the easterlies ensures the flux of moisture from the warm Indian Ocean. Air carried toward the west over central southern Africa is responsible for moving biomass burning and other emissions out over the Atlantic Ocean. The two systems interact at various times of the year. The link between the easterly and westerly transport results in large scale recirculation over the subcontinent.

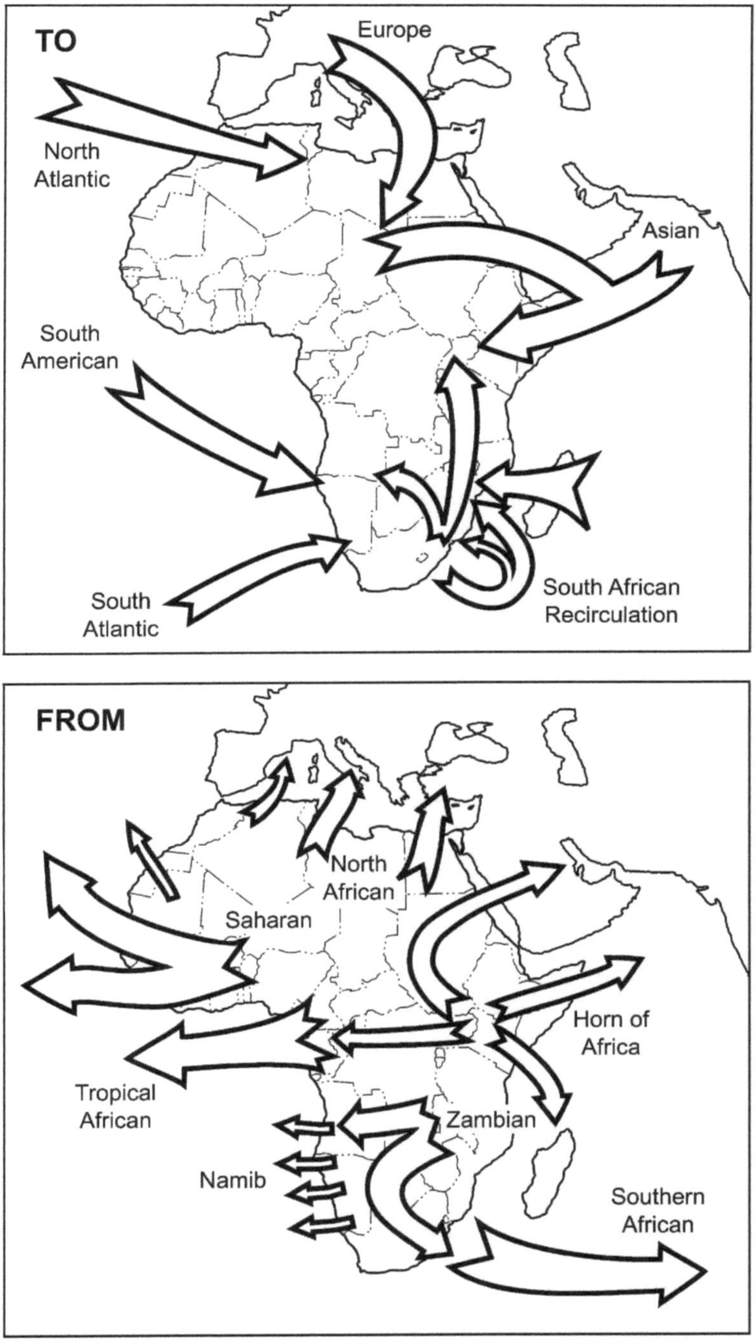

Fig. 14 Schematic diagram of the most important atmospheric transport pathway to and from the African continent

Africa is a large source of trace gas and aerosol emissions. The nature of the sources and size of the emissions vary greatly across the continent. Emissions from the African continent are transported to five other continents, namely Australia, South America, North America, Europe and Western Asia. Although Africa is not the most industrialized continent, it makes important contributions to the global atmospheric chemistry. Emissions from the African continent contribute both nutrients and pollution to the aforementioned continents as well as to the Indian and Atlantic Oceans.

8 References

1. Tegen I, Fung I (1994) J Geophys Res 99:22897–22914
2. Tegen I, Fung I (1995) J Geophys Res 100:18707–18726
3. Sokolik IN (1999) IG Activities Newsletter 17:12–14 May
4. Guerzoni S, Chester R (1996) The impact of desert dust across the Mediterranean, Kluwer, Dordrecht
5. Behrenfield MJ et al. (1996) Nature 383:508–511
6. Siversten B et al. (1995) SADC ELMS, Report 35, Lesotho
7. Wells RB, Lloyd SM, Turner CR (1996) National air pollution source inventory – Inventory of scheduled processes. In: Held G, Gore B, Surridge AD, Tosen GR, Turner CR, Walmsley RD (eds) Air pollution and its impacts on the South African Highveld. Environmental Scientific Association, Cleveland
8. Benkovitz CM (1996) J Geophys Res 101:29239–29253
9. Örn S et al. (1996) International Meteorological Institute, Report CM91, Stockholm, ISSN 0280-445X
10. Meter S et al. (1999) Nucl Inst Meth Phys B 150:433–438
11. Meter SL (1999) Unpublished MSc dissertation, University of the Witwatersrand, Johannesburg, South Africa
12. Andreae MO, Atlas E, Cachier H, Cofer WR, Harris GW, Helas G, Koppman R, Lacaux J, Ward DE (1996a) Trace gas and aerosol emissions from savanna fires, Levine JS (ed) Biomass Burning and Global Change. MIT Press, Cambridge 278–294
13. Lacaux JP, Cachier H (1993) Biomass burning in Africa: An overview of its impacts on atmospheric chemistry. In: Crutzen PJ, Goldammer J (eds) Fire in the Environment: The ecological, atmospheric and climatic importance of vegetation fires. John Wiley, New York 159–191
14. Piketh SJ, Annegarn HJ, Kneen MA (1996) Regional scale impacts of biomass burning emissions over southern Africa. In: Levine JS (ed) Biomass Burning and Global Change. MIT Press, Cambridge 320–326
15. Cahoon RC et al. (1992) Nature 359:812–815
16. Scholes RJ et al. (1996) J Geophys Res 101:23667–23676
17. Scholes RJ et al. (1996) J Geophys Res 101:23677–23682
18. Müller C (1992) Unpublished MSc Dissertation, University of the Witwatersrand, Johannesburg
19. Piccot SD et al. (1992) J Geophys Res 97:9897–9912
20. Guenther A et al. (1995) J Geophys Res 100:8873–8892
21. Swap RJ, Tyson PD (1999) S Afri J Sci 95:63–71
22. Novelli PC et al. (1994) Science 263:1587–1590
23. Zencker T, Thompson AM, Mcnamara DP, Kucsera TJ, Harris GW, Wienhold FG, LeCanut P, Andreae MO, Koppman R (1996) Regional trace gas distribution and air mass characteristics in the haze layer over southern Africa during the biomass burning season (September/October 1992): Observations and modelling from the STARE/SAFARI-92/DC3. In: Levine JS (ed) Biomass Burning and Global Change. MIT Press, Cambridge, Mass 296–308

24. Diab RD et al. (1996) J Geophys Res 101:23823–23833
25. Diab RD et al. (1996) J Geophys Res 101:23809–23821
26. Marufu l et al. (2000) J Geophys Res 105:14513–14530
27. Cosijn C, Tyson PD (1996) S Afri J Sci 92:381–386
28. Tyson PD et al. (1997) J Geophys Res 102:10623–10635
29. Annegarn HJ, Turner CR, Helas G, Tosen GR, Rorich RP (1996) Gaseous pollutants. In: Held G, Gore BJ, Surridge AD, Tosen GR, Turner CR, Walmsley RD (eds) Air pollution and its impacts on the Southern African Highveld. Environmental Scientific Association, Cleveland, chap 6, pp 25–34
30. Garstang M, Tyson PD (1997) Atmospheric circulation, vertical structure and transport. In: Wilgen BW, Andreae MO, Goldammer JG, Lindesay JA (eds) Fire in the southern African savanna: Ecological and Environmental Perspective. Witwatersrand University Press, Johannesburg, South Africa 57–88
31. Garstang M et al. (1996) J Geophy Res 101:23721–23736
32. Tyson PD et al. (1996) Int J Clim 16:265–91
33. Piketh SJ et al. (1999) Nucl Inst Meth Phys Res B 150:350–355
34. Tyson PD (1997) Prog Phys Geog 21:79–101
35. Tyson PD, Preston-Whyte RA (2000) The weather and climate of Southern Africa. Oxford University Press, Cape Town
36. Piketh SJ et al. (1999) J Geophys Res 104:1597–1607
37. Gatebe CK (1999) Unpublished PhD Thesis, University of the Witwatersrand, Johannesburg
38. Galpin A, Turner CR (1999) S Afri J Sci 95:223–225
39. Zunckel M et al. (2000) Atmos Environ 34:2797–2808
40. Fishman J et al. (1990) J Geophys Res 95:3599–3617
41. Herman JR et al. (1997) J Geophys Res 102:16911–16922
42. Prospero JM, Carlson TN (1972) J Appl Meteorol 11:283–297
43. Husar RB et al. (1997) J Geophys Res 102:16889–16909
44. Prospero JM (1999) J Geophys Res 104(D13):15917–15927
45. Prospero JM (1996) Saharan Dust Transport over the North Atlantic Ocean and Mediterranean: An Overview. In: Guerzoni S, Chester R (eds) The Impact of Desert Dust Across the Mediterranean, Kluwer, Dordrecht
46. Swap RJ et al. (1992) Tellus 44(B):133–149
47. Prodi F, Fea G (1979) J Geophys Res 84c:6951–6950
48. Hamonou E et al. (1999) J Geophys Res 104:22257–22270
49. Krishnamurti TN et al. (1993) J Geophys Res 98:10621–10641
50. Thompson AM et al. (1996) J Geophys Res 101:24251–24278
51. Salati E, Vose P (1984) Science 225:129–138
52. Vitousek PM, Sanford RL Jr (1986) Nutrient cycling in moist tropical forest. A Rev of Eco Syst 17:137–168
53. Terblanche DE et al. (1999) S Afr J Sci 96:15–21
54. Piketh SJ et al. (2000) S Afr J Sci 96:244–246
55. Wenig M, Spichtinger N, Stohl A, Held G, Beirle S, Wagner T, Jähne B and Platt U (2003) Intercontinental transport of Nitrogen Oxide pollution plumes. Atmospheric Chemistry and Physics 3:387–393
56. Takahashi K, Oba T and Koizumi L (1997) Quaternary charge of productivity and carbonate, dissolution process in coastal upwelling region, western Arabian Sea. Journal of Geological Society, Japan 103:1162–1174
57. Sturman AS, Tyson PD and D'Abreton PC (1997) Transport of air from Africa and Australia to New Zealand. Journal of the Royal Society of New Zealand 27:485–498
58. Rayner PJ and Law RM (1995) A comparison of modeled responses to prescribed CO_2 sources. CSIRO Division of Atmospheric Research Technical Paper No. 36
59. Cooper DJ, Watson AJ and Nightingale PD (1996) Large decrease in ocean-surface CO_2 fugactiy in response to in situ iron fertilization. Nature 383:511–513

The Handbook of Environmental Chemistry Vol. 4, Part G (2004): 197–223
DOI 10.1007/978-3-540-40037-0

Numerical Modelling of Long-Range Pollution Transport

Martin G. Schultz[1] · Isabelle Bey[2]

[1] Max Planck Institute for Meteorology, Bundesstrasse 55, 20146 Hamburg, Germany
E-mail: martin.schultz@dkrz.de
[2] Swiss Federal Institute of Technology (EPFL) ENAC-LMCA, 1015 Lausanne, Switzerland
E-mail: isabelle.bey@epfl.ch

Abstract A quantitative understanding of long-range transport of air pollution involves several aspects from different science areas: emissions of ozone precursors, greenhouse gases, aerosols and aerosol precursors must be known not only in their amount, but also with regard to their geographic distributions and seasonal variations, wind fields must be available in sufficiently high spatial and temporal resolution, vertical transport processes such as convection or frontal lifting must be captured, and the chemical transformations of pollutants must be considered. Three-dimensional numerical models of atmospheric chemistry and transport provide a means to address all of these issues simultaneously. In this chapter we discuss the state-of-the-art of such models and present a summary view of a number of issues related to long-range transport (i.e. transport pathways, regional and global impact of various source regions) as it emerges from numerical simulations of the atmosphere. We focus on tropospheric ozone, because it illustrates the interactions between transport and chemistry, and it is also of immediate relevance for air quality regulations. Global models have contributed significantly to the understanding of the long-range transport of ozone and its precursors, but there is still considerable uncertainty with respect to important terms of the ozone budget. This makes quantitative predictions of the impact of long-range transported air pollution difficult. We discuss the shortcomings of today's models and provide an outlook into future developments.

Keywords Tropospheric ozone · Aerosols · Eulerian models · Lagrangian models · Chemistry transport models · Advection · Diffusion · Convection · Pollutant transport · Trace gas emissions · Fossil fuel combustion · Biomass burning

List of Abbreviations

BIBLE	Biomass Burning and Lightning Experiment, a NASDA/GLACE aircraft experiment series
CTM	Chemistry Transport Model, a three-dimensional Eulerian computer model of the atmosphere
GOME	Global Ozone Monitoring Experiment, a UV/visible spectral scanning sensor on board the European ERS-2 satellite
GTE	Global Tropospheric Experiment, a NASA programme of airborne measurements of the tropospheric chemical composition
INDOEX	Indian Ocean Experiment
ITCT 2k2	Intercontinental Transport and Chemical Transformation experiment
MOPITT	Measurements Of Pollution In The Troposphere, a correlation spectroscopy sensor on board the TERRA satellite
PEM	Pacific Exploratory Mission, a series of NASA GTE experiments in the Pacific region with PEM West A and B in the North Pacific (1991, 1994) and PEM-Tropics A and B in the South Pacific (1996, 1999)
PHOBEA	Photochemical Ozone Budget of the Eastern North Pacific Atmosphere experiment
ppb (or ppbv)	parts per billion (by volume), SI unit: nmol/mol, common unit for trace gas mixing ratios
TRACE-P	Transport and chemistry experiment-Pacific, NASA GTE aircraft campaign in spring 2001

1 Introduction

Long-range transport (LRT) of airborne gas-phase and aerosol pollutants has reached the agenda of environmental authorities as it has become clear that air pollution is a problem of global, or at least of hemispheric scale. Violations of the European air quality standard for surface ozone[1] (O_3) still occur in several regions throughout the summer, despite the implementation of measures for controlling ozone precursor emissions. Even though there is only limited evidence for direct import of ozone in long-range transport episodes, global modelling studies have suggested that LRT plays a significant role for the air quality in different regions by increasing the number of violations of national standards and by lowering the amount of regional pollution necessary to exceed these thresholds [1–4].

Besides ozone, LRT of aerosols has recently received much attention due to the discovery of a quasi-permanent haze layer over much of southern Asia, the so-called 'Atmospheric Brown Cloud'. Inhabitants of western and central Europe are familiar with occasional transport of Saharan dust into these regions. Saharan and Asian dust have also been identified in several filter samples collected across the North American continent [5–8] (see also Chap. 11). Other pollutants, for which LRT is important, include mercury and polycyclic aromatic hydrocarbons

[1] 110 μg/m³ (equivalent to 55 ppb) for an 8-h average.

(so-called 'persistent organic compounds' or POPs). Mercury is transported from European, Asian, and North American sources into the Arctic, where it accumulates in the marine biosphere and in the local human population (see Chap. 2). The chemical and transport characteristics of POPs are still very uncertain, but it is clear that they have a long reaching impact due to their long lifetimes and the fact that they can be re-suspended even years after they were deposited onto the ground [9, 10]. In this chapter we will mainly discuss LRT of tropospheric ozone and its precursors (nitrogen oxides (NO_x), carbon monoxide (CO), hydrocarbons (VOC)). However, we note that the general relationships between source and receptor regions are often similar for other pollutants – at least if they are predominantly of anthropogenic origin.

Several aircraft field campaigns with detailed measurements of the atmospheric chemical composition have been conducted over the last decade to sample the outflow of pollution from various regions of the world. These measurements have documented the ubiquitous occurrence of intercontinental pollution transport events, as discussed in previous chapters. However, no suitable measurement strategy exists to quantify the amount of exported pollution and to assess its fate on the global scale. Carefully validated global 3-D (three dimensional) chemistry transport models are the only adequate tools available for this task. In this chapter, we show how global models of tropospheric chemistry have been used in combination with observations from various platforms (airborne field experiments, satellites, and surface sites) to answer a wide range of scientific questions related to LRT. These include:

- What are the pathways of pollution export from a given geo-political region? Aircraft campaigns are usually conducted over a limited period of time and a limited region of a world. However, their comprehensive set of measurements and fine-scale resolution provide strong constraints for numerical model simulations. A model which successfully reproduces the chemical and meteorological measurements obtained in the field can be used with some confidence to extend the temporal and spatial scale of a given aircraft campaign. Issues such as seasonality and interannual variability of pollution export can then be investigated.
- To what extent does pollution outflow from a given region contribute to the air quality of the regions downwind? The quantification of such contributions is rarely possible from observations alone, and global 3-D models are needed to capture the spatial and temporal variability of LRT. As an example, Fig. 1 displays the simulated impact of emissions from a number of sources on the global atmospheric distribution of carbon monoxide. Such studies have important political implications because pollution imported from upwind continents and regions can lead to an offset of emission reduction strategies that are implemented locally [11].
- What are the chemical transformations in LRT pollution plumes? The lifetime of primary atmospheric pollutants[2] depends on the efficiency of the sink processes (often reaction with the OH radical). The loss rate itself is controlled

[2] i.e. species which are directly emitted, e.g. from combustion of fossil fuel or biofuels.

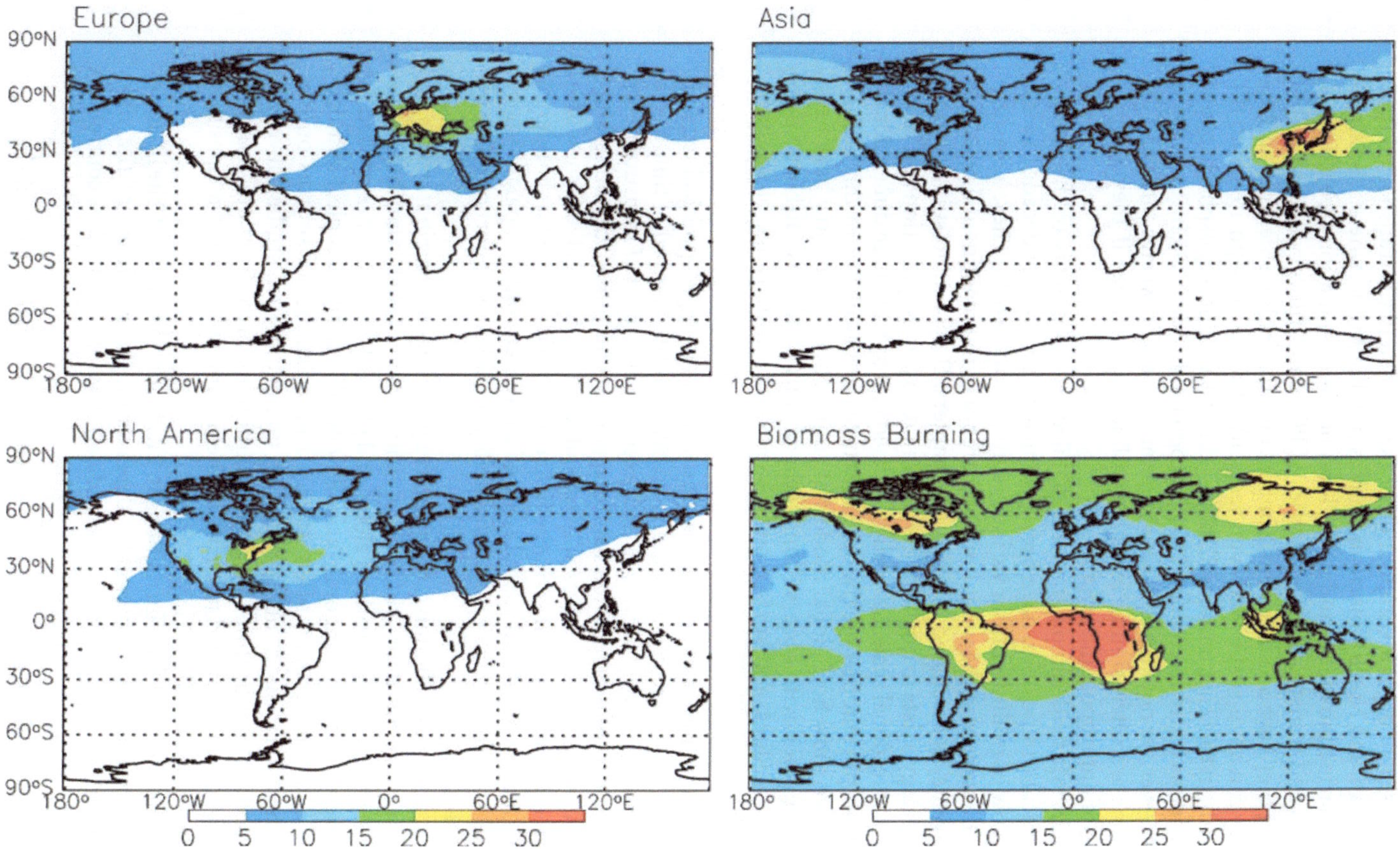

Fig. 1 Percentage of the total column of carbon monoxide from fossil fuel combustion sources in Europe, Asia, and North America and global biomass burning sources in the GEOS-CHEM model for July 1995. Westerly winds dominate transport in the extra-tropics of the Northern Hemisphere, and move pollution from the three main industrial regions eastwards. For instance, pollution from Asia impacts the troposphere over the Pacific Ocean and North America while pollution from North America impacts the Atlantic troposphere. Biomass burning in southern Africa and Brazil disperses in the trade winds polluting much of the tropics. Boreal forest fires in North America and Asia peak in summer and cause pollution over the Arctic, and sometimes over Europe, Asia, and North America. Note that boreal fires in North America were unusually intense in the summer of 1995. Courtesy of Bryan Duncan

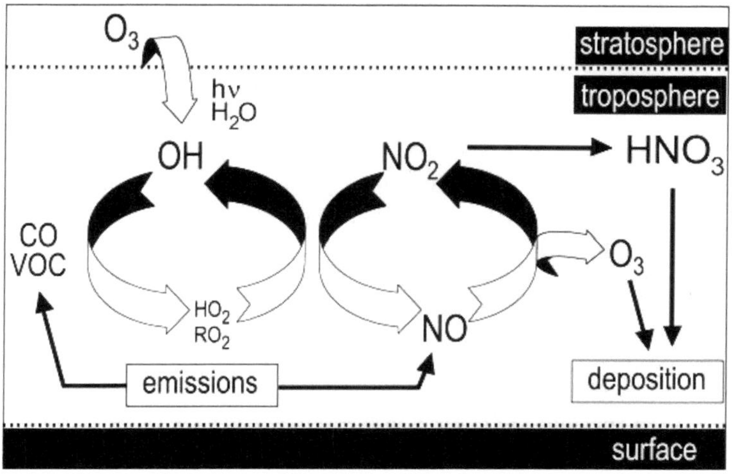

Fig. 2 Simplified schematic of chemical ozone formation in the troposphere. During the day, the ozone precursors NO_x, CO, CH_4, and VOCs, which are emitted from combustion sources, but also from natural processes, are oxidised by the hydroxyl radical (OH), and they react in a coupled catalytic cycle to form ozone. The dominant loss processes for tropospheric ozone are deposition at the surface, reaction with the HO_2 radical, and indirect losses due to the removal of NO_x

by meteorological parameters and the concentration of other trace compounds. Ozone is a secondary pollutant, which is formed in the atmosphere from photochemical reactions of its precursors NO_x and VOC (Fig. 2). Therefore, the long-range impact of ozone precursor emissions depends not only on the magnitude of the emissions, but also on the chemical composition of a pollution plume, and on the temperature, pressure, light intensity, and humidity during its transport.

- What are the global budgets of key airborne pollutant species? When ozone precursors are exported into the free troposphere or into the marine boundary layer, they perturb the chemical regime of the so-called "background atmosphere" and lead to enhanced photochemical ozone formation. It is not clear yet, to what extent there is net ozone production during long-range transport events.[3] Models must be able to yield a good representation of the global ozone budget, before such issues can be resolved. This mandates the use and continuous improvement of comprehensive schemes for emissions, transport, and chemistry.

Global models of tropospheric chemistry are now sufficiently advanced that they have already been used to assess past and future changes of the atmospheric chemical composition and its relation to climate change [12]. They have also been employed successfully to provide chemical forecasts before and during aircraft campaigns, where their results influenced the flight planning.

[3] atmospheric chemical reactions produce and destroy ozone in various chemical cycles. Net ozone production occurs if the formation rate of ozone exceeds its loss rate.

2
What Constitutes a State-Of-The-Art Chemistry Transport Model?

Since the late 1980s, several 3-D global models of transport and chemistry in the atmosphere have been developed (e.g. [13–21]). Current models of tropospheric chemistry typically include 50–100 chemical compounds and 100–300 photochemical reactions, and they are run at a resolution of about 2° latitude by 2° longitude with 20–60 vertical levels from the surface into the stratosphere. This amounts to about 300,000–1,000,000 individual Eulerian grid boxes, for which hundreds of numerical equations have to be solved at each time step. Evaluations with surface observations, vertical soundings, and large-scale field campaigns have shown that current models reproduce with reasonable success the mean global distributions of ozone and its precursors in the atmosphere. However, various detailed intercomparison exercises have also revealed severe differences and biases between models, indicating that our understanding of some important processes and how they should be represented in global models is still rather limited.

In a very general sense, the role of an atmospheric chemistry model is to integrate the continuity equation for trace constituents in space and time. The dominant terms in the mass budget equation of gas-phase species are: horizontal and vertical advection, turbulence (including convection), chemical formation and loss, emissions, dry deposition, and scavenging in clouds and precipitation. For aerosols, additional processes must be taken into account if one wants to monitor their size distribution, solubility, acidity, and chemical composition. For reasons of memory limitations, computational efficiency, and complexity, the individual terms of the continuity equation are solved sequentially rather than concurrently (so called 'operator splitting'). While large-scale physical processes (synoptic advection) are explicitly resolved in the models, smaller scale processes (such as convection) have to be parameterised using empirical relationships between large-scale model quantities.

Emissions of anthropogenic origin (e.g. related to fossil fuel use in power plants, surface and aircraft traffic) are typically prescribed as monthly mean fluxes, which are derived from extensive compilations of energy use and combustion efficiency data, or from extrapolations of field measurements probing the release of trace substances from different types of sources. Natural emissions, such as volatile organic compounds released from plants and also the source of NO_x from lightning, are often calculated online in parameterised form in order to capture their large variability. Dry deposition is parameterised as a first order loss process. The rate constant (deposition velocity) is computed using a so-called resistance analogue scheme [22]. Parameterisations for the scavenging of trace gases and aerosols in and below clouds are tied to the level of detail, which is used in the description of cloud microphysical processes of the physical atmosphere model (e.g. [23]). Further details on the equations and parameterisations employed in current chemistry transport models can be found for example in [24], or in the papers describing the individual models in use (see above).

3 Pollution Transport in the Northern Mid-Latitude Belt

Fossil fuel combustion in Asia, North America and Europe is clearly the main factor which determines the distributions of ozone and its precursors in the northern hemisphere. Most of the emission sources are located in the mid-latitude belt[4] where the population density is highest. It has only recently been recognized that air quality in a given region can be affected by compounds (such as ozone and aerosols) transported from far away sources. Thus, it is important for local environmental authorities to acquire a quantitative understanding of the amount of pollution that can be controlled by local emission reduction strategies and how much pollution is transported from outside. The reduction of the latter requires international agreements. This section analyses how global models have contributed to our understanding of export pathways and the regional and global impacts of Asian, North American, and European pollution, respectively.

3.1 Asian Pollution

The first studies of the impact of Asian continental outflow on the global tropospheric composition were motivated by the rapid economical growth in the countries of this region and the concomitant increase in ozone precursor emissions. For example, Berntsen et al. [15] examined the impact of doubling Asian emissions (relative to the 1980 level) and reported a significant ozone increase (up to 30%) in the upper troposphere over a large region, which was attributed to efficient convective transport. Since then, a number of modelling studies have investigated the chemical outflow from Asia with a special focus on the late winter and early spring season when the outflow from Asia to the Western Pacific is strongest, because cyclogenesis and convective activity (especially in Southeast Asia) and also biomass burning in southeast Asia are at their maximum. Most of the global simulations investigating Asian export have been constrained by observed concentrations of trace gases and aerosols provided by aircraft experiments conducted to sample Asian outflow on the western side (e.g. PEM-West B, TRACE-P) and the eastern side of the Pacific (e.g. PHOBEA, ITCT 2k2).

The earlier PEM-West B mission was an exploratory mission which gave a first characterization of the Asian outflow, while the TRACE-P mission was dedicated to sampling and characterizing the continental outflow from the Asian continent in a quantitative manner [25].

Numerical 3-D model simulations of the Asian continental outflow pointed out its complex chemical mixture and identified some of the export pathways. Anthropogenic sources and biomass burning were found to make comparable contributions to the chemical concentration enhancements in the outflow [26]. The presence of high aerosol concentrations including sulfate, nitrate and mineral aerosols, in this outflow was shown to complicate further its characterization and influence the chemical evolution of the pollution plumes over the Western Pacific

[4] i.e. the latitude range between about 30°N and 65°N.

[27]. During the PEM West B campaign, the maximum outflow was seen in the lower troposphere (2–5 km) and resulted from episodic lifting ahead of cold fronts [26, 28]. These preliminary results were confirmed by several simulations of the TRACE-P region for the campaign period in March 2001. Liu et al. [29] and Carmichael et al. [30] found that lifting of air ahead of cold fronts and boundary layer outflow behind the fronts are important mechanisms to transfer the anthropogenic pollution from the boundary layer to the free troposphere. Orographic lifting over central and eastern China also contributes to the transport of pollution to the free troposphere. Convective lifting is important for pollution export (especially that of biomass burning origin) but is confined to Southeast Asia (south of 35°N) [29].

Model simulations of the TRACE-P experiment also revealed interesting aspects of the chemical evolution of the plumes during the transport across the Pacific (see below) and of the global impact of Asian emissions. Wild et al. [31, 32] found that the impact of Asian emissions on global ozone strongly depends on the meteorological conditions during a given export event. Anticyclonic conditions favour regional ozone production over the emission region while only little export of ozone and its precursors occurs. Conversely, cooler and cloudier weather conditions reduce local ozone production but lead to stronger export from the boundary layer of ozone precursors, thus having a bigger impact on global ozone production because of the longer ozone lifetime in the UT.

The impact of Asian emissions extends well beyond the western Pacific. Jaeglé et al. [33] conducted a simulation of the PHOBEA II campaign (which took place at the western US coast around the same time as TRACE-P) to examine the origin of CO and ozone in the north east Pacific troposphere and to quantify the contribution of Asian pollution to their burden. They found that their global model is capable of reproducing well the observed CO concentrations (including the temporal variability and vertical gradients), and that it captures several of the observed cases of rapid intercontinental transport in their location, time, and magnitude. Based on these results, Jaeglé et al. [33] extended their analysis beyond the region and time interval of the campaign and found that Asian and European emission sources both contributed significantly to the total column CO in the NE Pacific (33% and 15%, respectively) while their contribution was smaller for ozone (12% and 5%, respectively).

These various modelling studies have shown that the Pacific region experiences frequent events of LRT. Even though export occurs only periodically, global model simulations show that these events significantly affect the mean distribution and the global burden of chemical compounds and that the chemical composition over the Pacific is characterized by a mix of pollution originating from Asian sources of fossil and biofuel combustion and from biomass burning. Global modelling studies have also indicated that pollution outflow from Asia could affect air quality over the United States, in particular due to their impact on the background ozone concentrations. Jacob et al. [11] estimated that tripling of Asian emissions (relative to 1985) would lead to an increase of monthly mean ozone concentrations of 2–6 ppb in the Western United States and 1–3 ppb in the Eastern United States.

3.2 North American Pollution

As discussed in Chap. 3, the pathways for pollution export from the United States and transport across the Atlantic Ocean to Europe include lifting to the free troposphere by warm conveyor belts associated with mid-latitude cyclones [34], lifting to the free troposphere (FT) by convection [35, 36], as well as low-level transport driven by the Icelandic low pressure system [1]. Using a global 3-D model of tropospheric chemistry and transport, Li et al. [1] found that transatlantic transport of ozone and CO pollution in the lower troposphere occurs year round (as the westerlies extend to the surface) while transport in the middle and upper troposphere is most important in summer after the pollution has experienced vertical transport through convection, especially over the central and eastern United States. Horowitz et al. [36] and Liang et al. [37] conducted a budget analysis for the United States and found that 25% to 30% of the NO_x emitted from fossil fuel combustion in the United States is exported out of the U.S. boundary layer as NO_x, PANs or HNO_3. Stohl et al. [38] used correlations between NO_y and CO measurements taken during the 1996 and 1997 NARE aircraft experiments to estimate NO_y export efficiencies. They found that most of the NO_y emitted at the surface was removed within the U.S. boundary layer, and that less than 10% of the NO_x emitted at the surface actually reached altitudes above 3 km. Based on the same observations, Li et al. [39] estimated an NO_y export efficiency out of the U.S. boundary layer of about 15% (20% in their global model). Discrepancies with previous global model studies were attributed to differences in the scavenging scheme that removes HNO_3 during the export.

Model studies have indicated that pollution export from North America and subsequent transport to Europe could have a significant impact on surface ozone concentrations in Europe. One should note however that so far, there are only few direct observations of such events (see Chap. 3). Li et al. [1] estimated that North American pollution contributes about 2–4 ppb on average to surface ozone over Europe in summer with a maximum of 5–10 ppb during transatlantic episodes. They estimated that 10–20% of the ozone associated with North American pollution actually goes into the European boundary layer. However, due to meteorological settings, these transport events occur when ozone concentrations over Europe are about 50–60 ppb, i.e. close to the European air quality threshold for ozone. Although the average amplitude of the North American pollution signal appears to be relatively small, the model of Li et al. [1] suggests that 20% of the air quality standard violations during the summer of 1997 would not have happened without fossil fuel emissions from North America. According to Wild and Akimoto [40], the United States emissions have potentially the greatest impact on surface ozone concentrations over continents downwind (compared to European or Asian emissions), because of their magnitude and the relatively short distance between the source and receptor continent.

Up to now, most of the modelling studies dealing with long-range transport from North America have only considered fossil fuel emissions although several atmospheric observations have indicated that boreal forest fires can also significantly impact the distributions of pollutants over the Atlantic Ocean and Europe

[41–43]. The emissions from such fires are estimated to be about 60% and 9% of the fossil fuel emissions in North America for CO and NO_x, respectively [12, 44], and they show a large interannual variability [45].

3.3 European Pollution

According to preliminary experimental and modelling studies the export of pollution from Europe in winter occurs via three primary pathways, 1. to the northeast in the prevailing westerlies (Russian Arctic and Siberia), 2. to the northwest (North Atlantic Ocean and European Arctic), and 3. to the south over the Mediterranean Basin (North Africa and the Atlantic Ocean) [44, 46–48]. The strength and location of these different pathways strongly depend on the North Atlantic Oscillation (NAO) [48], as discussed in more details below. In summer, predominant flows are to the east over Russia and to the south over the Mediterranean Basin and northern Africa [46, 48, 49]. Summertime convection lofts some of the European pollution into the upper troposphere. Hov and Flatøy [50] conducted short-term simulations of convective events over Germany and estimated that between 7% and 12% of NO_x emissions were vertically exported from the boundary layer as NO_x and 13–15% as PAN and HNO_3. Duncan and Bey [48] reported two important areas of convection centred over Germany and Poland (consistent with what was indicated by [50]), and a third area over Russia near the Ural Mountains. Furthermore, dry lofting of European pollution into the middle troposphere occurs over northwestern Africa in the quasi-permanent West Saharan Low after the pollution has travelled to North Africa through the Mediterranean basin [48]. Vertical pumping of air out of Alpine valleys could be another important pathway for the export of European pollution [51] but, due to the fine scale required by the orographic complexity of the Alps, there are no modelling studies yet to investigate the potential contribution of such processes.

Summarizing the results from various models, European pollution appears to affect ozone concentrations mainly in regions relatively nearby, such as the Mediterranean basin and Northern Africa. Roelofs et al. [49] found that ozone in the lower troposphere over the eastern Mediterranean region is strongly impacted by precursor emissions from eastern and western Europe. Duncan and Bey [48] report that O_3 concentrations are enhanced by 5 ppbv on average in the middle troposphere over the regions of most intense vertical lofting (i.e., Germany, Russia, and northwestern Africa) and that O_3 generated from European pollution regularly enhances O_3 levels in northern Africa and the Near East. Only a small contribution of European emissions is found in surface air over North America and Asia [1, 40]. Weak European pollution plumes were also simulated in the boundary layer over Asia [29] although it was difficult to identify such signatures in the observed plumes [25]. Wild and Akimoto [40] suggested that (compared to Asian and North American emissions) European fossil fuel emissions have the smallest potential to perturb ozone distributions in the middle and upper troposphere.

It is however important to note that, contrary to pollution export from Asia and North America, the export of European air pollution has not yet been studied in a large-scale, comprehensive field experiment. A quantitative understanding of the impact of European emissions is hampered by the lack of data, in particular with respect to the strong outflow to the east and north of the major industrial areas.

4 Pollution Transport in the Tropics and Subtropics

In what is probably the first global model study of tropospheric CO, Pinto et al. [52] pointed out that a large tropical source of CO would be required in order to explain observations of enhanced pollution levels in the tropics, which had been linked to the wide spread occurrence of biomass burning by Seiler and Crutzen [53]. Further evidence was found by Fishman et al. [54], who identified a strong seasonal enhancement of the tropospheric ozone column over the South Atlantic in their satellite data analysis (see Chap. 9 for a more detailed discussion of the observational findings). This feature was then also related to biomass burning during the 1992 TRACE-A field experiment [55]. Meanwhile, a substantial amount of research has been done in order to quantify the amounts of trace gases and carbonaceous aerosols from vegetation fires in different regions and biomes of the world (cf. [56–66]), and several modelling studies have confirmed the important role of these emissions for the global budgets of tropospheric CO, NO_x, and ozone (cf. [12]). More recently, the formation of NO_x in tropical thunderstorms has been identified as another important factor determining the fate of biomass burning pollution exported into the troposphere [67].

In this section, we concentrate on the export mechanisms and impact of fires in the tropical and subtropical regimes as they emerge from the use of different numerical models. The most important regions and seasons of tropical biomass burning are Sub Saharan Africa (Dec-Feb), Central America and South East Asia (Feb-Apr), Central and Southern Africa and tropical South America (Jul-Oct) [59, 64, 68, 69]. The influence of lightning NO_x is discussed in the next section.

During the 1990s, several large-scale airborne field experiments were conducted in the Southern Hemisphere focusing on the impact of biomass burning emissions on the tropical South Atlantic (TRACE-A, 1992 [55]), and on the tropical and subtropical South Pacific (PEM Tropics A, 1996 [70], BIBLE A and B, 1998 and 1999 [71]). All of these experiments have documented the pervasive influence of biomass burning pollution even in areas which have long been thought of as pristine and pollution-free. Earlier measurements (from the 1977 GASP and the 1984 MAPS experiments) had also shown CO pollution signatures in the tropics, but these were generally ignored because of insufficient measurement accuracy or too limited sampling [72]. Simulations with various 3-D models and for different years have shown that the recurring increase in the tropospheric ozone column over the South Atlantic and the elevated concentrations of CO, NO_x, and other species sampled during TRACE-A are primarily due to emissions from fires in South Africa, while burning in South America makes a comparatively smaller contribution [73–75]. These findings were corroborated by detailed meteorological analyses, which identified the interplay of moist convection, large-scale dry convection, and extratropical low pressure systems as the dominant mechanism for pollution export out of Africa into the middle free troposphere over the South Atlantic [76–78]. South American biomass burning pollution is found over the Atlantic if it has been lifted directly into the upper troposphere in deep convective cells over the continent, and then advected eastwards in the main geostrophic flow [73, 74]. Polluted air masses exported from the African continent are generally trapped between stable layers, and a large fraction of the air is recirculated over the southern tip of the continent [77, 78]. Export over the South Atlantic and the Indian Ocean occurs quasi-permanently and is episodically enhanced by the passage of subtropical weather systems [75, 77]. Chatfield et al. [73] created the term "Great African plume" to indicate the coherence and extent of this phenomenon.

The PEM Tropics A experiment conducted over the Southern Pacific in September/October 1996 [70] revealed a far more frequent and intense influence of biomass burning pollution onto this region than had previously been expected [79–81]. Observed concentrations of CO and ozone in the middle free troposphere over the subtropical South Pacific frequently exceeded 80 ppbv – at least 30 ppbv above background [81]. Photochemical point model calculations by Schultz et al. [81] led to the conclusion that about 50% of the ozone in this region must have been transported from outside. In addition, the in-situ photochemical production of ozone in the lower troposphere is also largely controlled by the advection of NO_x reservoir species (PAN, HNO_3) from biomass burning in Africa and South America. Using a 3-D model with assimilated meteorology, Staudt et al. [74] confirmed these results, and identified the transport of NO_x from lightning over Indonesia as another important term in the ozone budget over the subtropical South Pacific. According to Staudt et al. [74] biomass burning emissions from South America and Africa contribute about equal amounts of CO to the tropospheric column over the South Pacific, while emissions from fossil fuel and biofuel combustion and from fires in Indonesia and Australia appear to have a smaller influence. Pollution from Africa dominates in the middle and upper free troposphere. According to global model studies of Schultz et al. [81] and Staudt et al. [73], South American outflow in the trade winds across the Andes constitutes the major pollution signal in the lower troposphere over the eastern Pacific. This export occurs north of Chile and episodically as is shown by recent simulations with finer resolution models (K. Longo, personal communication, 2003). Further outflow onto the Pacific occurs over Ecuador after the passage of cold fronts. During the PEM-Tropics A field experiment large pollution signals were found west of Guayaquil [82], but there were no measurements along the South American West coast further south.

Photochemical point model analyses of the BIBLE experiments [71] confirmed some of the PEM-Tropics results, but also revealed a strong influence of Indonesian outflow in the free troposphere over Australia [83]. Air masses advected from Indonesia showed significantly enhanced NO_x concentrations, which were attributed to lightning as well as convective lifting of boundary layer pollution [84]. Correlating numerical back trajectories with satellite data of cloud top heights, Koike et al. [84] could show that upper tropospheric NO_x was increased in continental convective outflow, but decreased in marine convective outflow. The average ozone increase observed between air masses sampled over Indonesia and those sampled over Australia was found to be consistent with the model estimated net formation rate of 1–2 ppb/day. Biomass burning pollution originating from north western Australia was confined to the boundary layer due to strong subsidence in the region at the time of the experiments [83].

The atmospheric impact of fires on the southeast Asian continent is difficult to investigate, because of the relatively strong influence of emissions from fossil fuel combustion. Modelling studies in support of the aircraft measurements during the March 2001 TRACE-P experiment [25, 29, 85] have revealed that the dominant fraction of biomass burning pollution is exported south of 35°N, and at altitudes above 3 km, typically after convective lifting over the continent (see above).

Indonesian fires, while certainly a huge factor in regional air pollution, are generally of little significance for the global burdens of trace gases and aerosols. A noticeable exception was the well known fire episode during the 1997 El Niño period, when extreme drought caused many fires to get out of control and large areas of peatland burned for several months. Areas of deep convection lay to the north of the major burning regions, especially off the northwestern coast of Sumatra, and lofted the pollution into the middle and upper troposphere above Indonesia. The pollution was subsequently transported in the prevailing easterlies to the tropical Indian Ocean and to the tropical and subtropical South Pacific Ocean in the southern subtropical jet [64]. Hauglustaine et al. [86] and Duncan et al. [64] used global chemistry transport models to estimate the impact of these fires on the tropospheric composition, and they both found large-scale regional ozone column enhancements of 20–25 DU, consistent with satellite observations from TOMS [87], and GOME [88]. The model of Duncan et al. [64] predicted an increase of the tropospheric CO column by 10–20% between 30°N and 45°S. Furthermore, large amounts of aerosols released from the 1997 Indonesian fires significantly altered the photodissociation rates of reactive compounds, and led to a noticeable reduction of the global tropospheric oxidizing power [67].

The transport patterns affecting the atmosphere over the Indian Ocean have received some attention recently, after a series of predominantly ship based experiments (INDOEX) have revealed high pollution levels in the Northern hemisphere and a sharp gradient towards the Intertropical Convergence Zone (ITCZ). Export of pollution from southern Asia is controlled largely by the Asian monsoon circulation with intense convection and large rainfall during summer and a northeasterly near-surface flow in winter (see Chap. 6). De Laat et al. [89] attribute most of the pollution in the northern part of the Indian ocean to agricultural waste and biofuel burning in India; however, emissions from the Middle East, China, and Southeast Asia also contribute. During summer, the intense convection allows for significant pollution export into the upper troposphere, where it is transported in the prevailing easterly subtropical jet and influences the Mediterranean region (see Chap. 6). Simulations with a chemistry transport model [49] indicate that an Asian pollution plume should be observable over Crete in around 25% of the time during summer.

Fires in Central America have received little attention so far, although in this region a large number of fires occur quite regularly each year in spring. Most of the pollution produced in these fires is exported over the Atlantic and Pacific ocean; however, occasional advection to the north has also been reported. Rogers and Bowman [90] use a trajectory model to analyse one particular episode in 1998, where large draught-induced fires in Mexico affected large areas of the U.S. due to enhanced southerly flow.

Summarizing this section on tropical and subtropical long-range pollution transport, we identified three dominant factors controlling the long-range transport of air pollution: (1) most of the pollution in these regions originates from biomass burning and therefore exhibits a large seasonal and interannual variability; (2) tropical deep convection plays a major role, because of rapid lofting of pollution into the middle and upper free troposphere, where it can be efficiently transported in the geostrophic flow, and because of the additional reactive nitrogen input from (continental) thunderstorms, which leads to enhanced ozone formation in the plume; (3) long-range transport of biomass burning pollution plumes occurs frequently and leads to a significant perturbation of the ozone budget over otherwise pristine oceanic regions.

5
Interactions Between Chemistry and Transport

The time scales for chemical transformations of gas-phase and aerosol pollutants in the atmosphere range from fractions of a second (i.e. for atomic and radical reactions) to several years (e.g. for methane oxidation). Transport processes occur on similar time scales; therefore it is no surprise that chemistry and transport both play a role in the pollution LRT. The turnover of atmospheric chemical reactions generally increases with increasing precursor concentrations, UV intensity, water vapour concentration, and temperature. However, the photochemical system of the atmosphere is highly non linear, and several feedbacks exist, which can distort this relationship. In a very general manner, the key chemical processes with respect to long-range transport and global ozone can be summarized as follows. (1) Near the source regions fast photochemical reactions oxidize precursor species (NO_x, CO, and VOCs) to form ozone, PAN, and other oxidants. The precursor concentrations decrease due to these photochemical reactions and because of deposition at the surface. Depending on the local weather conditions, pollution is either advected laterally within the boundary layer, or it is lofted into the free troposphere through large-scale ascent (i.e. warm conveyor belts), or through local convection (i.e. thunderstorms). (2) During this transport, soluble species (e.g. HNO_3, H_2O_2, etc.) may experience washout and heterogeneous removal. If transport occurs in the boundary layer, precursor substances are further destroyed in photochemical reactions and by deposition, and the chemical formation of secondary pollutants competes with their photochemical destruction. There is a vast body of literature analysing the conditions of net ozone formation in the polluted and rural boundary layer (cf. [91–94]). (3) Polluted air parcels which are lofted into the free troposphere experience further chemical reactions and they may be affected by additional input of pollutants (NO_x, aerosols) from lightning or aircraft traffic. (4) The lifecycle of a pollution plume typically ends with dispersion (e.g. along a blocking high pressure system) often following adiabatic or diabatic descent, which leads to the release of NO_x

from the thermal decay of reservoir species such as PAN or pernitric acid (HO_2NO_2) (cf. [74, 81, 85, 95]).

Whether or not net ozone formation occurs during the LRT depends to a large extent on the altitude of the plume trajectory. Generally, the lower temperatures (and the smaller amount of water vapour) in the upper troposphere slow down the chemical reactions so that pollution plumes can be transported over longer distances. Among the critical parameters for determining the net ozone tendency are the abundance of reactive nitrogens (NO_x) and the NO/NO_2 ratio. The colder temperatures in the upper troposphere increase the NO_x lifetime so that more ozone can be formed per NO_x molecule consumed, and the NO/NO_2 ratio increases with altitude, which shifts the photochemical balance in favour of ozone production. Because information on free tropospheric NO_x is still rather limited, there is considerable debate on whether significant amounts of additional ozone are formed during the transport of pollution plumes in the free troposphere.

Two of the first studies investigating the net photochemical tendency of ozone during transport across the Atlantic arrived at contradicting conclusions: Wild et al. [96] and Schultz et al. [95] both used a photochemical point model to follow the chemical evolution of an air parcel along a typical trajectory from the US East coast to Europe. While the ozone concentration in the free troposphere increased by a factor of two in the Wild et al. [96] study, Schultz et al. [95] found no increase in ozone, but rather a small decrease. Much of the difference between these results can be attributed to the initial concentration of ozone assumed at the start of the trajectory: [96] used a typical tropospheric background concentration of 30 ppb as a starting value, while [95] assumed significant processing of the air in the boundary layer and initiated their model with 60 ppb as an estimated value for convective outflow. According to the 3-D model by Li et al. [97] some high ozone events at Bermuda in spring can be explained either by direct export of ozone from the boundary layer over the United States, or by secondary ozone production of 4–8 ppb/day during the transport over the western North Atlantic. Reeves et al. [98] found examples of net O_3 production over the North Atlantic during two aircraft field campaigns in the spring and summer of 1997 which were either associated with in situ sources of NO or long-range transport of pollution, and Koike et al. [84] determined that net ozone formation of 1–2 ppb/day after outflow from convection over Indonesia was needed in order to explain the enhancement of ozone observed over Australia a few days later. During one of the flights of the TRACE-P campaign, a succession of Asian pollution layers were sampled across the Pacific with different levels of ozone and nitrogen species. Chemical tracer and back-trajectories analysis revealed that they all originated from a single outflow event due to a warm conveyor belt ahead of a cold front. The initial plume was split into a southern and a northern component because of a blocking high pressure system in the mid-Pacific. Heald et al. [85] used a global 3-D model in combination with CO observations from the space-borne MOPITT instrument to show that most of the reactive nitrogen remains as PAN in the northern plume while PAN decomposes to nitrogen oxides in the southern plume as it subsides, thus inducing ozone production.

Thus a number of modelling studies indicate that secondary ozone production can take place during LRT events. However, this secondary production might be limited to the regions close to the sources [99]. DiNunno et al. [99] used observed data from a number of GTE aircraft campaigns to constrain box model calculations and examine ozone photochemistry in fall and spring over the central and eastern North Pacific, a region typically influenced by the continental outflow from Asia. They found significant net ozone destruction for altitudes below 4 km and a small net production above, with a resulting net destruction integrated over the tropospheric column. Over the western Pacific the ozone tendency was either neutral or exhibited net production (depending on the season). Further away from the source regions, dispersion and chemical loss controlled the net ozone tendency.

Using a global model, Li et al. [39] estimated that the total NO_y exported (as NO_x and PAN) from the U.S. boundary layer eventually leads to an ozone production comparable to the direct export of ozone from the North American boundary layer, and Wild et al. [31, 32] estimated that the gross chemical production in the free troposphere and the boundary layer outside of the region of precursor emissions exceed the production in the continental boundary layer during their study period (spring 2001).

In order to investigate the impact of pollution plumes on the tropospheric ozone budget, we have analysed results from the global chemistry transport model MOZART-2 [21] in different oceanic regions using the local CO concentration to distinguish between background and plume conditions (Fig. 3). According to the model, the effect of pollution on the ozone budget is quite differ-

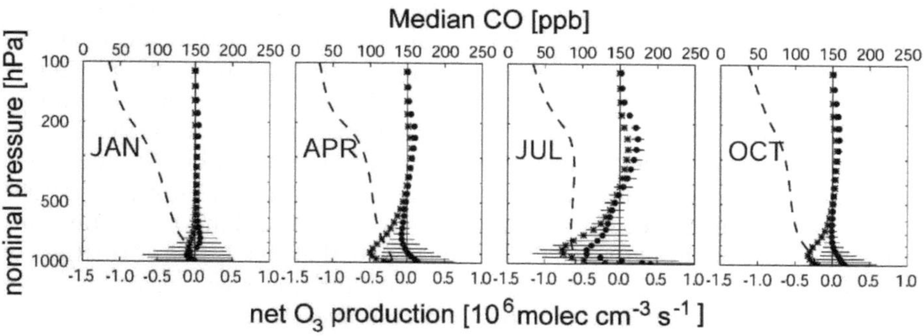

North Atlantic (45°W–15°W, 35°N–60°N)

Fig. 3 Simulated vertically resolved net photochemical ozone formation rate for different regions and seasons stratified by using the calculated CO concentration to determine "background" (*stars*) and "plume" (*bullets*) conditions.
The data used for this analysis are daily average values from a seven-year MOZART-2 chemistry transport model simulation. Background is defined as CO values below the multi-annual regional median value, and plume conditions were taken as median(CO)+1·σ(CO). The *error bars* denote one standard deviation of the net ozone formation in pollution plumes. The scatter around the background ozone formation rates is of similar magnitude but has been omitted for clarity. The *dashed line* denotes the median CO concentrations. Panels are for the North Atlantic (45°W–15°W, 35°N–60°N) (first row), North Pacific (150°E–135°W, 25°N–50°N), South Atlantic (32°W–0°, 30°S–0°), South Pacific (175°W–90°W, 30°S–0°), and Indian Ocean (55°E–95°E, 30°S–5°N). Columns are for January, April, July and October (from left to right)

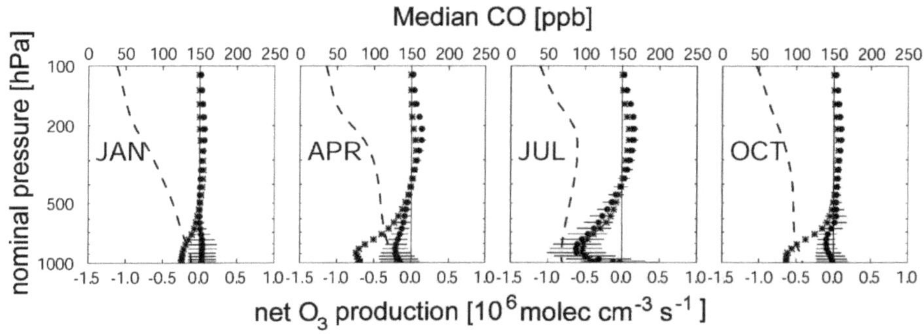

North Pacific (150°E–135°W, 25°N–50°N)

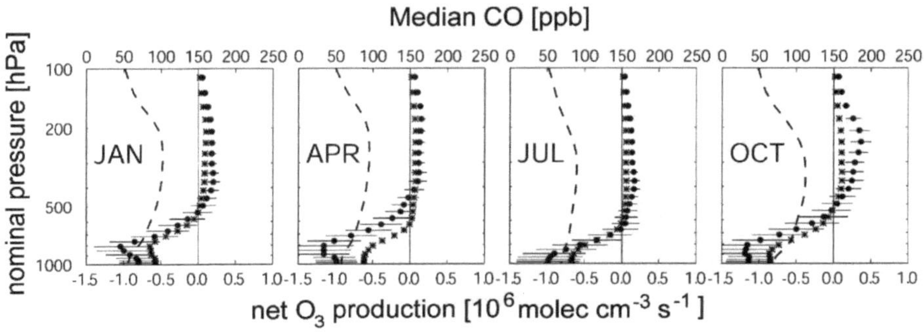

South Atlantic (32°W–0°, 30°S–0°)

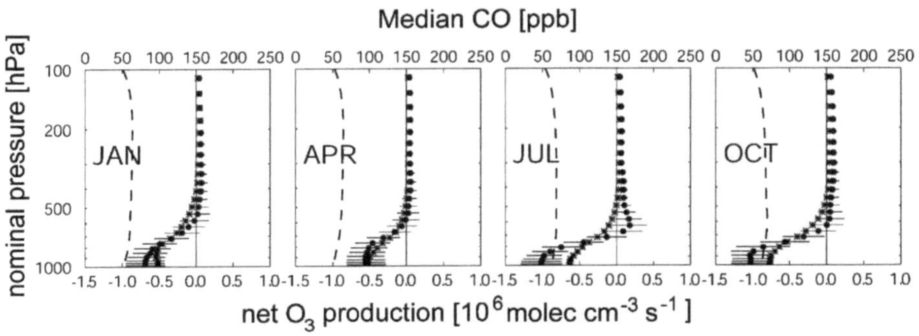

South Pacific (175°W–90°W, 30°S–0°)

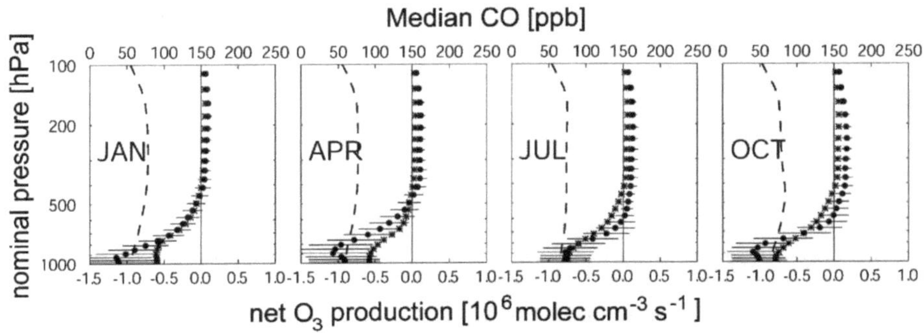

Indian Ocean (55°E–95°E, 30°S–5°N)

Fig. 3 (continued)

ent from region to region. Generally, one observes the same behaviour as discussed in [99], i.e. ozone tends to be destroyed in the lower troposphere, while there is some small net production above about 2–6 km altitude. In the northern hemisphere mid-latitudes, the pollution export from Asia onto the Pacific or from North America onto the Atlantic always increases the net formation rate of ozone (except for the northern Pacific in summer). In contrast, biomass burning pollution over the southern subtropical oceans tends to enhance the ozone destruction in the lower troposphere, whereas enhanced formation can be observed above. These differences point to differences in the pollution mix entering the different regions. Due to the somewhat lower temperatures and less available UV radiation (and possibly also due to faster export mechanisms), more precursors are conserved in the northern mid latitude plumes where they can contribute to the gross ozone production. Another important factor may be the higher NO_x/CO ratio predicted for the northern hemisphere marine boundary layer, which is in part due to the large amount of NO_x emissions from ships [44].

Ozone formation in the troposphere is controlled largely by the availability of NO_x (cf. [3, 100]). Emission source strengths of NO_x in the combustion of fossil fuels and from biomass burning are still very uncertain, and there are also significant surface emissions from soils and vegetation. In the middle and upper troposphere, aircraft emissions and the formation of NO_x in lightning flashes, which maximize in tropical latitudes [101, 102], provide additional NO_x input. Moxim and Levy [103], and Thompson et al. [67] attributed a significant fraction of the observed enhancements of the tropospheric ozone column over the Southern Atlantic to tropical lightning. Staudt et al. [75] highlight its importance for the Southern Pacific ozone budget, and Duncan et al. [104] suggest that NO_x from lightning was more important for ozone formation during the 1997 Indonesian fire episode than the release of NO_x from the fires themselves. While it is undisputed that lightning NO_x does have an influence on tropospheric ozone concentrations, there is still considerable uncertainty in the quantification of its source strength. Current estimates range from 2–12 Tg N/year [105, 106]. The intensity of lightning depends on cloud parameters such as the maximum updraft velocity, which also controls the tracer transport from the boundary layer into the middle and upper troposphere. The altitude of the convective outflow has important implications for the transport speed and direction, and it also determines the chemical lifetime of the lofted pollutant species.

6 Interannual Variability of Long-Range Transport

The examination of the interannual variability of the LRT is made difficult by the lack of long-term data. Most of the knowledge we have so far concerning interannual variability has come from modelling studies. Changes in the meteorology and changes in the emissions over a given region appear to be the key factors that both contribute to interannual variability of long-range transport of pollution [48, 107, 108].

In the Northern Hemisphere, the North Atlantic Oscillation (NAO) was found to contribute significantly to the year-to-year variability of long-range transport processes, especially in the North Atlantic region. Li et al. [1] conducted a five-year (1993–1997) simulation and found a significant correlation (0.57 for all seasons and 0.72 in spring) between the NAO index and the ozone concentrations originating from the North American boundary layer and arriving at the western tip of Europe. The NAO also influences the strengths and locations of the export pathways for European pollution. In their 11-year simulation (1987–1997), Duncan and Bey [48] found that export of European pollution is more frequent when the NAO is in a positive phase, as the winds are then stronger over Europe. Consequently, the CO burden from the European sources tends to be lower (higher) over the North Atlantic Ocean and higher (lower) over the Arctic, when the NAO is in a positive (negative) phase. Such a signal is also seen in the remotely-sensed NO_2 vertical columns from GOME, which show enhanced northward NO_2 transport and reduced NO_2 outflow into the North Atlantic from Central Europe during positive phases of NAO [109]. Year-to-year changes in pollution export from Asia seem to be influenced by the El Niño Southern Oscillation. Jones et al. [110] qualitatively discuss correlations between the long-range transport of CO and aerosols from biomass burning sources based on a five-year data set of observations at Lauder, New Zealand. Both species exhibit a seasonal cycle consistent with biomass burning emissions, and they show a strong dependence on the phase of the El Niño Southern Oscillation. It must be noted, however, that the frequency and intensity of tropical savanna and forest burning also vary with drought intensity associated with El Niño [68]. Liu et al. [29] used a global model to examine changes in the continental outflow from East Asia in four different years. They found that the dominant patterns for continental outflow are similar over various years but they observed a significant variability related to El Niño. For example, the Spring 2001 (corresponding to a La Niña period) was characterized by unusually frequent cold fronts passing over the Pacific Rim and by strong convection in South East Asia, which, in turn, leads to unusually strong boundary layer outflow and convective outflow into the upper troposphere.

Changes in emissions also induce some variability to LRT. Duncan and Bey [48] found for example that the decrease in the "European" CO burden follows nearly linearly the decrease in European CO emissions. This results in a significant downward trend in CO concentrations observed at a number of sites in the mid and high-latitudes of the Northern Hemisphere that are affected by pollution transport from Europe [111]. While fossil fuel emission changes tend to follow a smooth trend rather than showing much interannual variability, this is different for biomass burning emissions, in particular from temperate and boreal forest fires [45]. Several atmospheric observations have indicated that biomass burning from boreal fires significantly impact the distributions of pollutants over the Atlantic Ocean and even Europe [41,42] but quantification remains difficult. As of now, no modelling study exists, where the interannual variability of long-range pollution transport in the Southern hemisphere has been analysed in a systematic manner. In part, this reflects the still very large uncertainties in the estimates of biomass burning emissions and their interannual variability.

The interannual variability in transport patterns further complicates the analysis of the impact of biomass burning on tropospheric ozone. For example,

Chandra et al. [112] have found that 50% of the tropospheric ozone column enhancements during the 1997/98 Indonesian fire episode were due to El Niño related transport changes.

7
Conclusions: Current Limitations of Global Models and Future Developments

Chemical models, and especially global 3-D atmospheric chemistry transport models, have contributed significantly to our understanding of the processes related to the LRT of pollution. In particular, global models have provided insights into the pathways contributing to LRT, the potential impact of emissions from one continent on the pollutant concentrations over another region downwind, and on the influence of LRT on the global tropospheric composition. The frequency and intensity of LRT events shows great variability from year to year. Global modelling studies indicate that some of this variability is due to pressure gradient changes related to quasi-periodic climate variability patterns such as ENSO or NAO. On the other hand, some recent studies have also found a large interannual variability in the emission source strength, in particular for biomass burning emissions.

In the northern extra-tropics, the LRT of anthropogenic emissions from North America and Asia (fossil and biofuel combustion) and Europe dominates, although boreal forest fires can have a significant impact in certain years. Table 1 (adapted from [2]) gives an overview of current estimates of the impact of ozone precursor emissions from the main anthropogenic source regions (i.e. North America, Asia, Europe) on regions downwind. Typically, "foreign" emissions enhance ozone levels at the surface by about 5 ppb on average, but up to 15 ppb during events. If emissions continue to rise in the future, the intercontinental impacts would become even more severe. For example, Asian emissions could lead to occasional ozone enhancements of up to 30–40 ppb over the western U.S. in spring [28].

Pollution in the southern hemisphere originates mainly from savanna and forest fires in the tropical and subtropical latitudes. Practically all global models agree that tropical biomass burning sources have a large impact on primary pollutants such as tropospheric CO. Recent developments in emissions inventories allow for a better spatial and temporal distribution of fires by using satellite observations of smoke aerosols [64], active fires [85, 107, 120], or burnt areas [65]. These new data sets significantly improve the simulation of specific biomass burning related transport events [104, 121]. However, there is still great uncertainty about the role of vegetation fires for the global tropospheric ozone budget.

LRT has direct implications on the global tropospheric composition, and in particular on the global budget of ozone, although this still eludes robust quantification at present. The photochemical formation and destruction of ozone is closely linked to meteorological parameters. In general, atmospheric reactions are faster in the planetary boundary layer and slower in the middle and upper free troposphere, so that pollution signatures are preserved over longer distances at higher altitudes. While the mixing ratio of primary pollutants, such as CO, or

Table 1 Surface ozone enhancements at northern mid-latitudes from intercontinental transport of pollution (reproduced from [2] with permission)

Source region	Receptor region	O_3 enhancement (ppbv)	Method of estimate	Reference
Asia	Northwestern U.S., spring	4 (mean), 7.5 (max)	Sensitivity simulation with no anthropogenic emissions from source region	Berntsen et al. [113]
Asia	Western U.S., spring	3–10 (range during Asian pollution events)	Sensitivity simulation with no surface emissions from source region	Yienger et al. [28]
Asia	Europe, U.S.	1.0 (U.S.), 0.8 (Europe)	Annual mean enhancements from sensitivity simulations with 10% increases in emissions from source region; results were multiplied by 10 to estimate total effect of current anthropogenic emissions from the source continent	Wild and Akimoto [40]
Europe	Asia, U.S.	1.1 (Asia), 0.9 (U.S.)		
U.S.	Europe, Asia	2.0 (Europe), 0.8 (Asia)		
Asia+ Europe	U.S., summer	4–7 (typical afternoon range), 14 (max)	Sensitivity simulation with no anthropogenic NO_x and NMVOC emissions from source region	Fiore et al. [114]
Europe	East Asia, spring	3 (daytime mean)		Liu et al. [29]
North America	Europe, summer	2–4 (daytime mean), 5–10 (events)		Li et al. [1]
Europe	East Siberia	2 (annual) 3 (spring-summer)	Difference between median observed O_3 concentrations in 1997–1999 for air masses originating from Europe vs. from Siberia and high latitudes	Pochanart et al. [115]
North America	Mace Head, Ireland	0.4 (winter), 0.2 (spring), –0.3 (summer), –0.9 (fall)	Mean observed difference in O_3 concentrations in 1990–1994 for air masses originating from the U.S. and Canada vs. from Iceland and Greenland	Derwent et al. [116]
North America	Europe, yearly mean	18 (Atlantic fringes), 10–15 (central Europe)	Ozone produced in tropospheric column over source region	Derwent et al. [117]
Asia	Europe, yearly mean	9 (Atlantic fringes), 5–7 (central Europe)		

Table 1 (continued)

Source region	Receptor region	O_3 enhancement (ppbv)	Method of estimate	Reference
Background (anthropogenic methane)	U.S., summer	6 (afternoon mean)	Sensitivity simulation with anthropogenic CH_4 emissions reduced globally by 50%; we doubled O_3 enhancements from that simulation were doubled to estimate total enhancement from anthropogenic CH_4	Fiore et al. [114]
Background (1980–1998)	U.S.	3–5 (spring, fall)	Observed trend in the lower quantiles of the O_3 frequency distribution at rural sites	Lin et al. [118]
Background (1984–2002)	U.S. west coast	10	Observed trend at surface sites and from aircraft missions (1984–2002)	Jaffe et al. [119]
Asia (future)	U.S.	2–6 (western U.S) 1–3 (eastern U.S.) highest in April–June	Sensitivity simulation with tripled Asian NO_x and NMVOC emissions	Jacob et al. [11]
Asia (future)	Western U.S., spring	30–40 (max during Asian pollution events)	Sensitivity simulation with quadrupled Asian emissions	Yienger et al. [28] Adapted from[2]

All estimates are from global 3-D models except those of [115, 116, 118, 119]

VOCs, generally decreases during transport in the atmosphere, there are species, such as ozone, for which secondary formation occurs in the atmosphere. Many field studies have found ozone formation and loss rates close to photochemical equilibrium in LRT plumes, indicating that the enhanced production of ozone due to higher precursor concentrations balances the greater loss caused by the higher ozone concentrations themselves.

Much of the uncertainty in the global budget of tropospheric ozone is related to the global budget of NO_x and the estimates of its emission sources at the ground as well as from lightning flashes and aircraft. Deep convection plays a very important role for the vertical lofting of surface pollutants. Due to the high local variability and episodic nature of convection, and because of the correlation between thunderstorm intensity and lightning activity (and thus NO_x formation), it is extremely difficult to separate the contributions of surface emissions and thunderstorms from each other.

The physical parameterisations constitute probably the most important weakness of current global models, which can at present only afford grid sizes of about 100 km and thus require semi empirical formulations for subscale processes like convection and boundary layer diffusion. The relatively coarse resolution of today's models, and their inability to resolve sharp orographic features, could also lead to a bias for estimates of pollution export fluxes in certain regions of the world. For example it is unclear, whether the export of pollution due to orographic lifting [51] is of global significance. Wild et al. [31, 32] have thus raised the question whether current models are able to capture quantitatively the impacts of ozone precursor emissions. This question is related to the quality of the subgrid scale transport parameterisations, because the rate of precursor oxidation and ozone formation depends non-linearly on the chemical mix as well as on the rate of dilution and surface deposition. As a general statement, one can say that resolution matters mostly where strong gradients in temperature, winds, or trace gas and aerosol concentrations are observed. Besides the boundary layer, this is also of particular importance for the upper troposphere and lower stratosphere region, and the ozone exchange between the two. This process represents one of the major uncertainties in today's estimates of the global tropospheric ozone budget, and it also impedes the reliability with which we can assess the chemical impact of intercontinental pollution transport. With increasing computer resources it can be expected that the horizontal and vertical resolution of models will be further refined.

Another weakness of today's global three-dimensional models, which warrants further improvement, concerns the representation of the chemical processes. In reality, thousands of VOCs with different oxidation rates and reaction products are emitted from fossil fuel or biomass combustion, while only a very limited number of species are taken into account in the chemical mechanisms used in current global models. The level of detail with which global models represent VOC oxidation might thus not be adequate [4].

The global impact of aerosols (which act as sites for heterogeneous reactions and modify photolysis rates) on the distributions of trace gases is still poorly known and only poorly represented in global models even though there are some indications that they can significantly change the global and regional budgets of key tropospheric species such as OH and ozone. Dentener et al. [122] found for example that uptake of O_3, N_2O_5 and HO_2 radicals on mineral dust could lead to a 10% decrease of O_3 concentrations in regions with a high load of mineral dust. Martin et al. [123] incorporated both the radiative and the heterogeneous effects of aerosol dust in a simulation of tropospheric O_3 and determined that these effects have little impact on global O_3 distributions (because of competing effects) but can have significant regional impacts.

Current model development efforts are therefore directed towards a full interactive coupling between the atmospheric chemistry and the physical atmosphere models, and also between aerosol dynamics and chemistry and gas-phase chemistry. Atmospheric observations, both in-situ and from remote sensing, are of paramount importance to validate these models, and they are also increasingly used to improve the estimates of pollutant emissions. Global models play an important role for integrating and interpreting these observations, but their main task will remain the prediction of future changes in the climate system and atmospheric composition change, and the assessment of the consequences for mankind.

Acknowledgements The authors wish to thank Bryan Duncan and Yutaka Kondo for providing unpublished material, and Birgit Grodtmann for help with editing the manuscript. Martin Schultz acknowledges funding from the German BMBF under contract 07ATF39.

8 References

1. Li Q, Jacob DJ, Bey I, Palmer PI, Duncan BN, Field BD, Martin RV, Fiore AM, Yantosca RM, Parrish DD, Simmonds PG, Oltmans SJ (2002) J Geophys Res 107:10.1029/2001JD001422
2. Holloway T, Fiore AM, Hastings MG (2003) Intercontinental transport of air pollution: will emerging science lead to a new hemispheric treaty? Environ Sci Technol 37:4535–4542
3. Prather M et al. (2003) Geophys Res Lett 30:10.1029/2002GL016285
4. Volz-Thomas A, Beekman M, Derwent R, Law K, Lindskog A, Prevot A, Roemer M, Schultz M, Schurath U, Solberg S, Stohl A (2003) In: Eurotrac Synthesis and Integration Report, vol 1. Margraf, Weikersheim, chap 3
5. Prospero JM (1999) Proc Nat Acad Sci USA 96:3396
6. Prospero JM, Olmez I, Ames M (2001) Water Air Soil Pollut 125:291
7. Prospero JM, Savoie DL, Arimoto R (2003) J Geophys Res 108:10.1029/2001JD001524
8. Husar RB, Tratt DM, Schichtel BA, Falke SR, Li F, Jaffe D, Gassó S, Gill T, Laulainen NS, Lu F, Reheis MC, Chun Y, Westphal D, Holben BN, Gueymard C, McKendry I, Kuring N, Feldman GC, McClain C, Frouin RJ, Merrill J, DuBois D, Vignola F, Murayama T, Nickovic S, Wilson WE, Sassen K, Sugimoto N, Malm WC (2001) J Geophys Res 106:18317
9. Wania F, Mackay D (1993) Ambio 22:10
10. Klecka G, Boethling R, Franklin J, Grady L, Graham D, Howard PH, Kannan K, Larson RJ, Mackay D, Muir D, van de Meent D (eds) (2000) Evaluation of persistence and long-range transport of organic chemicals in the environment. SETAC Press, Pensacola, USA, pp 362
11. Jacob DJ, Logan JA, Murti PP (1999) Geophys Res Lett 26:2175
12. Houghton JT et al. (eds) (2001) IPCC: Intergovernmental Panel on Climate Change, 3rd Assessment Report. Climate change, the scientific basis. Cambridge University Press
13. Zimmermann PH, Feichter J, Crutzen PJ (1988) Chem Geol 70:105
14. Müller JF, Brasseur GP (1995) J Geophys Res 100:16445
15. Berntsen T, Isaksen ISA, Wang WC, Liang XZ (1996) Tellus 48B:13
16. Wang Y, Jacob D, Logan J (1998) J Geophys Res 103:10713
17. Brasseur GP, Hauglustaine DA, Walters S, Rasch PJ, Muller JF, Granier C, Tie XX (1998) J Geophys Res 103:28265
18. Bey I, Jacob DJ, Yantosca RM, Logan JA, Field B, Fiore AM, Li Q, Liu H, Mickley LJ, Schultz MG (2001) J Geophys Res 106:23073
19. Mickley LJ, Murti PP, Jacob DJ, Logan JA, Rind D, Koch D (1999) J Geophys Res 104: 30153
20. Lawrence MG, Crutzen PJ, Rasch PJ et al. (1999) J Geophys Res 104:26245
21. Horowitz et al. (2003) A global simulatin of tropospheric ozone and related tracers: description and evalation of MOZART, version 2. J Geophys Res 108(D24)4784, doi: 10.1029/2002JD002853
22. Wesley ML (1989) Atmos Environ 23:1293
23. Giorgi F, Chameides WL (1985) J Geophys Res 90:7872
24. Jacobson MZ (1999) Fundamentals of atmospheric modeling. Cambridge University Press
25. Jacob DJ, Crawford JH, Kleb MM, Connors VS, Bendura RJ, Raper JL, Sachse GW, Gille JC, Emmons L, Heald CL (2003) J Geophys Res 108:9000, doi: 10.1029/2002JD003276
26. Bey I, Jacob DJ, Logan JA, Yantosca RM (2001) J Geophys Res 106:23097
27. Zhang Y, Carmichael GR (1999) J Appl Meteorol 38:353
28. Yienger JJ, Galanter M, Holloway TA, Phadnis MJ, Guttikunda SK, Carmichael GR, Moxim WL, Levy HII (2000) J Geophys Res 105:26931
29. Liu H, Jacob DJ, Bey I, Yantosca RM, Duncan BN, Sachse GW (2003) J Geophys Res 108: 10.1029/2002JD003102

30. Uno I, Carmichael GR, Streets DG, Tang Y, Yienger JJ, Satake S, Wang Z, Woo JH, Guttikunda S, Uematsu M, Matsumoto K, Tanimoto H, Yoshioka K, Iida T (2003) J Geophys Res 108:10.1029/2002JD002845
31. Wild O, Sundet JK, Prather MJ, Isaksen ISA, Akimoto H, Browell EV, Oltmans SJ (2003) Chemical transport model ozone simulations for spring 2001 over the western Pacific: comparisons with TRACE-P lidar, ozonesondes, and total ozone mapping spectrometer columns. J Geophys Res 108 (D21), 8826, doi:10.1029/2002JD003283
32. Wild O, Prather MJ, Akimoto H, Sundet JK, Isaksen ISA, Crawford JH, Davis DD, Avery MA, Kondo Y, Sachse GW, Sandholm ST (2003) J Geophys Res (in press) doi: 10.1029/2003JD004041
33. Jaeglé L, Jaffe DA, Price HU, Weiss-Penzias P, Palmer PI, Evans MJ, Jacob DJ, Bey I (2004) Sources and budgets for CO and O_3 in the Northeastern Pacific during the spring of 2001: Results from the PHOBEA-II Experiment, J Geophys Res (in press)
34. Stohl A, Trickl T (2001) IGACtivities Newsletter NARE 24:10
35. Jacob DJ, Logan JA, Gardner GM, Yevich RM, Spivakovsky CM, Wosfy SC, Sillman S, Prather MJ (1993) J Geophys Res 98:14817
36. Horowitz LW, Liang J, Gardner GM, Jacob DJ (1998) J Geophys Res 103:13451
37. Liang J, Horowitz LW, Jacob DJ, Wang Y, Fiore AM, Logan JA, Gardner GM, Munger JW (1998) J Geophys Res 103:13435
38. Stohl A, Trainer M, Ryerson TB, Holloway JS, Parrish DD (2002) J Geophys Res 107:10.1029/2001JD000519
39. Li Q, Jacob DJ, Parrish DD, Munger JW (2003) J Geophys Res (submitted)
40. Wild O, Akimoto H (2001) J Geophys Res 106:27729
41. Forster C et al. (2001) J Geophys Res 106:22807
42. Spichtinger N et al. (2001) Geophys Res Lett 28:4579
43. Waibel AE, Fischer H, Wienhold FG, Siegmund PC, Lee B, Strom J, Lelieveld J, Crutzen PJ (1999) Chemosphere-Global Change 1:223
44. Olivier JGJ, Berdowski JJM (2001) Global emission sources and sinks. In: Berdowski JR, Guicherit R, Heij BJ (eds) The climate system. Swets & Zeitlinger Publishers, Lisse, pp 33–77
45. Wotawa G, Novelli P, Trainer M, Granier C (2001) Geophys Res Lett 28:4575
46. Stohl A, Eckhardt S, Forster C, James P, Spichtinger N (2002) J Geophys Res 107:10.1029/2001JD001396
47. Lelieveld J et al. (2002) Science 298:794
48. Duncan B, Bey I (2003) J Geophys Res (submitted)
49. Roelofs GJ, Scheeren HA, Heland J, Ziereis H, Lelieveld J (2003) Atmos Chem Phys 3:1199
50. Hov O, Flatoy F (1997) J Atmos Chem 28:319
51. Prévôt ASH, Dommen J, Bäumle M (2000) Atmos Env 34:4719
52. Pinto JP, Yung YL, Rind D, Russel L, Lerner JA, Hansen JE, Hameed S (1983) J Geophys Res 88:3691
53. Seiler W, Crutzen OJ (1980) Clim Change 2:204
54. Fishman J, Brackett VG, Fakhruzzanman K (1992) J Atmos Terr Phys 54:589
55. Fishman J, Hoell JM, Bendura RD, McNeal RJ, Kirchhoff VWJH (1996) J Geophys Res 101:23865
56. Hao WM, Liu MH, Crutzen PJ (1990) In: Goldammer JC (ed) Fire in the tropical biota, pp 440–462
57. Andreae MO et al. (2001) Geophys Res Lett 28:951
58. Hao WM, Liu MH (1994) Global Biogeochem Cycl 8:495
59. Cooke WC, Wilson JJN (1996) J Geophys Res 101:19395
60. Levine JS (1996) Biomass burning and global change, vols 1 and 2. MIT Press, Cambridge, London
61. Liousse C, Penner JE, Huang CC, Walton JJ, Eddleman H, Cachier H (1996) A global three-dimensional model of carbonaceous aerosols. J Geophys Res 101:19411–19432
62. Galanter M, Levy H II, Carmichael GR (2000) J Geophys Res 105:6633
63. Lavoué D, Liousse C, Cachier H, Stocks B, Goldammer J (2000) J Geophys Res 105:26871

64. Duncan BN, Martin RV, Staudt AC, Yevich R, Logan JA (2003) J Geophys Res 108: 10.1029/2002JD002378
65. Hoelzemann JJ, Brasseur GP, Granier C, Schultz MG, Simon M (2003) J Geophys Res (submitted)
66. Liousse C, Andreae MO, Artaxo P, Barbosa P, Cachier H, Grégoire J-M, Hobbs P, Lavoué D, Mouillot D, Penner J, Scholes M, Schultz MG (2004). In: Granier C, Artaxo P, Reeves CE (eds) Emissions of atmospheric trace compounds, chap 3. Kluwer Academic Publishers, Dordrecht, The Netherlands (in press)
67. Thompson AM, Doddridge BG, Witte JC, Hudson RD, Luke WT, Johnston JE, Johnston BE, Oltmans SJ, Weller R (2000) Geophys Res Lett 27:3317
68. Barbosa PM, Stroppiana D, Grégoire J-M, Pereira JMC (1999) Global Biogeochem Cycl 13:933
69. Schultz MG (2002) Atmos Chem Phys 2:387
70. Hoell et al. (1999) J Geophys Res 104:5567
71. KondoY, Ko M, Koike M, Kawakami S, Ogawa T (2002) J Geophys Res 107: 10.1029/2001JD000820
72. Chatfield RB, Guo Z, Sachse GW, Blake DR, Blake NJ (2002) J Geophys Res 107: 10.1029/2001JD000497
73. Chatfield RB, Vastano JA, Li L, Sachse GW, Connors VS (1998) J Geophys Res 103:28059
74. Staudt AC, Jacob DJ, Logan JA, Bachiochi D, Krishnamurti TN, Sachse GW (2001) J Geophys Res 106:32571
75. Staudt AC, Jacob DJ, Logan JA, Bachiochi D, Krishnamurti TN, Poisson N (2002) J Geophys Res 107:10.1029/2000JD000296
76. Diab RD, Jury MR, Combrink J, Sokolic F (1996) J Geophys Res 101:23809
77. Garstang M, Tyson PD, Swap R, Edwards M, Kallberg P, Lindesay JA (1996) J Geophys Res 101:23721
78. Tyson PD, Garstang M, Thompson AM, D'Abreton P, Diab RD, Browell EV (1996) J Geophys Res 102:10623
79. Fenn et al. (1999) J Geophys Res 104:16197
80. Gregory et al. (1999) J Geophys Res 104:5677
81. Schultz MG, Jacob DJ, Wang Y, Logan JA, Atlas E, Blake DR, Blake NJ, Bradshaw JD, Browell EV, Fenn MA, Flocke F, Gregory GL, Heikes BG, Sachse GW, Sandholm ST, Shetter RE, Singh HB, Talbot RW (1999) J Geophys Res 104:5829
82. Blake NJ et al. (1999) J Geophys Res 104:16213
83. KondoY, Koike M, Kita K, Ikeda H, Takegawa N, Kawakami S, Blake D, Liu SC, Ko M, Miyazaki Y, Irie H, Higashi Y, Liley B, Nishi N,Zhao Y, Ogawa T (2002) J Geophys Res 107: 10.1029/2001JD000820
84. Koike M, Kondo Y, Akutagawa D, Kita K, Nishi N, Liu SC, Blake DR, Kawakami S, Takegawa N, Ko M, Zhao Y, Ogawa T (2002) J Geophys Res 107:10.1029/2001JD000823
85. Heald CL, Jacob DJ, Fiore AM, Emmons L, Gille JC, Sachse GW, Browell EV, Avery MA, Vay SA, Crawford JH, Westberg DJ, Blake DR, Singh HB, Sandholm ST, Talbot RW, Fuelberg HE (2003) J Geophys Res (2004) (submitted)
86. Hauglustaine DA, Brasseur GP, Levine JS (1999) Geophys Res Lett 21:3305
87. Chandra S, Ziemke JR, Min W, Read WG (1998) Geophys Res Lett 25:3867
88. Burrows JP, Weber M, Buchwitz M, Rozanov VV, Ladstätter-Weißenmayer A, Richter A, deBeek R, Hoogen R, Brahmstedt K, Eichmann KU, Eisinger M, Perner D (1999) J Atmos Sci 56:151
89. De Laat ATJ, Lelieveld J, Roelofs GJ, Dickerson RR, Lobert JM (2001) J Geophys Res 106:28481
90. Rogers CM, Bowman KP (2001) J Geophys Res 106:28357
91. Lin X, Trainer M, Liu SC (1988) J Geophys Res 93:15879
92. Chameides WL et al. (1992) J Geophys Res 97:6037
93. Sillman S, Logan JA, Wofsy SC (1990) J Geophys Res 95:1837
94. Sillman S, He DY (2002) J Geophys Res 107:10.1029/2001JD001123
95. Schultz M, Schmitt R, Thomas K, Volz-Thomas A (1998) J Geophys Res 103:13477

96. Wild O et al. (1996) J Geophys Res
97. Li Q, Jacob DJ, Fairlie TD, Liu H, Yantosca RM, Martin RV (2002) J Geophys Res 10.1029/2002JD00213
98. Reeves CE et al. (2002) J Geophys Res 107:10.1029/2002JD002415
99. DiNunno, Davis BD, Chen G, Crawford J, Olson J, Liu S (2003) J Geophys Res 108:8237
100. Schultz MG, Diehl T, Brasseur GP, Zittel W (2003) Air pollution and climate impacts of a global hydrogen economy. Science 302:624–627
101. Orville RE, Henderson RW (1986) Mon Weather Rev 114:2640
102. Christian HJ et al. (2003) J Geophys Res 108:10.1029/2002JD002347
103. Moxim WJ, Levy HII (2000) J Geophys Res 105:17393
104. Duncan BN, Bey I, Chin M, Mickley LJ, Fairlie TD, Martin RV, Matsueda H (2003) J Geophys Res (submitted)
105. Price C, Penner J, Prather M (1997) J Geophys Res 102:5929
106. Lawrence MG, Chameides WL, Kasibhatla PS, Levy H II, Moxim W (1995) In: Volland H (ed) Handbook of atmospheric electrodynamics. CRC Press, Boca Raton, pp 189–202
107. Schultz MG, Brasseur GP, Feichter J, Granier C, Hoelzemann J, Rast S (2002) Analysis of factors controlling the variability and trends of the tropsopheric chemical composition. International Global Atmospheric Chemistry Conference, Hersonisos, Crete, September 18–25, 2002
108. Karlsdottir S, Isaksen I, Myhre G, Berntsen T (2000) J Geophys Res 105:28907
109. Eckhardt S, Stohl A, Beirle S, Spichtinger N, James P, Forster C, Junker C, Wagner T, Platt U, Jennings SG (2003) Atmos Chem Phys Discuss 3:3223
110. Jones NB, Rinsland CP, Liley JB, Rosen J (2001) Geophys Res Lett 28:709
111. Novelli PC, Masarie KA, Lang PM, Hall BD, Myers RC, Elkins JW (2003) J Geophys Res 108:10.1029/2002JD003031
112. Chandra S, Ziemke JR, Bhartia PK, Martin RV (2002) J Geophys Res 107: 10.1029/2001JD000447
113. Berntsen TK, Karlsdottir S, Jaffe DA (1999) Geophys Res Lett 26:2171
114. Fiore AM, Jacob DJ, Bey I, Yantosca RM, Field BD, Wilkinson JG (2002) J Geophys Res 107:10.1029/2001JD000982
115. Pochanart P, Akimoto H, Kajii Y, Potemkin VM, Khodzher TV (2003) J Geophys Res 108:10.1029/2001JD001412
116. Derwent RG, Simmonds PG, Seuring S, Dimmer C (1998) Atmos Environ 32:145
117. Derwent RG, Stevenson DS, Collins WJ, Johnson CE (2002) EPA/EMEP Workshop on Hemispheric Pollution, Bad Breisig, Germany
118. Lin CY, Jacob DJ, Munger JW, Fiore AM (2000) Geophys Res Lett 27:3465
119. Jaffe D, Price H, Parrish D, Goldstein A, Harris J (2003) Geophys Res Lett 30: 10.1029/2003GL017024
120. Generoso S, Bréon F-M, Balkanski Y, Boucher O, Schulz M (2003) Atmos Chem Phys 3:1211
121. Chin M, Ginoux P, Kinne S et al. (2002) J Atmos Sci 59:461
122. Dentener FJ, Carmichael GR, Zhang Y, Lelieveld J, Crutzen PJ (1996) J Geophys Res 101: 22869
123. Martin RV, Jacob DJ, Yantosca RM, Chin M, Ginoux P (2003) J Geophys Res 108: 10.1029/2002JD002622

The Handbook of Environmental Chemistry Vol. 4, Part G (2004): 225–253
DOI 10.1007/978-3-540-40037-0

Intercontinental Transport of Ozone from Tropical Biomass Burning

Anne M. Thompson

NASA Goddard Space Flight Center, Laboratory for Atmospheres, Greenbelt, Maryland 20771, USA
E-mail: thompson@gator1.gsfc.nasa.gov

Abstract Photochemically reactive gases released by fires (e.g., NO, CO, hydrocarbons) interact as they do in an urban environment to form ozone. There has been interest in the impacts of tropical biomass fires on ozone since about 1980. Evidence for intercontinental transport of ozone from tropical burning appeared in the mid-1980s. Tropical meteorology plays a part in tropospheric ozone distributions in the tropics through interactions of pollution with large-scale circulation, deep convection, regional phenomena (e.g., Atlantic subsidence, the West African and Asian monsoons). Characteristics of dynamical patterns affecting tropical tropospheric ozone are described. Selected observations, mostly based on satellite views and soundings, illustrate regional influences and intercontinental-scale transport over the Indian Ocean region, the Pacific and the Atlantic.

Keywords Tropospheric ozone · Intercontinental transport · Biomass burning · Africa · Indian Ocean · Atlantic Ocean · Pacific Ocean · Convection

1 Introduction

1.1 Chemistry

The atmospheric impacts of tropical fires came to attention in the 1970s, largely through the work of Crutzen, in relationship to anthropogenic carbon dioxide and other long-lived greenhouse gases [1,2]. It was realized that photochemically reactive gases released by fires (e.g., NO, CO, hydrocarbons [HC]) would interact as they do in an urban environment to form ozone, a greenhouse gas. Several sets of observations stimulated research into the relationship between biomass fires and ozone. A Brazilian field campaign in the 1980 dry season measured ozone and emissions of reactive gases [3,4]. Surface measurements made in west Africa were among the first to demonstrate a seasonality in ozone [5] that could be explained by savanna fires. Table 1 shows the "smog"-type reactions involved in ozone formation. Italicized are those species measured during various experiments (Table 2) that contributed to an understanding of ozone photochemistry and biomass burning sources.

Table 1 Photochemical reactions linking methane, NMHC, CO, and NO with ozone

OH forms from ozone photolysis

$O_3 + h\nu \rightarrow O(^1D) + O_2$
$O(^1D) + H_2O \rightarrow OH + OH$

Methane oxidation

$OH + CH_4\,(+ O_2) \rightarrow CH_3O_2 + H_2O$
$CH_3O_2 + NO \rightarrow NO_2 + CH_3O$
(Form formaldehyde: $CH_3O(+ O_2) \rightarrow HCHO$)
$HCHO + h\nu \rightarrow H_2 + CO \leftarrow$ formation of CO
$HCHO + h\nu(+ O_3) \rightarrow HO_2 + CHO$

NMHC oxidation

$OH + NMHC\,(+ O_2) \rightarrow RO_2 + H_2O$
$RO_2 + NO \rightarrow NO_2 + RO$

CO oxidation by OH produces HO_2

$OH + CO(+ O_2) \rightarrow CO_2 + HO_2$

Conversion of NO to NO_2 by HO_2, RO_2, CH_3O_2

$CH_3O_2 + NO \rightarrow NO_2 + CH_3O$
$RO_2 + NO \rightarrow NO_2 + RO$
$HO_2 + NO \rightarrow NO_2 + OH$

Formation of O_3

$NO_2 + h\nu \rightarrow O + NO$
$O + O_2(+ M) \rightarrow O_3 + M$

Table 2 Ozone observations (campaigns, networks, datasets)

Aerosols99 (cruise)	
ABLE2	– Amazon Boundary Layer Experiment (A=1985, B=1987)
BIBLE	– Biomass Burning and Lightning Experiment (A=1998, B=1999)
CARIBIC	– Civil Aircraft for Regular Investigation on the Atmosphere based on an Instrument Container (1997–2001)
DECAFE	– Dynamique et Chimie Atmosphère des Forêts Equatoriale (1988)
EXPRESSO	– Experiment for Regional Sources and Sinks of Oxidants (1996)
FOS	– Fires in Savannas (1991)
GASP	– Global Atmospheric Sampling Project (1977)
INDOEX	– Indian Ocean Experiment (1999)
MOPITT	– Measurements of Pollution in the Troposphere (2000–)
MOZAIC	– Measurements of Ozone and water vapor by Airbus In-service Aircraft (1995–)
PEM	– Pacific Exploratory Mission (1991, 1994, 1996, 1999)
SAFARI-1992	– Southern African Fire Atmospheric Research Initiative
SAFARI-2000	– Southern African Atmospheric Research Initiative
SHADOZ	– Southern Hemisphere Additional Ozonesondes (1998–)
STRATOZ	– Stratospheric Ozone
TRACE-A	– Transport and Atmospheric Chemistry near the Equator-Atlantic (1992)
TROPOZ	– Tropospheric Ozone

1.2 Early Observations

It was not right away that the long-range impact of biomass burning was apparent. The 1980 Brazilian campaign, the 1985 ABLE-2A in Brazil, and early African ozone measurements were performed near source regions. Aircraft measurements of ozone during TROPOZ and STRATOZ, taken over the Atlantic, pointed to transport of ozone and precursors like CO and other biomass burning byproducts, e.g., carboxylic acids in tropical precipitation [6]. Along the Brazilian coast, the ozone sounding record at Natal (6°S, 35°W) displayed a seasonal enhancement in free tropospheric ozone consistent with transport from fire activity in southern Africa [7]. How could this happen? Through a combination of convective mixing of ozone and/or its precursors followed by advection in relatively stable air parcels. Chatfield and Delany [8] describe two distinct processes that enhance ozone as a consequence of convection. In a mechanism referred to as "cook-then-mix," ozone is formed in the boundary layer near sources (e.g., savanna fires, urban areas) before entrainment into a convective cell. In "mix-then-cook," as depicted in Fig. 1, ozone precursors are transported out of the boundary layer through injection into a convective cell. Whereas soluble gases like nitric acid and hydrogen peroxide are removed in cloud systems, essential species contributing to ozone formation, NO, CO, and predominant hydrocarbons, are insoluble. Indeed, the photochemical lifetimes of NO and ozone itself are enhanced. Once transported above the boundary layer, they may last for several weeks to a month. Additional ozone forms downwind of convection during the transport process [8–11]. Lightning also affects tropospheric ozone by introducing NO directly into the upper troposphere.

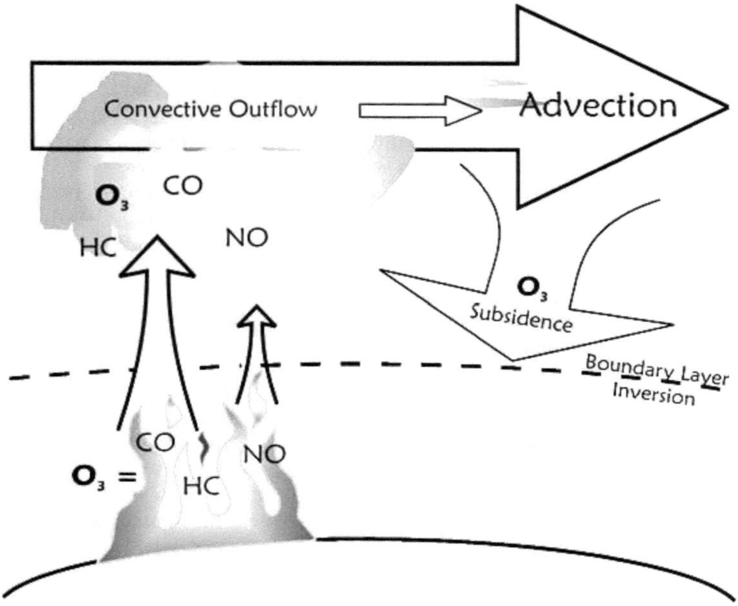

Fig. 1 Schematic of ozone formation in the free troposphere made possible by convective transport of ozone precursors NO, HC (hydrocarbons), CO

Although some of the early 1980s observations were suggestive, satellites and aircraft experiments remote from sources (e.g., FOS/DECAFE, TROPOZ II, SAFARI/TRACE-A, PEM-Tropics A) revealed the far-ranging impact of biomass burning on ozone and its precursor carbon monoxide (CO). Measurements from space (an instrument aboard the Space Shuttle) in October 1984 and again in October 1994 showed widespread enhancements in CO over large oceanic regions in the tropics [12, 13]. More recently, a three-year record of continuous CO measurements from the MOPITT instrument aboard NASA's Terra Spacecraft have confirmed locations and seasonality of the Shuttle snapshots of pyrogenic CO [14]. Chatfield and coworkers [15] showed that commercial airliner (GASP=Global Atmospheric Sampling Project) CO measurements pointed to tropical impacts of biomass burning in the 1970s. More recent commercial airline sampling programs (MOZAIC, CARIBIC, CO and ozone measurements aboard a Japanese jetliner) have confirmed widespread impacts of biomass fires [16–18].

The big push to connect ozone and long-range transport came from Fishman and coworkers [19, 20], who estimated the tropical tropospheric ozone (TTO) column amount by taking the difference between total ozone and stratospheric column ozone measured by two satellites. The resulting climatological maps gave a global picture of the tropics, showing a large seasonal maximum over the south tropical Atlantic occurring in austral Spring when southern hemisphere fire activity is high. The south Atlantic tropospheric ozone feature represents a geographical as well as seasonal maximum in that the Atlantic tropospheric

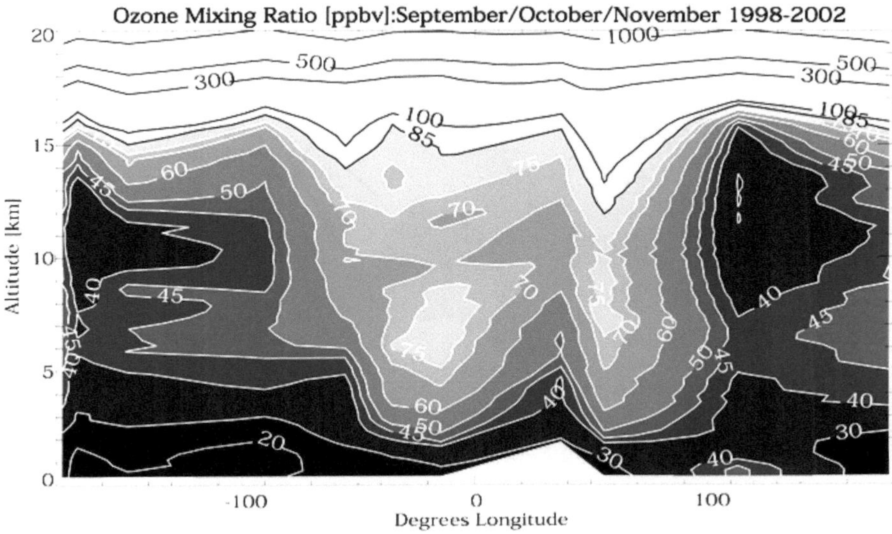

Fig. 2 Structure illustrating "wave-one" pattern due to more tropospheric ozone over the Atlantic and adjacent continents than over the Pacific. Based on averaged soundings taken in 1998–2002 during September–October–November

ozone column appears to be greater than the Pacific ozone column at all times of year.

The Atlantic-Pacific difference is one component contributing to a global feature in total ozone first seen by satellite in the early 1990s. Shiotani [21] pointed out that a thicker total ozone column over the Atlantic compared to the Pacific produces a standing wave-one pattern in the equatorial zone. Although satellite data [22] implied that the wave was in the troposphere, satellite vertical resolution is insufficient to determine whether the wave is a stratospheric or tropospheric feature [23, 24]. A new compendium of ozone sounding data, taken from the SHADOZ (Southern Hemisphere Additional Ozonesondes) network, and resolved to 5 km in most cases, has provided the first definitive identification of the wave as a tropospheric feature [25, 26]. A cross-section of the tropospheric wave-one ozone feature appears in Fig. 2. Upper tropospheric ozone over Pacific and Indian Ocean sites often shows the imprint of convection, with lower ozone injected upward whereas the upper troposphere over Ascension or Natal is enriched by ozone over 65 ppbv. A model interpretation of tropospheric ozone column depth for September 1992 appears in Fig. 3, where various chemical sources of ozone are illustrated as a function of longitude.

During the southern hemisphere burning season (July–November), southern African pollutants may head toward the Indian Ocean (Fig. 4). A large multi-national, multi-platform (6–8 aircraft and numerous ground-based experiments in half a dozen countries) called SAFARI-92 (Southern African Fire Atmospheric Research Initiative) was assembled to study ecological aspects of fires, trace gas emissions from savanna fires, and near and far impacts of fire activity on

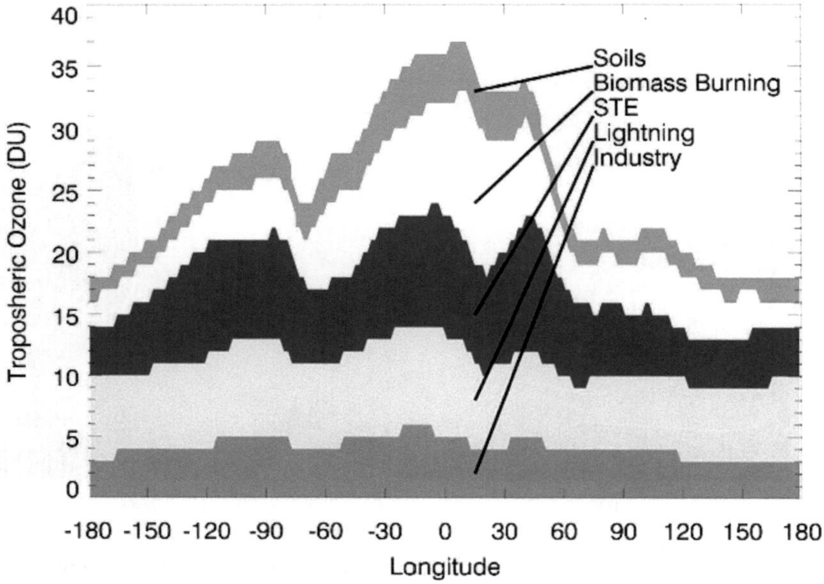

Fig. 3 Absolute concentrations of various processes contributing to the tropical tropospheric ozone column amount determined from a model simulating September 1992 [92]. Lightning and soils provide NO, that reacts to form ozone. Climatological emissions are used in the model but the meteorological fields are based on ECMWF analyses for 1992

atmospheric composition [27–29]. A partner experiment to SAFARI-92, TRACE-A (Transport and Atmospheric Chemistry near the Equator-Atlantic) tested the hypothesis that biomass burning is responsible for the south Atlantic ozone maximum. SAFARI/TRACE-A, like TROPOZ and FOS/DECAFE, confirmed the photochemical origins of much of the Atlantic ozone. Compendia of studies related to biomass burning and the atmosphere prior to and following SAFARI-92 and TRACE-A are given by Levine [30, 31].

1.3 Meteorological Factors Affecting Ozone

With increasingly extensive observations of ozone and other chemical constituents, a more complex picture of chemical and dynamical interactions affecting tropical ozone is emerging. Regional meteorological patterns (Fig. 4) and synoptic-scale systems introduce a high degree of variability into ozone distributions. For example, Thompson and coworkers [25] report that integrated tropospheric ozone column varies by a factor of three at ten tropical stations. Randriambelo and coworkers [32] demonstrate that "seasonal pollution" at Réunion is in fact a series of polluted episodes alternating with "background" ozone (25–30 ppbv in the lower troposphere).

Figure 5 shows the typical pattern of general circulation. Interaction with these large-scale meteorological patterns and regional dynamics affects what might

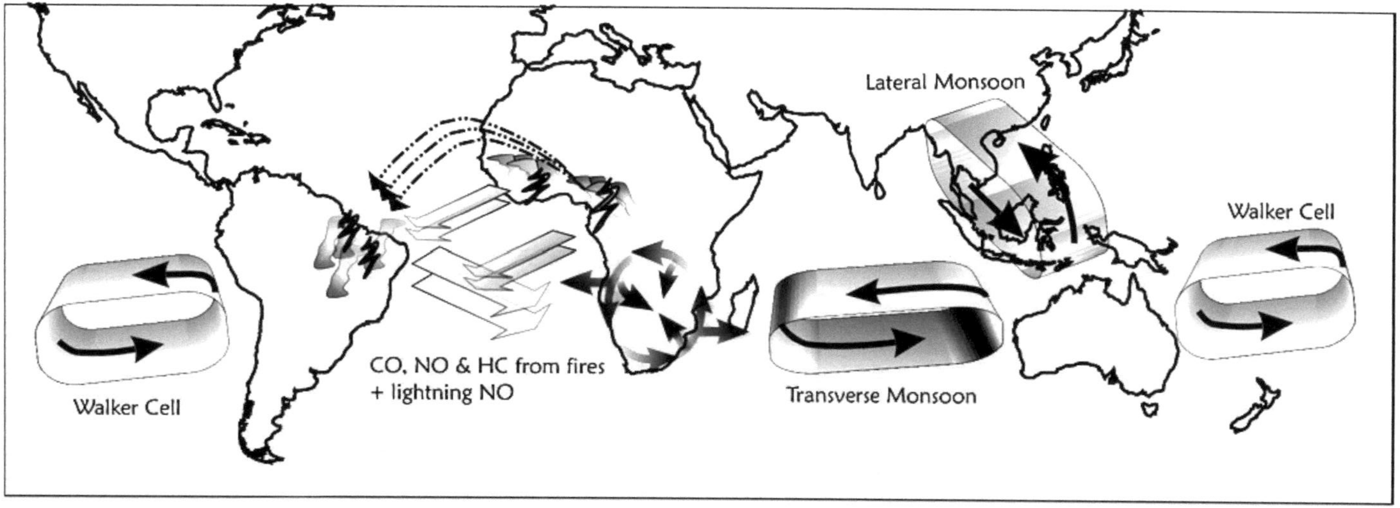

Fig. 4 Regional circulation patterns affecting tropical ozone. Southern African recirculation that tends to concentrate ozone over the continent with sporadic outflow to the Atlantic and Indian Oceans. The Atlantic is further enriched in ozone from convective outflow from South America and cross-hemispheric mixing at the ITCZ. In the latter case, ozone from north of the ITCZ adds to the south Atlantic burden. Monsoonal patterns (winter monsoon) that affect ozone over the Indian Ocean

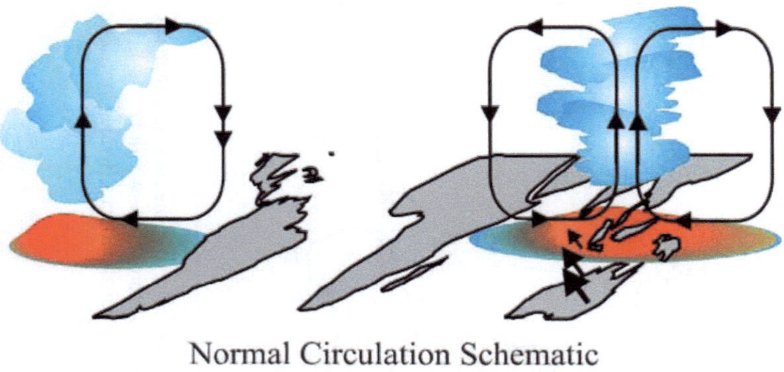

Fig. 5 Schematic of zonal circulation in the tropics. Rising regions are more convective than areas of subsidence. Flows are perturbed during an El Niño-Southern Oscillation (ENSO) and Indian Ocean Dipole event

otherwise be straightforward intercontinental flows. Note the convective tendency over the Pacific and subsidence over parts of the Indian and Atlantic Oceans. During the winter monsoon period the cells shown in Fig. 4 interact with ozone. Flows from South America are shown interacting with the mid-to-upper tropospheric westerlies to transport ozone toward the Atlantic. Subsidence over the Atlantic adds to the ozone column burden in this region [33]. Further augmentation comes from advection into the lower and middle troposphere from African sources.

Chemical-dynamical interactions are captured to varying degrees in coupled chemical-transport models, as described in chapters by Schultz and Bey (p. 197) and by Lawrence (p. 131). An alternative modeling approach uses forward trajectories initialized at source regions over the major continents to assess intercontinental impacts. In Stohl's analysis [34], about half the burden of tracer parcels received over Africa originated from Africa. One-quarter of air parcels received over Australia also originated over Africa; 46% of Australia's imported air parcels came from Asia. In the following sections, one should bear in mind the conclusions of the Stohl analysis and the other model studies as well as tropical observations reported by Pochanart and coworkers (East Asia, p. 99), Lawrence (South Asia, p. 131), and Piketh (Africa, p. 173). We attempt here to synthesize observations about the intercontinental transport of ozone from biomass fires by examining the large impact regions: Indian Ocean, Pacific, and Atlantic.

2 Indian Ocean Region

Pollution over the tropical Indian Ocean is affected principally by four source regions: Africa, south central Asia (mostly India), southeast Asia, and the maritime continent (major components – Indonesia, New Guinea, Australia). In the following sections the relationship between each of these regions to trans-oceanic and intercontinental transport over the Indian Ocean is illustrated.

2.1 African Pollution

Examples of export transport pathways from southern Africa to the Indian Ocean appear in the works of Chatfield and Piketh and their coworkers [15, 35]. In [15] the pattern indicated by a simulated CO plume coincides with elevated smoke aerosol observed in September 1996 by the TOMS satellite. The corresponding satellite fire count data [36] from the AVHRR (Advanced Very High Resolution Radiometer) instrument for the PEM-Tropics A period (28 August to 6 October 1996) shows the highest distribution of fires in the region shaded in Fig. 6. Note coincidence of fire locations with the high CO plume.

From various indicators (Southern Oscillation Index, Outgoing Longwave Radiation, Indian Dipole Mode Index, precipitation and wind fields), as well as from satellite tropospheric ozone and aerosol measurements [37], it appears that 1996 was a typical year for circulation affecting the Indian Ocean. Ozone pollution traveled to the western Pacific where it was observed from aircraft during PEM-Tropics A [38–41]. What did ozone over the Indian Ocean look like in September 1996? Are the patterns suggested by the CO simulation of Chatfield and coworkers [15] applicable to ozone observations? We address this question with ozone profile data from Réunion Island (21°S, 55°E) and Watukosek (Java, Indonesia, 7.5°S, 113°E).

Figure 7 shows ozone profiles from soundings taken over La Réunion on 29 August and 9 October 1996. Radiosondes launched at the same time (not shown) reveal a capped marine boundary layer below 800 hPa. No pollution appears in the marine boundary layer (ozone mixing ratio is 20–30 ppbv, Fig. 7). Air-parcel back trajectories (800 hPa, red, Fig. 8a) are located over the ocean south of Africa. Between 400 and 700 hPa ozone mixing ratios from 50 to 115 ppbv sug-

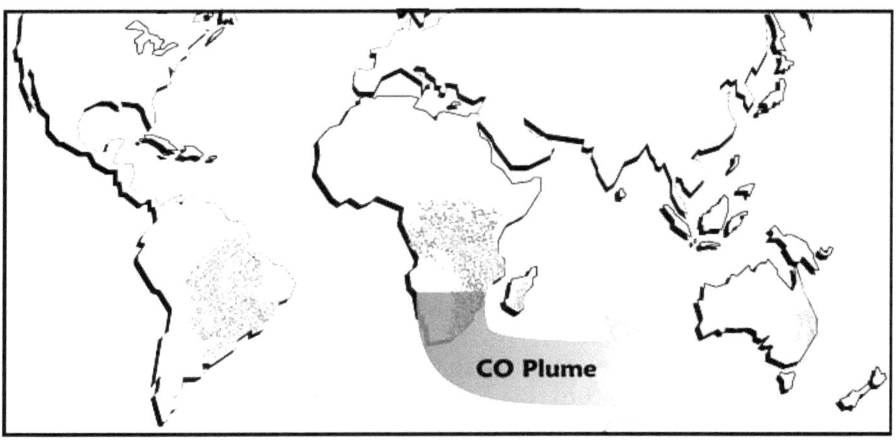

Fig. 6 Schematic of simulated CO plume for September 1996 [15] heading from Africa toward the Indian Ocean superimposed over fires observed from AVHRR at the same time [36]. PEM-Tropics-A field mission [59] took place during this period

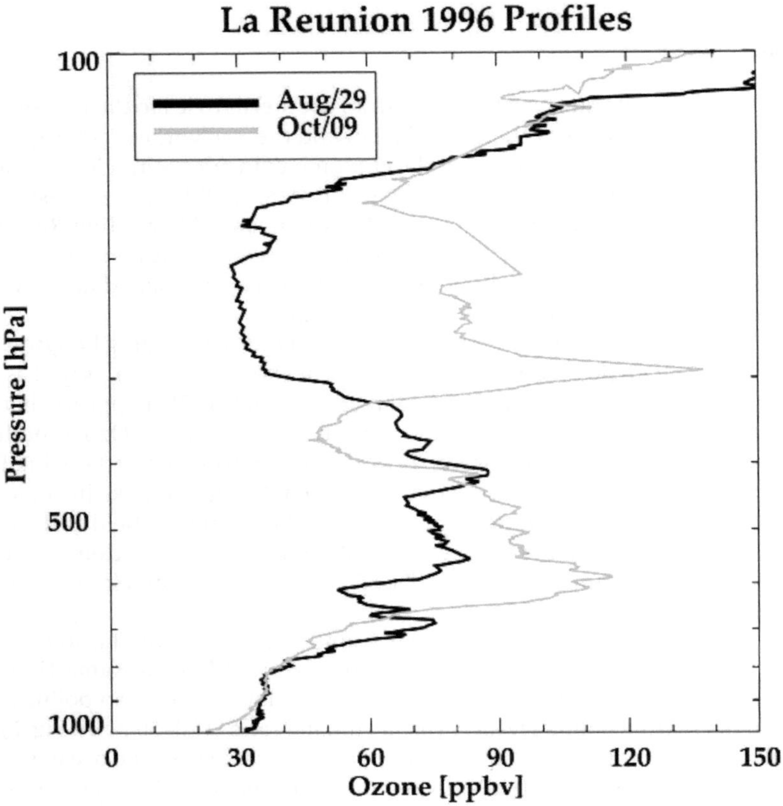

Fig. 7 Two profiles from ozone soundings taken over La Réunion during the dry season in 1996

gest pollution (Fig. 7). This is confirmed by air parcel back trajectories (blue, purple in Fig. 8) that originate over regions of active fires (compare fire locations in Fig. 6).

Figure 9 shows profiles over Watukosek in July and September 1996. Similar back-trajectory analysis shows pollution in the free troposphere (ozone >50 ppbv at 300–600 hPa) appears to originate from eastern Africa and the western Indian Ocean. Note that on 17 September 1996 there is also relatively high ozone near the surface. This might be due to localized biomass fires and/or transport from northern Australian fires (trajectories at 700, 800 hPa, not shown; fires in Fig. 6).

How typical are these results? Ozonesonde observations at these stations, based on the SHADOZ record (Fig. 5 in [25]) confirm that this is the time of year of greatest tropospheric column ozone. For Réunion, column-integrated tropospheric ozone amounts computed by Baldy, Taupin and their coworkers appear in Fig. 9 in [42] and Fig. 2 in [43]. Watukosek climatology shows more sporadic influences of fires and lower column amounts than over Réunion, on average [44].

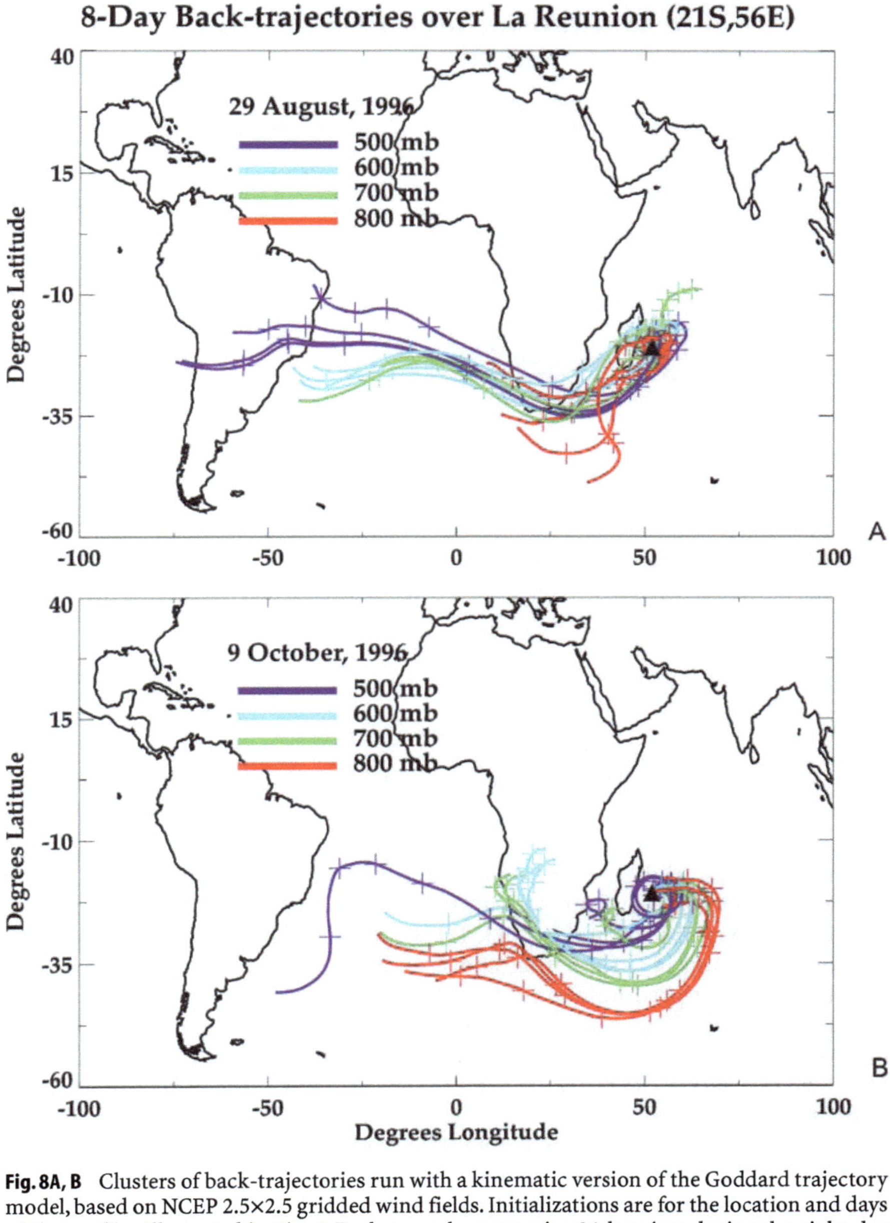

Fig. 8A, B Clusters of back-trajectories run with a kinematic version of the Goddard trajectory model, based on NCEP 2.5×2.5 gridded wind fields. Initializations are for the location and days of the profiles illustrated in Fig. 7. Each + marks successive 24-h points during the eight-day period

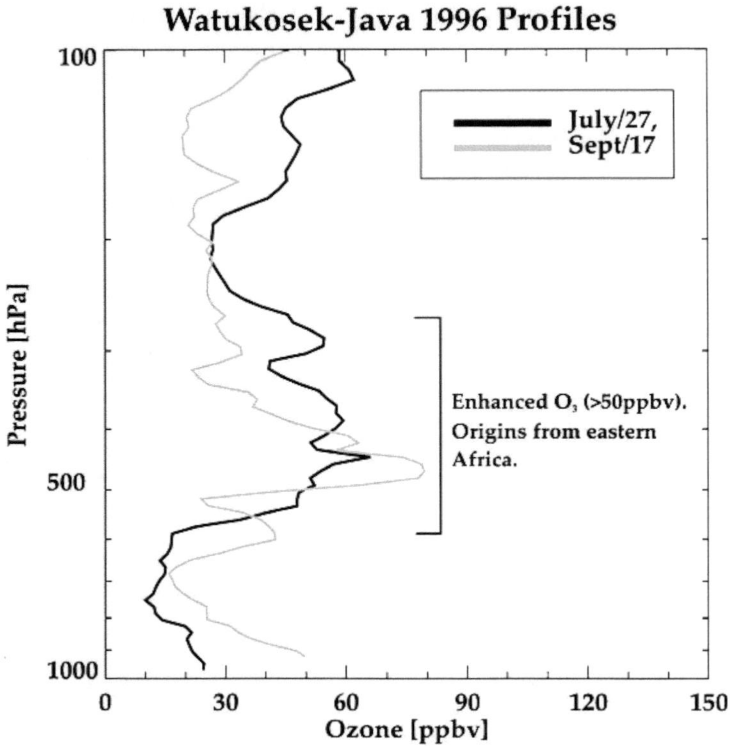

Fig. 9 Ozone profiles taken from soundings over Watukosek during the 1996 dry season

The reason for highest ozone over Réunion in October is transport of ozone and/or ozone precursors from African and Madagascar fires [42], although there are also stratospheric influences [43, 45]. Trajectories [42] and wind climatologies over Réunion [43] show that above 4 km (~700 hPa) during September–October, flows are predominantly from the southwest. Transit times from Madagascar and southeastern Africa are typically 1–2 and 4–5 days, respectively. A cross-section of mean ozone (mixing ratio as function of month and altitude from several years of SHADOZ data, Fig. 9 in [26]) shows elevated concentrations in September–October–November where flows are usually African (3–10 km). Some of this ozone could come from nitric oxide injected by lightning because the ITCZ (Intertropical Convergence Zone) is approaching Réunion at this time of year. At Watukosek, in addition to ozone exported from Africa, there may be influences of biomass burning in Southeast Asia, the Indonesian maritime continent (see below) and Australia. Figure 4b in Thompson et al. [25] shows relatively high surface ozone at Watukosek in September–October–November.

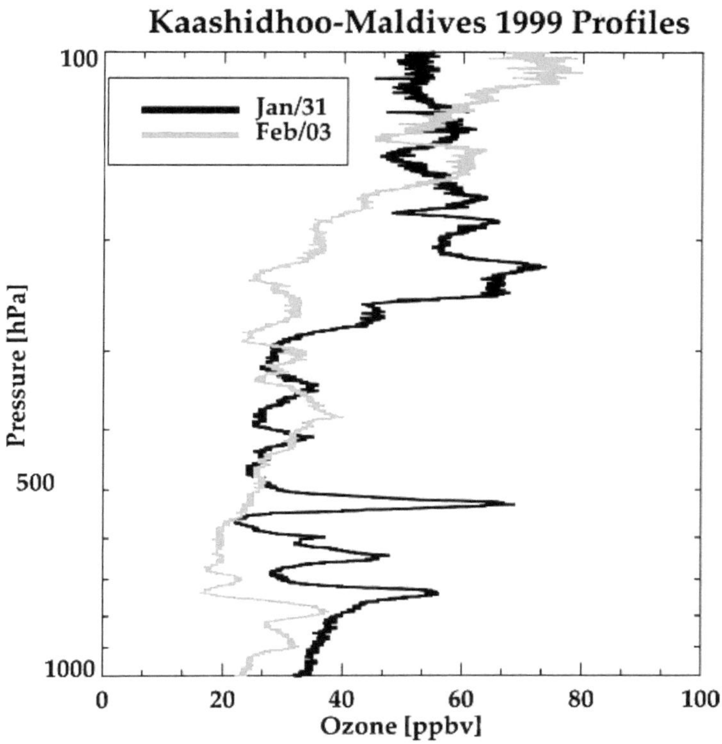

Fig. 10 Contrasting ozone soundings over the Kaashidhoo Observatory (Male, Maldives) taken during the INDOEX period in early 1999. The 31 January profile shows more evidence of pollution than the profile on 3 February. Back-trajectories indicate origins of the latter are marine whereas air parcels contributing to the 31 January profile were from southern India

2.2 South and Southeast Asian Pollution

Intercontinental transport relative to South Asian and southeast Asian sources was explored during INDOEX (Indian Ocean Experiment), a dedicated campaign that studied many processes with aircraft and ships working downwind of India. Working during January–March 1999, during the period of winter monsoon-to-spring transition, INDOEX was a multi-platform, multi-national experiment. The reader is referred to a collection of INDOEX papers in the Journal of Geophysical Research [46] and to an overview paper [47]. Highlights of intercontinental transport from south and southeast Asia during INDOEX are covered by Lawrence in Chap. 6.

Figures 3 and 4 in Chap. 6 show flows governing pollution transport from south Asia toward the central Indian Ocean region. The large-scale circulation (Fig. 4) which interacts with the pollutants includes downslope flow of ozone from the Himalayas. Although ozone pollution episodes during INDOEX were frequent, discriminating between pyrogenic and industrial sources requires analysis with tracers and modeling. In Phadnis et al. [48], evaluation of typical ozone budgets is made with a coupled chemistry-transport model driven with winds from a GCM. In the free troposphere (represented by 685 hPa, Fig. 10 in [48]), maximum ozone occurs over northern India. The fraction of pollution ozone, however, is greatest over SE India and the tropical Indian Ocean. This is due to a combination of subsidence with low-level outflow away from the continent. Similarly, the region of highest fractional pollution ozone in the lower-free troposphere due to SE Asian pollution extends over Indonesia and the Indian Ocean (Fig. 11 in Phadnis et al. [48]). Based on the model, approximately half the ozone pollution is from biomass/biofuels burning and half from fossil fuel combustion.

The INDOEX experiment was timed to take advantage of maximum impact of south Asian emissions on the Indian Ocean. In ozone soundings taken from the Research Vessel Ronald H. Brown, pollution appeared as layers of ozone >50 ppbv in the free troposphere (Fig. 1c in Zachariasse et al. [49] is an example). Figure 10 presents contrasting ozone profiles from Male (Maldives Islands), an Indian Ocean site that intermittently received south Asian pollution during INDOEX. The two soundings (cf. Fig. 11b in Thompson et al. [25]) correspond to contrasting flows near the southern Indian pollution source region. The ozone-enriched layers at 750 hPa and 500 hPa on 31 January 1999 represent dehydrated continental air, traceable to south India and a localized region of pollution recirculation, respectively. In contrast, on 3 February, the lower free troposphere is moist with ozone mixing ratios of 20 ppbv; the flows are marine.

2.3 Indonesian Maritime and Australian Continents

deLaat and Lelieveld [50] showed that southeast Asian pollution tends to be transported more strongly toward the western Pacific than to the Indian Ocean. However, biomass burning in Indonesia and Australia, which often follows the El-Nino-La-Nina and Indian Ocean Dipole cycles, may affect pollution transport over the Indian Ocean [37, 51–53]. The impact of biomass burning on ozone formation has been evaluated with satellite, aircraft and sounding data. Immediately adjacent to fires, ozone was depleted [54] whereas in the lower troposphere, thin layers of ozone pollution were detected. The ozone loss might have been due to a decreased photochemical formation rate in the presence of heavy aerosol loading or it may have been due to heterogeneous removal.

Figure 11 displays derived satellite tropospheric ozone and smoke aerosol products for 10 October 1997 during the period of intense biomass fires over Kalimantan and New Guinea, the latter signified by gray shaded regions. The smoke is greatest near regions of burning. However, the ozone, most concentrated above the boundary layer [37], clearly shows intercontinental transport. Ozone from Africa extends east to Australia and ozone from the Indonesian fires reaches southern India and beyond.

The BIBLE-A campaign [55] conducted in late September–early October 1998 was in a building La Nina when biomass burning pollution was relatively low over

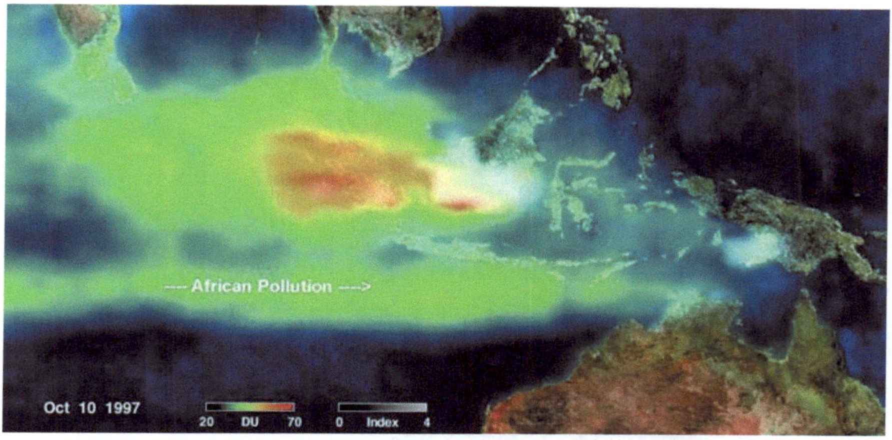

Fig. 11 Satellite view (Earth-Probe TOMS instrument) of absorbing aerosols (*gray scale*) and tropospheric ozone (*color scale*) during intense biomass burning in the wake of the ENSO-induced drought (August–November 1997). By the end of October 1997 the two distinct plumes had merged into one [37]

Indonesia [56]. Sampling over the Indonesian maritime continent, New Guinea and Australia gave evidence of intercontinental transport from Indonesia toward the central Indian Ocean. Convection introduced ozone precursors (from non-biomass burning pollution and lightning) into the upper troposphere where ozone formed at a 1–2 ppbv/day rate. Convective outflow toward the south introduced ozone into the westerlies (Fig. 12 in [56]) so the net flow was from Indonesia to northern Australia. The lightning input was observed to be greater in 1998 than during BIBLE-B in September-October 1999. Subsidence brought ozone over Australia and prevented northern Australian biomass burning from reaching the altitudes required for long-range transport.

3
Pacific Region

The Pacific is the largest oceanic basin affected by intercontinental-scale transport of ozone from biomass burning. The PEM (Pacific Exploratory Mission) series of experiments [57–60] turned up evidence of pollution transport throughout the tropical and mid-latitude Pacific in all seasons. Oltmans et al. [61] presents a climatology of tropospheric ozone over three Pacific stations (American Samoa, Fiji, San Cristóbal; cf. Fig. 12 for San Cristóbal) that receive ozone from biomass burning sources. The highest concentrations of free tropospheric ozone (5–10 km) occur in September–October–November (SON) when outflow from southern hemisphere biomass burning is greatest.

The greatest source of biomass burning emissions at Samoa and Fiji is from Africa, although ozone from SE Asian and Australian fires can also be detected. Individual ozone profiles with mean mid-tropospheric (5–10 km) ozone greater

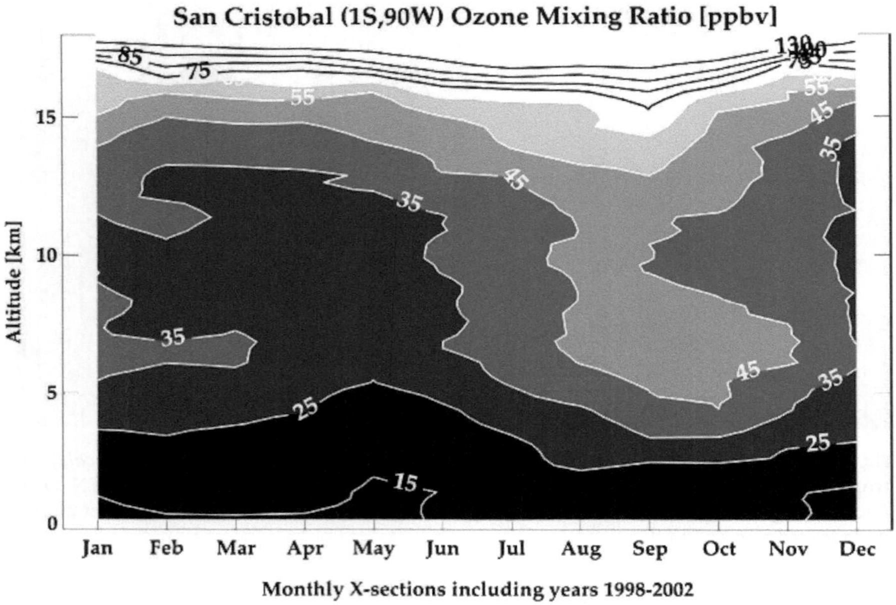

Fig. 12 Contour of ozone mixing ratio based on monthly averages from ozonesondes launched over San Cristóbal (Galapagos) from 1998 to 2002

than 55 ppbv at Samoa and Fiji, taken from 1998–2002 SHADOZ observations appear in Fig. 13. All but four of the profiles depicted are from October and November. Two features are noteworthy.

First, as Thompson and coworkers [25] have noted, ozone is highly variable in the southern tropical troposphere. Many profiles display very clean (mixing ratio <20 ppbv) layers, as well as polluted layers even during the time of year that pollution is most prevalent. The persistent and widespread polluted ozone layers over the Pacific have been noted by Newell et al. [62] and Stoller et al. [41]. The PEM-Tropics A experiment conducted in September–October 1996 [59, 60] measured a number of tracers (CO, hydrocarbons, PAN, formaldehyde, organic acids) to show the biomass burning origins of ozone.

Second, Fiji receives more pollution influence than Samoa (Figs. 8 and 9 in Thompson et al. [25]). There is a clustering of high values (~70 ppbv) at 4–6 km over Fiji. This contributes to the generally higher ozone contour in SON at Fiji compared to Samoa (compare Fiji and Samoa contours in Plate 3 in [61]). Back trajectories showing air parcel origins for the ozone profiles in Fig. 13 (not illustrated) show that both Australian and African fires may contribute to the ozone layers at Fiji and Samoa (cf. fire locations in Fig. 6).

At San Cristóbal (Fig. 12) the July-October period is characterized by a broad ozone feature with a mean mixing ratio contour of 45 ppbv. This appears at somewhat greater altitudes (6–8 km) than the corresponding features over Fiji and Samoa (Fig. 13). Trajectory analyses allow us to ascribe as the most likely origins of the high ozone layers at San Cristóbal (6–8 km) to be South American biomass

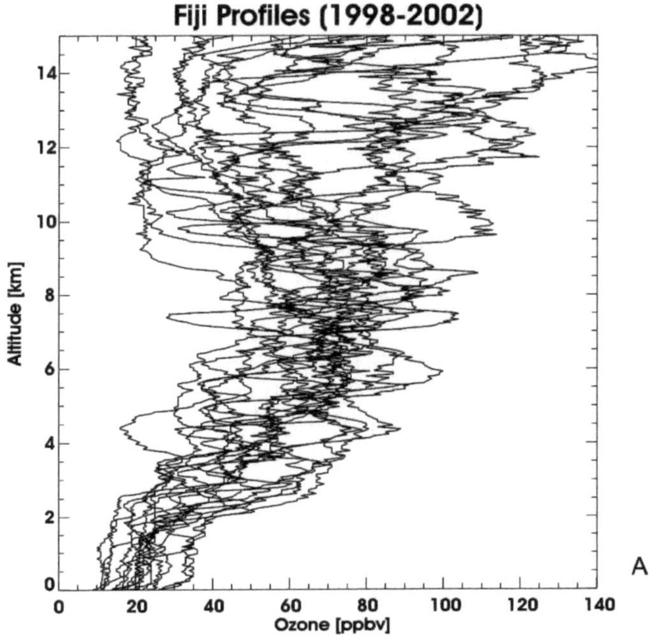

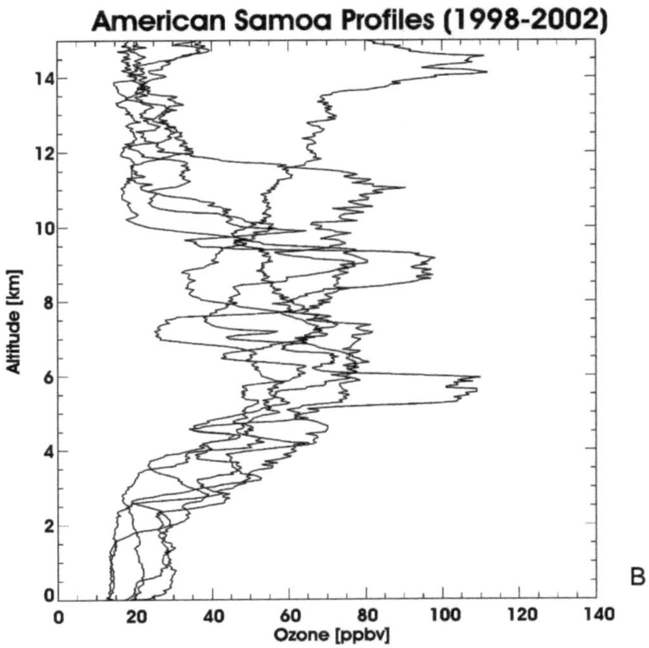

Fig. 13A, B Profiles from the SHADOZ ozonesonde record (1998–2002) over: **A** Fiji; **B** Samoa; for which the mean ozone mixing ratio in the 5–10 km layer exceeded 55 ppbv. Most of these profiles were taken during the September–November period

burning. During the latter part of the dry season, ozone and ozone precursors near fires are entrained into developing convective systems, in either a cook-then-mix or mix-then-cook mode ([8]; Fig. 1). These air parcels are then advected east toward the Pacific or into the westerlies toward the Atlantic (Fig. 4) or Africa [63, 64]. A case study of high ozone over San Cristóbal that is traced to high-ozone, biomass burning regions of South America is described by Oltmans et al. [61] (Plate 6).

Part of the motivation for the 2001 TRACE-P (Transport and Chemical Evolution – Pacific) aircraft experiment was to evaluate impacts of intercontinental transport during the northern hemisphere springtime. Although the focus was on mid-latitudes, where models indicated maximum flow, some of the Pacific regions sampled were expected to be affected by biomass fires [65]. Southeast Asian biomass fires influenced some flights out of Hong Kong [66]. In addition, some ozone soundings at Hong Kong (22°N, 114°E) were affected by African biomass burning (Figs. 5, 7, 11–14 in Oltmans et al. [67]).

4 Atlantic Region

4.1 Intercontinental Transport from Africa and South America

As described in the Introduction, much of our knowledge about intercontinental transport of ozone pollution is based on observations over the tropical Atlantic and adjacent continents. Flows from biomass burning regions over Africa and South America (Fig. 4) can supply ozone in long-lived layers. The lifetime of ozone (and some ozone precursors) above the boundary layer (Fig. 1) is up to several weeks. For example, peroxyacetyl nitrate (PAN) is a reservoir for the essential ozone precursor NO [68]. During transit from Africa, transformation of PAN to NO allows ozone to form far from source regions. The region of African recirculation (Fig. 4) shows the patterns of ozone (and ozone precursor) transport toward the south tropical Atlantic and the photochemical source from biomass fires schematically when the fires are south of the ITCZ. This applies to the July through October period. As the dry season advances south [69–71], so do the regions of most intense fire activity. During TRACE-A (September–October 1992), the highest ozone formation rates were found in the boundary layer near fires (10–15 ppbv/day ozone) and above 8 km where NO, presumably enriched by lightning [72], maintained a positive photochemical regime [63]. The upper tropospheric lightning influence on ozone occurred over the south tropical Atlantic and both adjacent continents.

In the middle troposphere (5–10 km) during TRACE-A (September–October 1992) high ozone layers (mean >65 ppbv) were associated with recirculation about the south tropical Atlantic gyre. SAFARI-92 (also conducted during September–October 1992), SAFARI-2000 (August–September 2000) and related climatological analyses demonstrated how persistent are some "absolutely stable layers" of ozone pollution [73–75]. A good example of ozone pollution in an extremely stable layer appears in Fig. 2 of Thompson et al. [76].

Recirculation of ozone over the persistent south tropical Atlantic high-pressure region allows ozone column amounts >35 Dobson Units (one Dobson Unit, DU, is equivalent to 0.1 cm ozone thickness at one standard atmosphere) to remain intact for weeks at a time. Analysis with air parcel back trajectories showed that the high ozone over Ascension Island and Natal [63] usually originated over Africa. Figure 14, with August and October ozone profiles at Natal and Ascension, respectively, from the SHADOZ record (2001 shown), is similar to the SAFARI-92/TRACE-A period [63]. Analysis with air parcel back-trajectories illustrates African origins for these soundings (Fig. 15). These trajectory transport patterns persist throughout the season of maximum burning in southern Africa (August–November).

Besides photochemical activity, a several-week lifetime and advection contributing to ozone, there is a tendency for ozone to accumulate over the Atlantic due to subsidence associated with sinking motions of the Walker circulation ([33]; also Fig. 18 in [63]). The result is sometimes a tropospheric ozone column more than 50 DU thickness. Atlantic subsidence was apparent in the ozone curtains from the UV-DIAL aircraft instrument during TRACE-A [77]. It also shows in the cross-section of mean monthly ozone mixing ratio vs altitude from the SHADOZ record at Ascension, Natal and Paramaribo. For Ascension and Natal (Fig. 16) intercontinental transport is signified by the localized ozone maximum (55 ppbv contour in June–August; 85 ppbv contour in September–October) between 5 and 10 km. At Ascension a secondary maximum (65 ppbv) appears in February. At this time of year, burning north of the ITCZ in western Africa is prevalent (Fig. 12 in Thompson [78]) but some interhemispheric transport of ozone to Ascension, south of the ITCZ, is probably occurring (see the next section).

North of the ITCZ, long-range ozone transport patterns are distinct from south of the ITCZ. This shows in several ways. First, the analysis of the total ozone wave-one pattern by Kim et al. [79] and Hudson and Thompson [80] reveals that it is largely a southern hemisphere phenomenon. Second, consider the ozone vertical and seasonal distribution over Paramaribo (Fig. 16c; also Fig. 2 in Peters et al. [81]). In SON, there is relatively high ozone in the free troposphere over Paramaribo but not as pronounced as the 5–10 km high-ozone feature observed over Natal (10° south of Paramaribo, and 20° east). During MAM relatively low ozone in the upper troposphere at Paramaribo (a localized 35 ppbv minimum at 12 km) looks as if convective redistribution of boundary layer ozone to the upper troposphere occurs as the ITCZ passes over. Where do the high-ozone mid-tropospheric ozone layers over Paramaribo come from? Analysis of a high ozone pollution episode in early August 2000 ([81]; Fig. 17) show wind fields at 500 hPa and linkage to African biomass fires (solid circles in Fig. 17 based on ATSR fire counts). Note that the transport to Paramaribo is an example of intercontinental and cross-hemispheric transport. Other episodes of South American biomass burning affecting Africa occurred during SAFARI-92/TRACE-A. Westerly flow of high ozone (especially post-convective) toward southern Africa and on to the Indian Ocean, affected ozone soundings over Irene, South Africa (Fig. 13 in [63]).

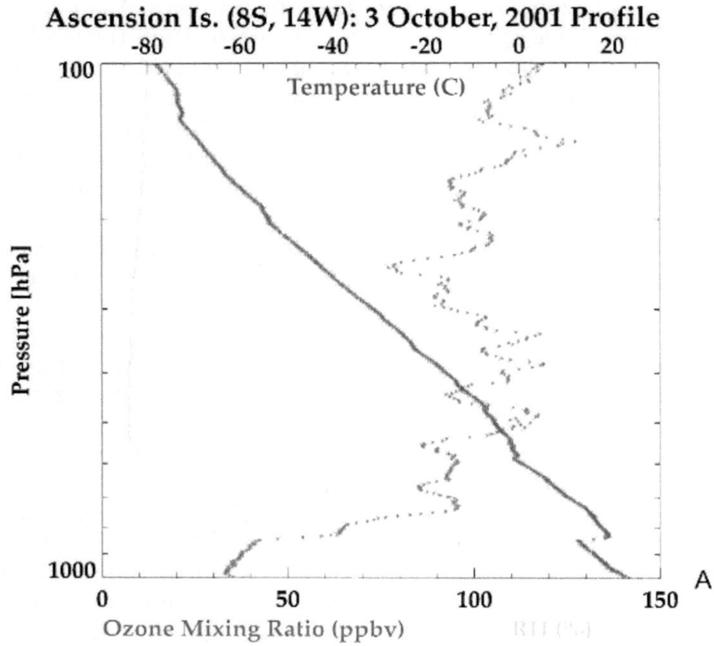

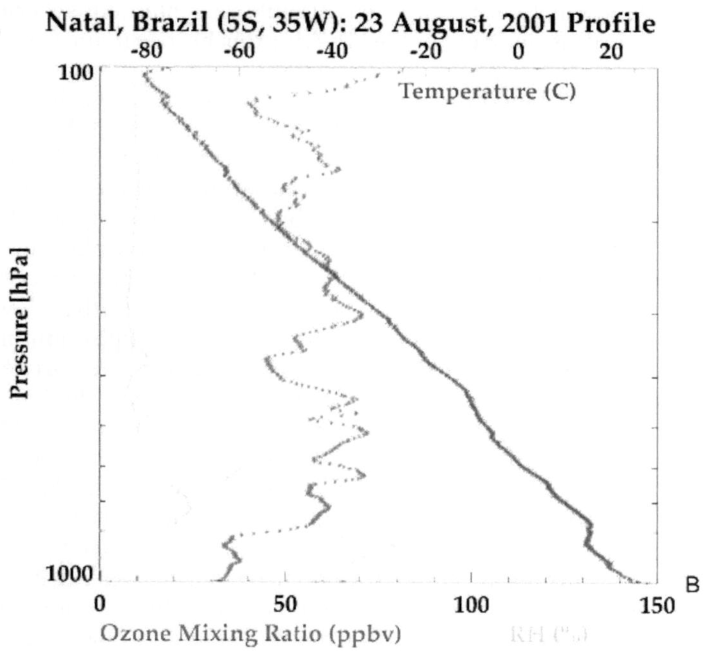

Fig. 14 **A** Ascension profiles with ozone layers associated with biomass burning. **B** Natal profiles of same type

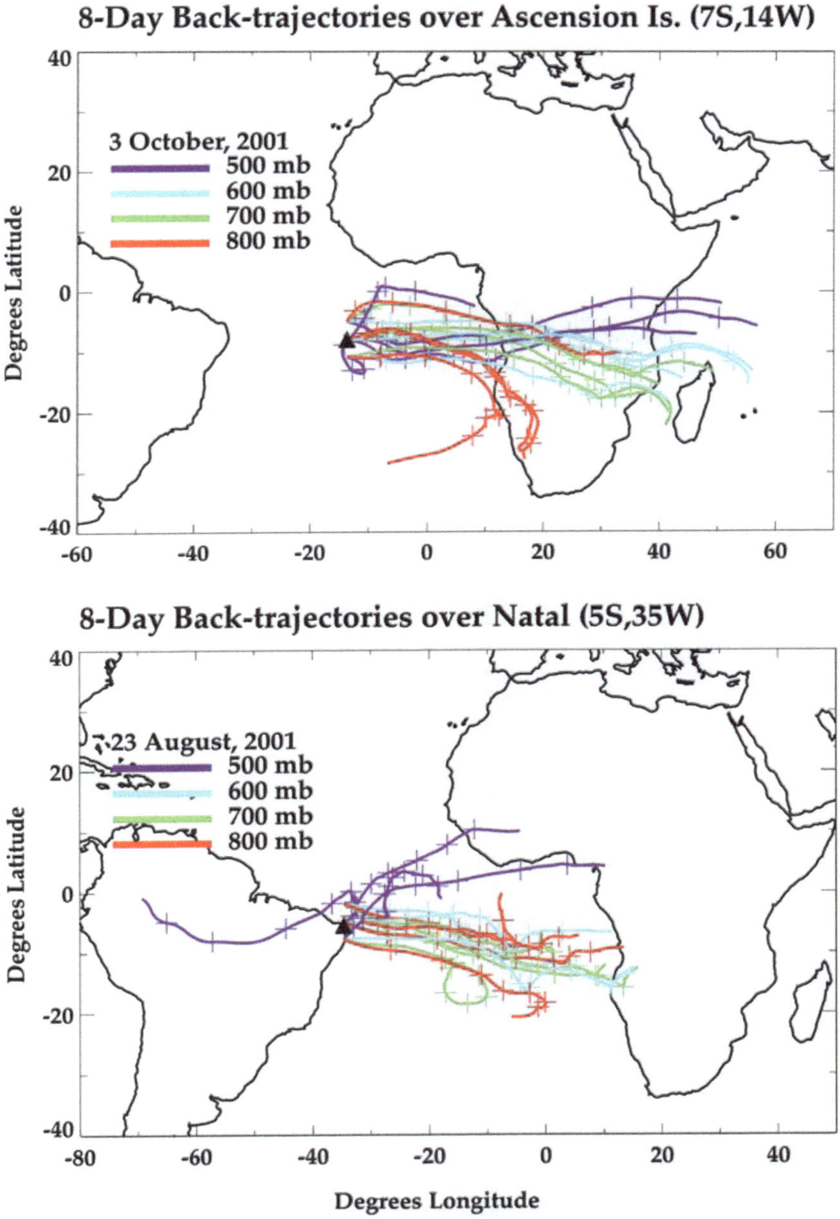

Fig. 15 Trajectories as in Fig. 8 for Ascension and Natal profiles shown in Fig. 14, respectively

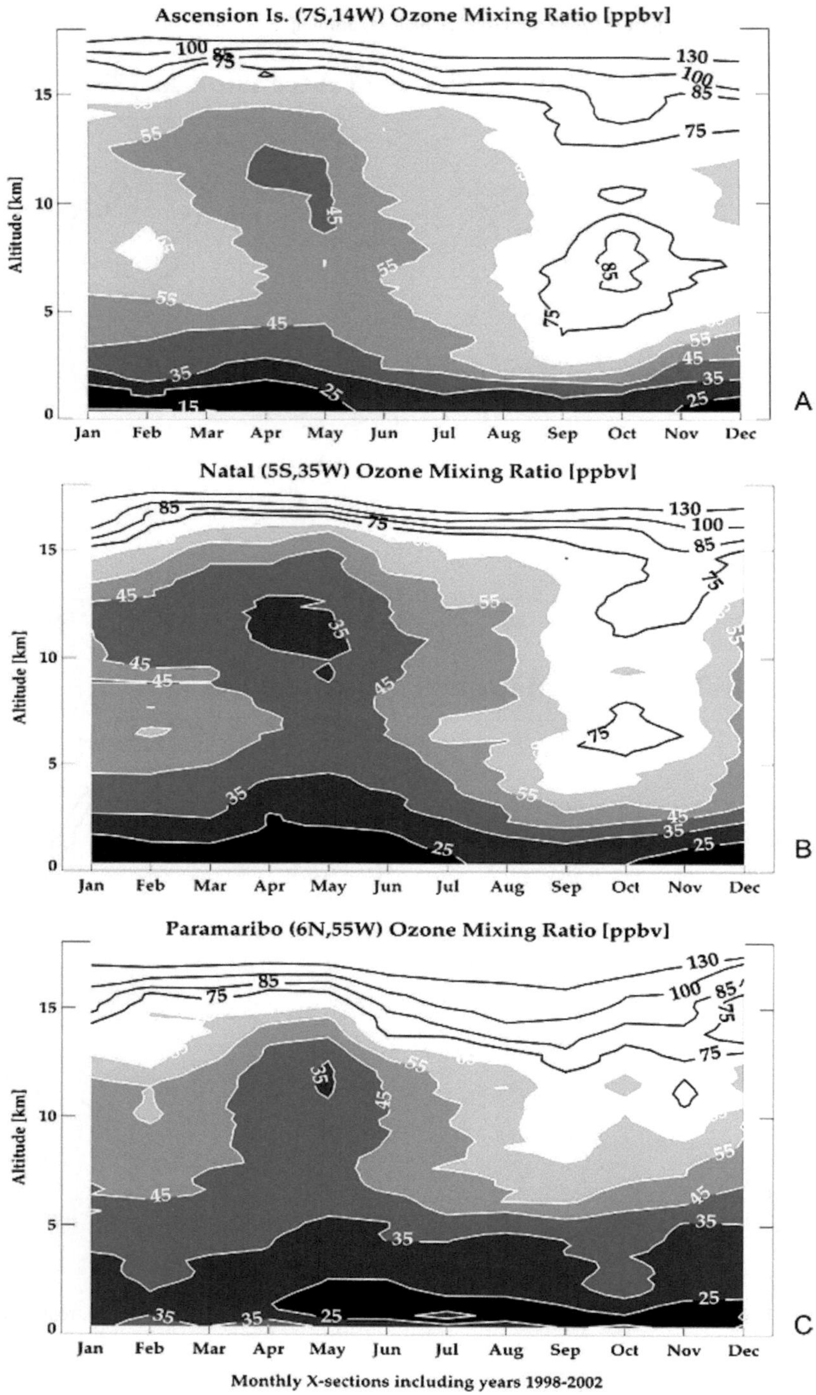

Fig. 16A–C Contoured ozone mixing ratio based on monthly averages from ozonesondes in 1998–2002 [25] launched over: **A** Ascension Island (8°S, 14°W); **B** Natal, Brazil (6°S, 35°W); **C** Paramaribo, Surinam (5°N, 55°W)

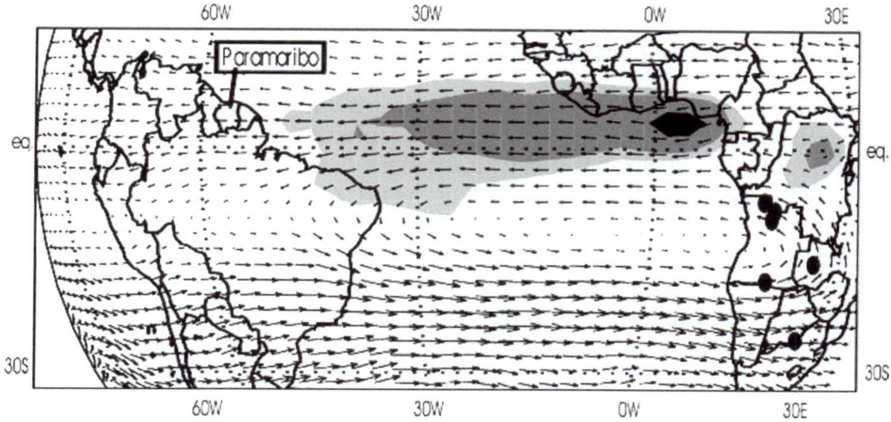

Fig. 17 A period of enhanced tropospheric ozone in early August 2000 [81] with 500 hPa wind fields shown. *Shading* represents regions of greatest concentration of air parcels transported from active fire areas (*solid circles*) and transported to Paramaribo by convection and advection

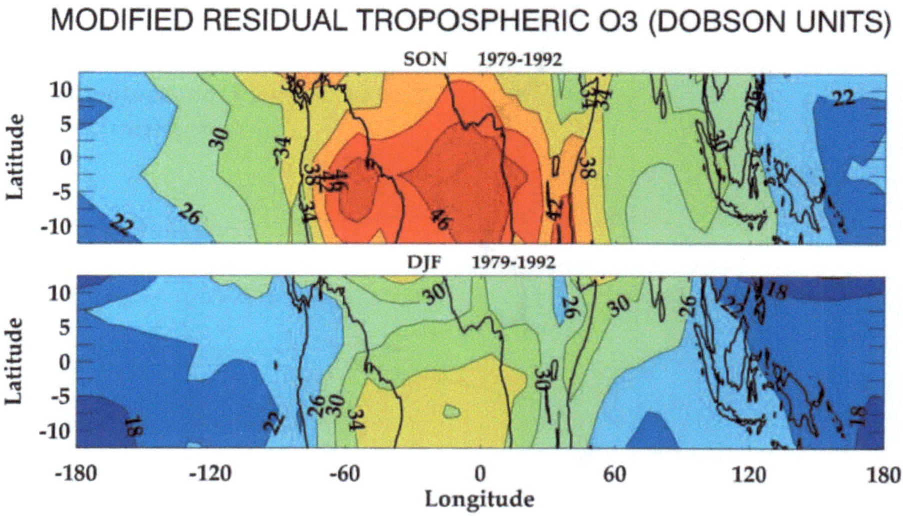

Fig. 18 The tropical Atlantic "ozone paradox" as seen from the TOMS satellite [22, 85]. During December–January–February (DJF), when northern equatorial biomass burning is at its peak (averaged absorbing aerosols, measured during the 1979–1992 Nimbus 7 observing period, *upper panel*), tropospheric column ozone is greater in the southern hemisphere (*lower panel*)

4.2 Cross-Hemispheric Intercontinental Transport and the "Atlantic Paradox"

As pointed out in the previous section, the wave-one pattern, signifying a relatively fixed longitudinal gradient in tropospheric ozone (maximum ozone column over the Atlantic), is mostly a southern hemisphere phenomenon. This appears to be due to the Atlantic north of the ITCZ being lower in ozone sources and more affected by vertical mixing of low-ozone boundary layer air by convection. A good depiction of the north-south ozone contrast was assembled from TRACE-A aircraft observations by Browell and coworkers (Fig. 17 in [77]). The result is that the highest tropospheric column ozone amount is between 10 and 20°S. Satellite observations (Fig. 18) show the north-south contrast is present

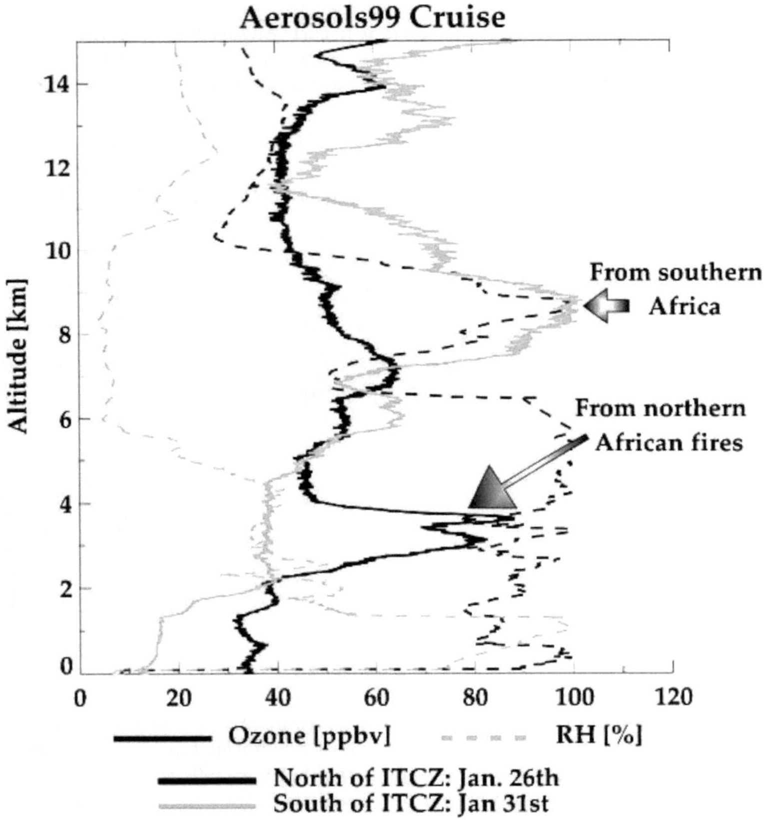

Fig. 19 Soundings taken in early 1999 during the Aerosols99 cruise on the *Research Vessel Ronald H. Brown.* Ozone profiles north (26 January, 5.5°N, 29.5°W) and south of the equator (31 January, 10.9°S, 10.3°W) illustrate the ozone paradox. The layers directly traceable to biomass burning from northern Africa are concentrated at 2–4 km in the 26 January sounding. In the 31 January sounding the highest ozone mixing ratios are in the middle troposphere (7–10 km). Origins are from southern Africa

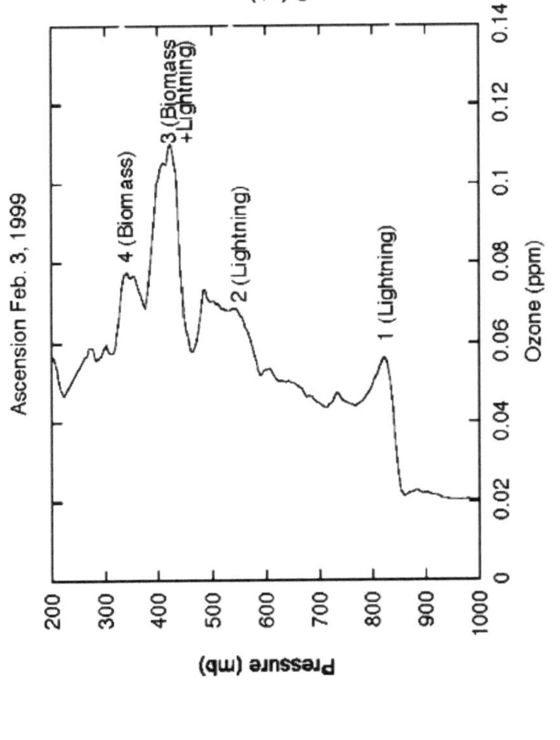

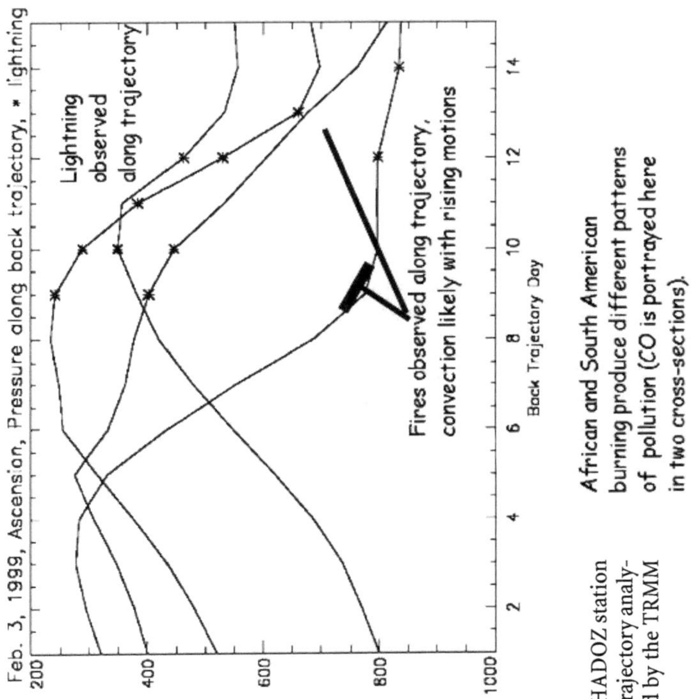

Fig. 20 Peaks of high ozone during the Aerosols99 cruise, taken from the SHADOZ station at Ascension Island (8°S, 14°W, 3 February 1999). Attribution made through trajectory analysis and coincidence of air parcel locations with fires and lightning detected by the TRMM (Tropical Rainfall Measuring Mission) satellite [90]

year-round [78, 82]. Note that there is more ozone south of the ITCZ in December–January–February even though biomass fires this time of year lead to regionally high-ozone regions in the northern tropics. As suggested by the annual cycle in tropospheric ozone at the Atlantic sounding stations (Fig. 16), the north-south gradient is most pronounced when south Atlantic subsidence and pyrogenic ozone sources are strongest, in September–October-November (upper panel in Fig. 18).

The latitudinal cross-section of tropospheric ozone structure over the Atlantic was first determined from ozone soundings launched during oceanographic transects between South America and Europe [83, 84]. Weller et al. [84] noted the relationship between ozone and water vapor on several Polarstern cruises, especially low ozone in the upper troposphere in the presence of higher water vapor when convection was present. The January–February 1993 ozonesonde record of Weller et al. [84] was nearly identical to that measured during the first North America-South Africa cruise ("Aerosols99") with ozonesondes (January-February 1999; Fig. 1 in [85]; Fig. 19). Thompson et al. [85] designated the appearance of more tropospheric ozone south of the ITCZ when biomass burning was maximizing photochemical ozone production north of the ITCZ as an "ozone paradox" and suggested that the cause was a combination of interhemispheric transport of ozone, greater subsidence and lightning frequency in the southern hemisphere. The contrast in ozone origins (northern Africa vs southern Africa) appears in their air parcel back-trajectories (not shown).

Transport of ozone precursors and other tracers across the ITCZ in the December–January period was observed during the TROPOZ II aircraft experiment [86, 87]. Several studies [88–90] have explored the Atlantic "ozone paradox" with additional data. Edwards et al. [88] show evidence for cross-hemispheric transport in the satellite CO data (MOPITT) for January 2001. Their mechanism (similar to that in Fig. 6a in [78]) is shown as the north African-to-Atlantic flow in Fig. 4. Interpretations of the ozone paradox using radiation data and satellite observations point to important roles for lightning, convection and biomass [89, 90]. Chatfield et al. [2003] relate a profile over Ascension (3 February 1999) that was taken during the Aerosols99 cruise [85] to lightning and convective redistribution of ozone that occurred during transport from the Indian Ocean as well as from Africa. It is possible that ozone layers originating from burning regions occur above those attributed to lightning (Fig. 20). More general interpretations of tropical Atlantic ozone origins based on climatology and models appear in [82, 91–93].

Acknowledgments Thanks to Jacquelyn Witte (SSAI at NASA/Goddard) for the trajectory analysis and graphics expertise. Thanks to Robert Chatfield (NASA/Ames) for comments on the manuscript and for communication of work in progress.

5
References

1. Crutzen PJ, Heidt LE, Krasnec JP, Pollack WH, Seiler W (1979) Nature 282:253
2. Seiler W, Crutzen PJ (1980) Climatic Change 2:207
3. Crutzen PJ, Delany AC, Greenberg J, Haagenson P, Heidt L, Lueb R, Pollock W, Seiler W, Wartburg A, Zimmerman P (1985) J Atmos Chem 2:233
4. Delany AC, Haagensen P, Walters S, Wartburg AF, Crutzen PJ (1985) J Geophys Res 90:2425
5. Cros B, Delmas R, Nganga D, Clairac B, Fontan J (1988) J Geophys Res 93:8355
6. Lacaux JP, Servant J, Baudet JGR (1987) Atmos Environ 21:2643
7. Logan JA, Kirchhoff VWJH (1986) J Geophys Res 91:7875
8. Chatfield RB, Delany AC (1990) J Geophys Res 95:18473
9. Pickering KE, Thompson AM, Wang Y, Tao WK, McNamara DP, Kirchhoff VWJH, Heikes BG, Sachse GW, Bradshaw JD, Gregory GL, Blake DR (1996) J Geophys Res 101:23993
10. Chatfield RB, Vastano JA, Singh HB, Sachse GW (1996) J Geophys Res 101:24279
11. Thompson AM, Tao WK, Pickering KE, Scala JR, Simpson J (1997) Bull Am Meteorol Soc 78:1043
12. Reichle HG, Connors VS, Holland JA, Sherrill RT, Wallio HA, Casas JC, Condon EP, Gormsen BB, Seiler W (1990) J Geophys Res 95:9845
13. Connors VS, Gormsen BB, Nolf S, Reichle HG (1999) J Geophys Res 104:21455
14. Lamarque JF, Edwards DP, Emmons LK, Gille JC, Wilhelmi O, Gerbig C, Prevedel D, Deeter MN, Warner J, Ziskin DC, Khattatov B, Franci GL Yudin V, Ho S, Mao D, Chen J, Drummond JR (2003) Geophys Res Lett 30:1688, doi: 10.129/2003GL017503
15. Chatfield RB, Guo Z, Sachse GW, Blake DR, Blake NJ (2002) J Geophys Res 107:D16, doi: 10.129/2001JD000497, 2002
16. Matsueda H, Inoue HY (1999) Geophys Res Lett 26:2413
17. Thouret V, Cho JYN, Newell RE, Marenco A, Smit HGJ (2000) J Geophys Res 105:20497
18. Zahn A, Brenninkmeijer CAM, Asman WAH, Crutzen PJ, Heinrich G, Fischer H, Cuijpers JWM, van Velthoven PFJ (2002) J Geophys Res 107:4337
19. Fishman JC, Watson CE, Larsen JC, Logan JA (1990) J Geophys Res 95:3599
20. Fishman J, Fakhruzzman K, Cros B, Nganga D (1991) Science 252:1693
21. Shiotani M (1992) J Geophys Res 97:7625
22. Thompson AM, Hudson RD (1999) J Geophys Res 104:26961
23. Shiotani M, Hasebe F (1994) J Geophys Res 99:14575
24. Newchurch MJ, Sun D, Kim JH (2001) Geophys Res Lett 28:3151
25. Thompson AM, Witte JC, Oltmans SJ, Schmidlin FJ, Logan JA, Fujiwara M, Kirchhoff VWJH, Posny F, Coetzee GJR, Hoegger B, Kawakami S, Ogawa T, Fortuin JPF, Kelder HM (2003) J Geophys Res 108:8241
26. Thompson AM, Witte JC, Oltmans SJ, Schmidlin FJ (2004) Bull Am Meteorol Soc (in press)
27. J Geophys Res (1996) 101, 30 October issue
28. Andreae MO, Fishman J, Garstang M, Goldammer JG, Justice CO, Levine JS, Scholes RJ, Stocks BJ, Thompson AM, van Wilgen B (1994) Biomass burning in the global environment. First results from the IGAC/BIBEX field campaign STARE/TRACE-A/SAFARI-92. In: Prinn RG (ed) Global atmospheric-biospheric chemistry. Plenum Press, New York
29. van Wilgen BW, Andreae MO, Goldammer JG, Lindesay JA (1997) Fire in the Southern Africa savannas: ecological and atmospheric perspectives. Witwatersand University Press, Johannesburg
30. Levine JS (1991) Biomass burning: atmospheric, climatic and biospheric implications. MIT Press, Massachusetts
31. Levine JS (1996) Biomass burning and global change. MIT Press, Massachusetts
32. Randriambelo T, Baray J-L, Baldy S, Thompson AM, Oltmans SJ, Keckhut P (2003) Annales Geophysiques (in press)
33. Krishnamurti TN, Sinha MC, Kanamitsu M, Oosterhof D, Fuelberg H, Chatfield R, Jacob DJ, Logan J (1996) J Geophys Res 101:889

34. Stohl A, Eckhardt S, Forster C, James P, Spichtinger N (2002) J Geophys Res 107:4684
35. Piketh SJ, Swap RJ, Maenhaut W, Annegarn HJ, Formenti P (2002) J Geophys Res 107:4817, doi: 10.129/2002JD002056
36. Olson JR, Baum BA, Cahoon DR, Crawford JH (1999) J Geophys Res 104:5865
37. Thompson AM, Witte JC, Hudson RD, Guo H, Herman JR, Fujiwara M (2001) Science 291:2128
38. Fenn MA, Browell EV, Butler CF, Grant WB, Kooi SA, Clayton MB, Gregory GL, Newell RE, Zhu Y, Dibb JE, Fuelberg HE, Anderson BE, Bandy AR, Blake DR, Bradshaw JD, Heikes BG, Sachse GW, Sandholm ST, Singh HB, Talbot RW, Thornton DC (1999) J Geophys Res 104:16197
39. Gregory GL, Westberg DJ, Shipham MC, Blake DR, Newell RE, Fuelberg HE, Talbot RW, Heikes BG, Atlas EA, Sachse GW, Anderson BE, Thornton DC (1999) J Geophys Res 104: 5677
40. Hoell JM, Davis DD, Jacob DJ, Rodgers MO, Newell RE, Fuelberg HE, McNeal RJ, Raper JL, Bendura RL (1999) J Geophys Res 104:5567
41. Stoller P, Cho JYN, Newell RE, Thouret V, Zhu Y, Carroll MA, Albercook GM, Anderson BE, Barrick JDW, Browell EV, Gregory GL, Scahse GW, Vay S, Bradshaw JD, Sandholm S (1999) J Geophys Res 104:5745
42. Baldy S, Ancellet G, Bessafi M, Badr A, Lan Sun Luk D (1996) J Geophys Res 101:835
43. Taupin FG, Bessafi M, Baldy S, Bremaud PJ (1999) J Geophys Res 104:8057
44. Fujiwara M, Kita K, Ogawa T, Kawakami S, Sano T, Komala N, Saraspriya S, Suripto A (2000) J Geophys Res 105:1879
45. Randriambelo T, Baray JL, Baldy S, Bremaud P, Cautenet S (1999) Geophys Res Lett 26:1287
46. J Geophys Res 106, 27 November 2001 issue
47. Lelieveld J, Crutzen PJ, Ramanathan V, Andreae MO, Brenninkmeijer CAM, Campos T, Cass GR, Dickerson RR, Fischer H, de Gouw JA, Hansel A, Jefferson A, Kley D, de Laat ATJ, Lal S, Lawrence MG, Lobert JM, Mayol-Bracero OL, Mitra AP, Novakov T, Oltmans SJ, Prather KA, Reiner T, Rodhe H, Scheeren HA, Sikka D, Williams J (2001) Science 291:1031
48. Phadnis M, Levy H II, Moxim WJ (2002) J Geophys Res 107:4790, doi:10.129/2002JD002190
49. Zachariasse M, Smit HGJ, van Velthoven PFJ, Kelder HJ (2001) Geophys Res 106:28441
50. de Laat, ATJ, Lelieveld J (2002) J Geophys Res 107:4739
51. Komala N, Saraspriya S, Kita K, Ogawa T (1996) Atmos Environ 30:1851
52. Fujiwara M, Kita K, Kawakami S, Ogawa T, Komala N, Saraspriya S, Suripto A (2000) Geophys Res Lett 26:2
53. Kita K, Fujiwara M, Kawakami S (2000) Atmos Environ 34:2681
54. Tsutsumi Y, Sawa Y, Makino Y, Jensen JB, Gras JL, Ryan BF, Diharto S, Harjanto H (1999) Geophys Res Lett 26:595
55. Kondo Y, Koike M, Kita K, Ikeda H, Takegawa N, Kawakami S, Blake D, Liu SC, Ko M, Miyazaki Y, Irie H, Higashi Y, Liley B, Nishi N, Zhai Y, Ogawa T (2002) J Geophys Res 108:8402
56. Kita K, Kawakami S, Miyazaki Y, Higashi Y, Kondo Y, Nishi N, Koike M, Blake DR, Machida T, Sano T, Hu W, Ko M, Ogawa T (2002) J Geophys Res 108:8400
57. (1996) J Geophys Res 101 20 January issue
58. (1997) J Geophys Res 102 20 December issue
59. (1999) J Geophys Res 104 20 July issue
60. (2001) J Geophys Res 106, 16 December issue
61. Oltmans SJ, Johnson BJ, Harris JM, Vömel H, Thompson AM, Koshy K, Simon P, Bendura RJ, Logan JA, Hasebe F, Shiotani M, Kirchhoff VWJH, Maata M, Sami G, Samad A, Tabuadravu J, Enriquez H, Agama M, Cornejo J, Paredes FJ (2001) J Geophys Res 106:32503
62. Newell RE, Thouret V, Cho JYN, Stoller P, Marenco A, Smit HG (1999) Nature 398:316
63. Thompson AM, Pickering KE, McNamara DP, Schoeberl MR, Hudson RD, Kim JH, Browell EV, Kirchhoff VWJH, Nganga D (1999) J Geophys Res 101:24251
64. Longo KM, Thompson AM, Kirchhoff VWJH, Remer LA, de Freitas SR, Dias MAFS, Artaxo P, Hart W, Spinhirne JD, Yamasoe MA (1999) J Geophys Res 104:12113
65. Liu H, Jacob DJ, Bey I, Yantosca RM, Duncan BN, Sachse GW (2003) J Geophys Res 108:8786, doi:10.129/2002JD003102

66. Chan CY, Chan LY, Chang WL, Zheng YG, Cui H, Zheng XD, Qin Y, Li YS (2003) J Geophys Res, 8800, doi:10.129/2003JD003127
67. Oltmans SJ, Johnson BJ, Harris JM, Thompson AM, Liu HY, Vömel H, Chan CY, Fujimoto T, Brackett VG, Chang WL, Chen JP, Kim JH, Chan LY, Chang HW (2004) J Geophys Res (in press)
68. Schultz MG, Jacob DJ, Wang YH, Logan JA, Atlas EL, Blake DR, Blake NJ, Bradshaw JD, Browell EV, Fenn MA, Flocke F, Gregory GL, Heikes BG, Sachse GW, Sandholm ST, Shetter RE, Singh HB, Talbot RW (1999) J Geophys Res 104:5829
69. Cahoon DR, Stocks BJ, Levine JS, Cofer WR III, O'Neill KP (1992) Nature 359:812
70. Justice CO, Kendall JD, Dowty PR, Scholes RJ (1996) J Geophys Res 101:23851
71. Duncan BN, Martin RV, Staudt AC, Yevich R, Logan JA (2003) J Geophys Res 108:2003
72. Smyth SB, Sandholm ST, Bradshaw JD, Talbot RW, Blake DR, Blake NJ, Rowland FS, Singh HB, Gregory GL, Anderson BE, Sachse GW, Collins JE, Bachmeier AS (1996) J Geophys Res 101:24165
73. Cosijn C, Tyson PD (1997) S Afr J Sci 92:381
74. Garstang M, Tyson PD, Swap RJ, Edwards M, Kållberg P, Lindesay JA (1996) J Geophys Res 101:23721
75. Tyson PD, Garstang M, Thompson AM, D'Abreton P, Diab RD, Browell EV (1997) J Geophys Res 102:10623
76. Thompson AM, Witte JC, Freiman MT, Phahlane NA, Coetzee GJR (2002) Geophys Res Lett 29:1976
77. Browell EV, Fenn MA, Butler CF, Grant WB, Clayton MB, Fishman J, Bachmeier AS, Anderson BE, Gregory GL, Fuelberg HE, Bradshaw JD, Sandholm ST, Blake DR, Heikes BG, Singh HB, Talbot RW (1996) J Geophys Res 101:24043
78. Thompson AM (2003) Biomass burning. In: Potter TD, Colman BR (eds) Handbook of weather, climate, and water: atmospheric chemistry, hydrology, and societal impacts, chap. 14. Wiley, New York
79. Kim JH, Hudson RD, Thompson AM (1996) J Geophys Res 101:24317
80. Hudson RD, Thompson AM (1998) J Geophys Res 103:22129
81. Peters W, Krol MC, Fortuin JPF, Kelder HM, Becker CR, Thompson AM, Lelieveld J, Crutzen PJ (2004) Tellus B 56:21
82. Martin RV, Jacob DJ, Logan JA, Bey I, Yantosca RM, Staudt AC, Li Q, Fiore AM, Duncan BN, Liu H, Ginoux P, Thouret V (2002) J Geophys Res 107:4351, doi:10.129/2001JD001480
83. Smit H, Kley D, McKeen S, Volz A, Gilge S (1989) The latitudinal and vertical distribution of tropospheric ozone over the Atlantic Ocean in the southern and northern hemispheres. In: Bojkov RD, Fabian P (eds) Ozone in the atmosphere. A Deepak, Hampton, p 419
84. Weller R, Lilischkis R, Schrems O, Neuber R, Wessel R (1996) J Geophys Res 101:1387
85. Thompson AM, Doddridge BG, Witte JC, Hudson RD, Luke WT, Johnson JE, Johnson BJ, Oltmans SJ, Weller R (2000) Geophys Res Lett 27:3317
86. Jonquières I, Marenco A, Maalej A, Rohrer F II (1998) J Geophys Res 103:19059
87. Jonquières I, Marenco A (1998) J Geophys Res 103:19075
88. Edwards DP, Lamarque J-F, Attié J-L, Emmons LK, Richter A, Cammas J-P, Gille JC, Francis GL, Deeter MN, Warner J, Ziskin DC, Lyjak LV, Drummond JR, Burrows JP (2003) J Geophys Res 108:4237
89. Jenkins GR, Ryu J-H, Thompson AM, Witte JC (2003) J Geophys Res 108:4745, doi:10.129/2002JD003297
90. Chatfield RB, Guan H, Thompson AM, Witte JC (2004) Geophys Res Lett (in press), doi:10.129/2003GL018866
91. Moxim WJ, Levy H II (2000) J Geophys Res 105:17393
92. Peters W, Krol M, Dentener F, Thompson A, Lelieveld J (2002) Atmos Chem Phys 2:103
93. Roelofs GJ, Lelieveld J, Smit, HGJ, Kley D (1997) J Geophys Res 102:10637

The Handbook of Environmental Chemistry Vol. 4, Part G (2004): 255–276
DOI 10.1007/978-3-540-40037-0

Intercontinental Transport of Trace Substances from Boreal Forest Fires

Nicole Spichtinger-Rakowsky · Caroline Forster

Department of Ecology, Technical University of Munich, Am Hochanger 13, 85354 Freising-Weihenstephan, Germany
E-mail: spichtinger@forst.tu-muenchen.de; forster@forst.tu-muenchen.de

Abstract Forest fire emissions have strong impact on the composition of trace gases and aerosols in the atmosphere. They are not only a local disturbance factor but are transported on regional and even hemispheric scales, and they lead to radiative forcing and modify tropospheric and stratospheric chemistry.

Recently, due to strong burning events in Canada and Siberia, boreal forest fires have attracted scientific interest; as for the last decades the main concern has been with tropical biomass burning. Boreal forest fires differ from tropical ones in the composition of the emissions and the spatial and temporal variation of their occurrence. Higher fuel consumption rates in the boreal region compared to the tropics lead to higher intensity of burning and larger quantities of combustion products with respect to climatic impact. Detailed measurements combined with atmospheric transport models are essential to characterize the extent of the forest fire impact downwind of the source regions in terms of atmospheric pollution and its health effects.

In this chapter we give an overview of the occurrence and nature of boreal forest fires and their emissions. Ground- and satellite-based monitoring of active fires and emissions originating from these fires are discussed, as they are the basis for quantitative transport modeling. The mechanisms which are involved in the transport processes associated with boreal fire emissions, and the modeling of the transport is described. Finally, cases of long-range transport events of forest fire emissions are documented.

Keywords Forest fire · Biomass burning · Long- range transport · Transport modeling · Boreal region

1 Introduction

The boreal region covers Alaska and the northern parts of Canada and Eurasia and contains about 29% of the world's forested area and 73% of global coniferous forests. More than one-third of the total terrestrial carbon is stored there [1, 2]. Thus, the boreal ecozone is a huge factor for the environment at local, regional, and global dimensions. As a net sink of atmospheric carbon – a consequence of plant photosynthesis and plant and microbial respiration – the boreal forest is an important factor in the global carbon cycle with strong impact on the global climate system. Not considering the effect of fires, the extended boreal forest stands were thought to be the main sink of carbon emissions from anthropogenic fossil fuel consumption and destruction of tropical forest [3]. However, the boreal forest is converted from a distinct carbon sink to a large carbon source if it is destroyed by fire [4]. That means, in addition to ecological modifications, there is a potential positive feedback mechanism between climate change and burning in the boreal region. The total biomass burned in the boreal region contributes around 1.2–11.5% of the global total burned biomass [5] and 15–20% of total global emissions from biomass burning [6].

In particular boreal forest fires affect not only the local ecosystem. Fishman et al. [7] have already mentioned that biomass burning emissions can be transported over large distances in the tropics. Meanwhile, a number of studies have shown that also boreal fires influence tropospheric chemistry on regional [8, 9] and even continental [10, 11] and hemispheric scales [12–14]. Recently even transport of forest fire emissions to the upper troposphere/lower stratosphere has been documented using POAM III and SAGE II satellite aerosol extinction data [15, 16] and Lidar measurements [17].

1.1 A Comparison of Tropical and Boreal Forest Fires

The input of biomass burning emissions into the atmosphere is often discussed as a phenomenon of the tropical rain forest and savannas and their climatic impact [18–20]. Because of the greater amount of carbon released from tropical biomass burning, the higher population density in the tropics and the complex issues of land use, tropical fires have attracted scientific and popular interest much more than boreal forest fires. Until the last decade boreal forest fires have almost gone unnoticed. However, the boreal forest contains one-third of the terrestrial carbon storage, emits up to 20% of the global biomass burning CO [21, 22] and is situated in the ecologically sensitive northern hemispheric high latitudes [1, 4, 23, 24]. In particular, effects of climate change in terms of increasing temperatures are expected to be more pronounced in the boreal regions compared with the tropics [1, 25, 26]. Furthermore, the lifetimes of many trace gases and aerosols in the atmosphere are much longer in the boreal region than in the tropics. As most fires in the tropics occur away from the Inter-Tropical Convergence Zone (ITCZ) during the dry season, the vertical transport of the emissions is limited by the trade wind inversion [20]. Only if low-level transport processes carry biomass burning smoke towards convergent regions do they become subject to deep convection, as was shown recently (e.g., [19]; see Chap. 8 of this book). In the boreal region the fire products can directly reach the free troposphere [1], where the lifetimes of the constituents are longer than in the lower troposphere, e.g., [27].

Comparatively low annual average temperatures and dryness at northern latitudes lead to accumulation of ground fuel through slow decomposition. Consequently, boreal fires burn with high intensity, consuming huge amounts of fuel, and spread particularly fast. The average fuel consumption is 2000 to 5000 kg/ha in the tropics, but up to 25,000 kg/ha in the boreal region [25, 28]. Sometimes the fires trigger and enhance deep convection in the atmosphere. Once the emissions have reached the free troposphere within high towering convection columns, they are picked up by long-range transport processes, as described below. The fire products travel over long distances and influence the atmospheric composition over hemispheric scales.

In contrast to tropical fires, which are mostly ignited by human activity, boreal forest fires are more often started by lightning flashes. These wildfires occur more frequent during warm and dry weather conditions and with increasing flash frequency and thus are characterized by a strong inter-annual variability (see below). Wildfires are thus susceptible to climatic change, while human made fires are solely intensified by weather conditions favoring fire [29]. This is also reflected by the return interval of fires, which is 3–5 years in savannas, where fire are ignited by humans and on average every 50–200 years in boreal regions [30, 31], where fires are more often ignited naturally by flashes. In the Canadian boreal region the contribution of lightning to fire ignition is higher than in Russia. In Russia only around 15% of the fires are caused by lightning [32], while in Canada this fraction is larger than 35%. Furthermore, lightning-triggered fires in unpopulated terrain are larger on average than those started by humans. They consume more than 85% of the total burned area in Canada [30].

1.2
Spatial and Temporal Distribution of Boreal Forest Fires

"Wildfire is a natural part of the life cycle of the boreal forest" [31]. The boreal region consists of large almost unpopulated territories. Therefore, many fires are not actioned because they pose no threat to human property. For example, in Canada these unrestricted fires contribute 50% of the total area burned [33].

Table 1 Areas burned in the boreal region, UN-ECE [34] averaged 1991–1998, Conard et al. [35] averaged 1990–1999

Country	Area burned	Reference
Canada	3.3 million ha	UN-ECE [34]
Russia	5–12 million ha	Conard et al. [35]
Scandinavia	1.900 ha	UN-ECE [34]

Without human intervention normally fires occur every 50 to 200 years. This so-called fire cycle has strong influence on the age structure, the species distribution, and the carbon storage of the forest stand. In general, forest fires are the major natural disturbance factor in boreal forests, because of huge amounts of highly flammable coniferous stands. Litter, consisting of slowly decompostable material, piles up on the floors and becomes extremely flammable.

Boreal forest fires occur mainly from May to September [5] with a maximum of burned area in June and July. Most fires occur in Canada and Russia, with two-thirds situated in Russia (Table 1).

The seasonal occurrence of boreal fires correlates well with summertime dryness, high temperatures and strong lightning activity. Notably the continentality, especially of the Siberian part of the boreal region, leads to strong dryness during the burning season [36]. Because of their dependence on meteorological conditions and anthropogenic influences, boreal forest fires have a high degree of interannual variability [37]. From 1959 to 1999 fires contributed from 2–75% (on average 18%) to Canadian CO_2 emissions [38]. There are some years with extremely high fire activity in the boreal region (e.g., 1987 and 1998 a total of 14.4 million ha and 17.9 million ha respectively) but there are also years with relatively little burning (e.g., in 1992 only 2–3 million ha). The years with high fire activity are associated with persistent dryness [39].

Forest fire occurrence and severity in Canada has increased since the 1970s [31, 40]. In Russia, fire occurrence had been nearly constant since 1960, but increased during the last decade [24]. These observations might be traced back to a changing climate. Due to higher temperatures, extended dryness and enhanced convective activity in a warmer climate increased lightning and fire occurrence is expected [24]. Stocks et al. [1] calculated a doubled CO_2 climate scenario with a General Circulation Model and found an earlier start of the burning season and an increase in area burned in Canada due to a fire favoring warmer climate. Furthermore, the fire-cycle was shortened under this scenario. Similarly, in Russia the severity and geographical extent of high to extreme fire danger conditions will increase by at least 50% over the next three decades [41].

1.3 Types of Boreal Forest Fires

Depending on the local vegetation structure [4] and the kind of fire ignition, different types of forest fires occur, which influence the emission structure and the subsequent transport. In the boreal region mainly three types of fire appear:

crown fires, surface fires, and peat fires. Surface fires consume grass, shrubs and litter, while crown fires burn the crowns of coniferous trees.

Canadian forest fires are often crown fires (around 80%) with high fire intensity and energy release. Siberian fires are 75% surface fires [42], although during severe fire years crown fires could contribute 50% [21]. The type of fire influences the injection height of emissions into the atmosphere. The high temperature- and energy-levels of crown fires cause a strong buoyancy-driven vertical uplifting of the burning products. Lavoue et al. [5] calculated an average injection height of Canadian forest fire emissions of 7600 m, and of Siberian fires of 3000 m. Spichtinger et al. [14] determined the injection height of Canadian crown fire emissions by comparing air mass trajectories of different altitudes starting at large fire locations with the smoke plumes visible in AVHRR images. The trajectories at 3000 to 5000 m corresponded most closely to the directions in which the smoke plumes traveled. The injection height is an important factor to consider in atmospheric transport models (see below).

Peat fires (also called ground fires) burn at the surface and even underground where they can burn over long periods. The combustion is incomplete, resulting in high emissions of CO and particles. This type of fire is mostly manmade, as naturally peat lands would not burn, because peats are normally saturated with water. Even during the dry season only some centimeters at the surface are drying out. However, if drained by humans, peat lands have a high fire danger and once peat fires are ignited they burn the highest amounts of organic matter per ha. Peat fires are a common tropical type of fire. For instance, the Indonesian fires of 1997/1998 were dominated by peat fires producing huge amounts of smoke and haze [29]. In the boreal region such fires are still uncommon. During the strong fires in the far east of Russia in 1998 only 1% of the fires were peat fires. However, such as during the fire season of the year 2002, severe peat fires occurred in the European part of the Russian federation. Smoke plumes of these fires were transported from Moscow towards Sweden and even Germany in 2002 [43]. Due to extended severe droughts and drainage by humans this fire type will likely occur more often in the future.

2 Fire Detection and Emissions

2.1 Fire Detection and Monitoring of Areas Burned

The location of and the area burned by fires are documented in official fire inventories of several burning regions, but their quality differs with region. For example, the Government of the Northwest Territories (2001) of Canada keeps a detailed record of all fires that burned more than 100 km^2. For all Canadian districts the Canadian Forest Service provides fire reports and maps containing areas burned and numbers of fires. The data are provided weekly and also for the whole burning period since the year 1998. For all fires larger than 200 ha a so-called Large Fire Database has been developed for Canada, including the location, starting and ending time of the large fires and other information [33, 37].

The situation in Siberia is more complex. For political and forest administrative reasons a large number of forest fires remains undetected in Siberia [28]. Therefore, official data for Siberia strongly underestimate areas burnt [21, 22, 44]. Most times the governmental data are cumulative yearly or monthly burned areas for the respective region or summary statistics, but no spatially resolved inventories similar to those of Canada are available.

In remote regions, ground-based fire documentation frequently underestimates the real extent of burning. There, the use of satellite techniques to detect fires and estimate the areas burned, considerably improved the data situation in recent years. Global fire maps with different quality and resolution are available. Satellite instruments record fire locations with globally consistent data quality and some of the data are available in near real-time. Major limitations are the sensitivity to cloud effects and other atmospheric parameters which influence the measurements [45].

From several satellite instruments data of hot spot locations (e.g., Along Track Scanning Radiometer (ATSR)), burned areas and scars (e.g., Landsat Thematic Mapper (TM), Moderate Resolution Imaging Spectrometer (MODIS), AVHRR), and recently of the fire energy (e.g., MODIS, Bi-spectral Infrared Detection (BIRD)) and the burning phase (e.g., MODIS) are available. These fire relevant parameters are derived from raw satellite data by applying sophisticated algorithms.

The AVHRR was one of the first instruments applied to fire monitoring [46–48]. Originally it was designed for meteorological observations. With series of threshold tests potential fire pixels, so-called hot spots, are marked due to their brightness temperature and reflectance. Falsely detected fire pixels are eliminated by taking into account land cover, cloud cover, and background temperatures. The low spatial resolution of the AVHRR sensor of 1 km^2 leads to overestimation of the area burned by small fire events, but this error is only small in Canada and Siberia, where large fires dominate [33, 41]. AVHRR is limited by its low saturation temperature of 315 K. In contrast, MODIS has a much higher saturation temperature of 400 to 500 K. Further developed fire products are provided globally in near real-time (Fig. 1). MODIS instruments fly on two different spacecraft, each of them delivering two observations daily with spatial resolutions of 250 m, 500 m, and 1 km. From these four daily observations, fire and smoke distributions are derived by using thermal signatures to separate fire signals from background signals [45, 49]. MODIS data are often used together with Geostationary Operational Environmental Satellites (GOES) to get information on the fire activity at the time between the single MODIS crossings. The GOES at a height of 36,000 km is fixed at a point above the equator and senses fires in the western hemisphere with high temporal resolution. MODIS can distinguish between ground and surface fires; even a rough estimate of smoldering-to-flaming ratio will be developed [49] by accounting for different structure of emitted products during the phases of burning. Additionally, MODIS determines the emitted radiant energy released by combustion processes to estimate the rate of atmospheric pollutant production.

Another sensor to analyze fire energy is BIRD. It has been specifically designed to detect hot spots and to analyze infrared radiation emitted from fires and vol-

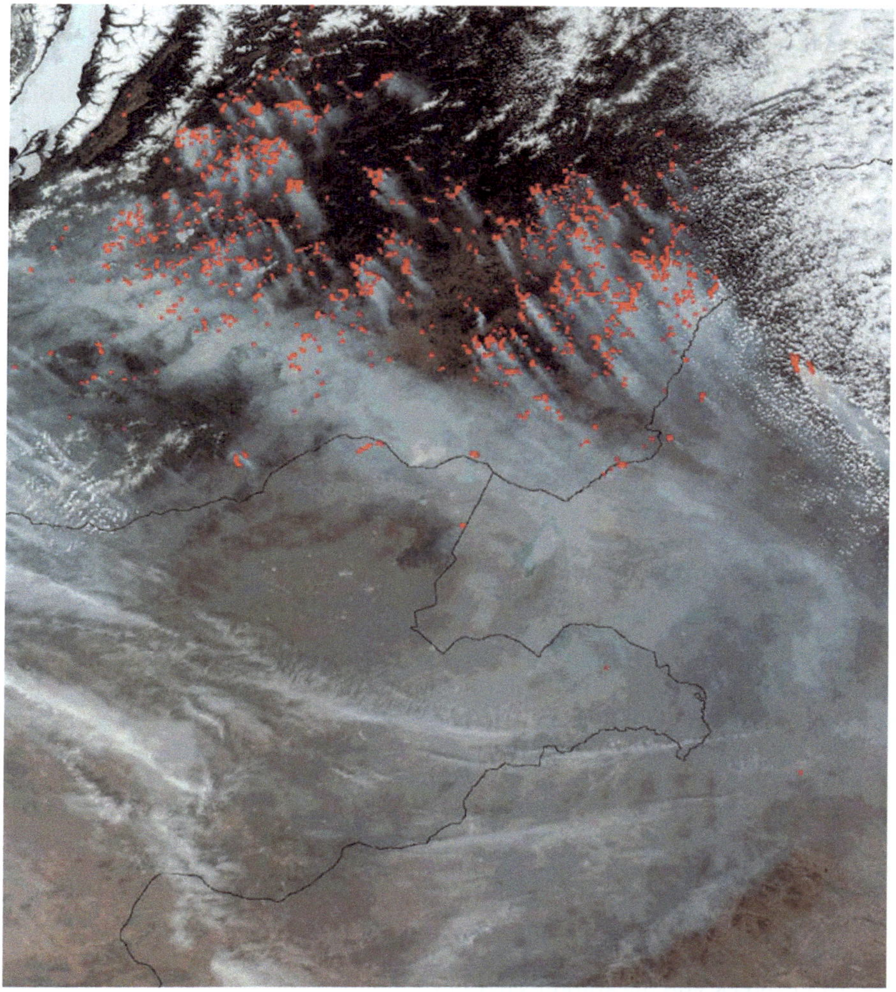

Fig. 1 Fire hot spots and smoke plumes east of Lake Baikal detected by MODIS on May 7th, 2003, onboard of the Aqua satellite with a pixel size resolution of 1 km

canoes [50]. Higher spatial resolution is obtained with the Landsat TM. Page et al. [29] estimated the burned area and material of the Indonesian fires of 1997 with remote sensing techniques by using TM images before and after fire detection and combined them with field measurements. However, TM's temporal coverage is limited and the revisit frequency is rather low (up to 16 days) [45].

Other satellite instruments are applied for fire detection in addition to the examples discussed above. Both the instruments itself and the fire detection algorithms are in a status of rapid development, such that improved fire products can be expected soon.

2.2
Trace Gas and Aerosol Emissions

The emissions from forest fires influence the atmosphere in various ways. Biomass burning emits large quantities of carbon dioxide (CO_2), carbon monoxide (CO), methane (CH_4), nitrogen oxides (NO_x), and aerosols. Amiro et al. [38] estimates that in extreme fire years, forest fire produced carbon emissions could arise to amounts equivalent to 75% of the total carbon emissions from fossil fuel consumption in Canada. In the Russian part of the boreal zone, the fires contribute 30–40% of the total natural and anthropogenic carbon flux into the atmosphere [36].

The largest part (80–90%) of forest fire carbon emissions consists of CO_2 [20] (IPCC, 1995). CO_2 and CH_4 are important greenhouse gases, which lead to global warming. The solar radiation budget is strongly affected by aerosols emitted from forest fires, too [51–53]. O'Neill et al. [54] found that optical depth variations in Canada can be linked to forest fire indices to a rate of 80%. Directly, aerosols backscatter sunlight and absorb sunlight and thermal radiation. Indirectly the cloud formation and cloud physics are affected by increasing numbers of cloud condensation nuclei (CCN) e.g. [42, 55, 56] and decreasing droplet sizes within the cloud, thus producing less rain [57, 58].

Tropospheric chemistry is influenced by the emission of CO, CH_4 and NO_x as precursors of tropospheric ozone (O_3) – an effective greenhouse gas itself – and nitric acid (HNO_3), a product which is part of the nitrogen chemical reaction cycle, leading to acid rain [59]. Additionally, other trace gases are emitted like methyl bromide and methyl chlorine influencing stratospheric ozone chemistry as they are involved in catalytic cycles of stratospheric O_3 destruction. Finally, forest fires are a source of stratospheric aerosols [15–17].

Forest fire emissions strongly depend on the phase of burning: smoldering or flaming. These two phases of burning occur at different temperatures and have different combustion efficiencies [60], defined as the amount of CO_2 relative to total carbon emitted. Flaming occurs at higher temperatures up to 1200 K with higher combustion efficiency and large amounts of CO_2. In contrast, smoldering is a very low-intensity incomplete combustion process. Smoldering temperature is around 600 K and the amounts of CO, CH_4, and particles emitted are relatively high [20, 61].

2.3
Inventories for Boreal Forest Fire Emissions

Information on the fire emissions of the different trace gases and aerosols are usually given as emission factors or emission ratios. The emission factor is a measure for the amount of species released per fuel mass burned (examples are shown in Table 2). The emission ratio (examples are shown in Table 3) is the mass of species released per mass of a reference species, which is normally CO_2 or CO. The emission ratio can be converted to an emission factor by multiplying with an appropriate conversion factor, typically the carbon content of the fuel burned [18].

Table 2 Emissionfactors are given in gram species per kilogram dry matter burned (numbers takenfrom Andreae and Merlet et al. [18])

Species	Extratropical forest	Tropical Forest
CO_2	1569±131	1580±90
CO	107±37	104±20
CH_4	4.7±1.9	6.8±2.0
NO_x(as NO)	3.0±1.4	3.9±2.4

Table 3 Emissionratios in % (numbers taken from Cofer et al. [62])

Canadian fires	CO/CO_2	CH_4/CO_2
Flaming	6.7±1.2	0.6±0.2
Smoldering	12.3±1.9	1.2±0.3

Data on the emissions by forest fires are especially important for the transport modeling of forest fire emissions, as these data are the basis for a quantitative representation of the atmospheric impact of forest fires in transport models (see below). They must best be provided in an emission inventory as emission per unit area and time at a specified spatial and temporal resolution, e.g.:

$$DFE = FC \times PA \times EF$$

where *DFE* are the daily fire emissions (kg), *FC* the daily fuel consumption (kg/m^2), *PA* the pixel or unit area (m^2), and *EF* the emission factor (g/kg) [46]. In addition, transport simulations need information on the emission height. This parameter has a great impact on the model results due to the variability of transport and wet deposition with altitude [63].

A first global estimation of the burned biomass budget was given by Seiler and Crutzen [64]. Since then a number of observational and modeling studies have been performed to estimate emission heights and emission factors for trace gases and aerosols released from wildfires and prescribed fires. Delmas et al. [65] and Andreae and Merlet [18] summarized these data on emissions from various types of biomass burning obtained in the last couple of decades including tropical, e.g., [25, 66] and boreal, e.g., [25, 60, 62, 67] fire emissions. A first spatial and monthly inventory for boreal and temperate wildfire emissions for the years 1960–1997 was given by Lavoue at al. [5]. They developed a methodology to construct 1°×1° forest fire emission maps for various regions in the northern hemisphere by accounting for the different vegetation types and associating the fuel loads and consumption efficiencies of these vegetation types to their location on the globe. Recently, the study of Duncan et al. [68] provided a global estimate of seasonal and interannual variability of biomass burning CO emissions for use in chemistry transport models (CTMs) based on satellite observations. Moreover, several recent studies on emission estimations for severe fire years, e.g., [6, 69] and different boreal forest fire regions e.g. [21, 42, 70, 71] are available.

Although a number of estimates for boreal forest fire emissions exist, these emission data are still highly uncertain for several reasons. First, boreal wildfires mostly occur in solitary regions, especially in Russia, where burned area statistics are often incomplete (see above). Second, the severity of the fires and the injection heights of the emissions are highly uncertain, as the convection associated with the fire is highly variable and unpredictable and depends on the type of fire (see below). Third, emissions from residual smoldering combustion, which can be produced for a long time after the passage of the fire front, are still largely uncharacterized [61, 72, 73]. Bertsci et al. [73] proposed a method to estimate emissions from residual smoldering combustion and obtained large changes in estimates of previous biomass burning emissions with the inclusion of residual smoldering combustion emissions.

3 Transport Processes

Once injected to the free troposphere, boreal forest fire emissions can be subject to substantial vertical and horizontal transport. Basically, the transport processes involved in the long-range transport of boreal forest fire emissions can be the same as for anthropogenic emissions. These are warm conveyor belts (WCBs), convection, and slow low level transport, which are described in more detail in Chap. 2 of this book. However, a great difference between anthropogenic emissions and boreal forest fire emissions is that fires strongly affect radiative processes through absorption by soot, produce high temperatures, and can therefore develop their own dynamics and convection, which injects aerosols and trace gases to the free troposphere [1]. Injection heights of gases and aerosols can reach several kilometers (see above), where some climatically relevant species like O_3 are chemically more stable and thus depleted more slowly during transport than in the boundary layer. Plant material like conifer needles, pollen cones, bracts, and wood fragments can also be lifted, then horizontally transported and deposited at least 20 km away from the fire [74].

In wildland firefighting two kinds of fires are distinguished: the two-dimensional surface fires and the three-dimensional fires, in which the fire itself generates the winds that drive the fire [75]. The surface fires are reasonably predictable. They are driven by the winds and temperatures of the surrounding. Emissions released by such fires can thus passively be taken up by a WCB or by convection similar to anthropogenic emissions. However, when winds in the lower 2–3 km of the troposphere are light, the buoyancy in the convective column over the fire can produce strong vertical growth of the convective column and the fire is no longer a surface, but a three-dimensional phenomenon [75]. The strong updrafts and downdrafts can drive strong horizontal outflow and inflow, which feeds fresh oxygen to the fire, allowing it to intensify rapidly [76]. The behavior and the spread of such fires are highly unpredictable. In addition, the convective column over the fire can evolve into a dry storm, which has the potential to ignite new fires by lightning.

If deep convection occurs in the vicinity of a fire, the convection associated with the fire itself can intensify the deep convective activity, which in turn can increase the fire activity. A number of recent studies where forest fire aerosols and trace gases were observed relatively deep in the lower stratosphere already suggested that convective activity is a key factor for the transport of forest fire emissions to high altitudes, e.g., [15, 17, 77]. However, one of the first studies that documented a case, where strong winds associated with a thunderstorm likely intensified a forest fire in Canada, which in turn intensified the convective activity in the thunderstorm, was published recently by Fromm and Servranckx [16]. A super cell developed, and within the over shooting top of the super cell aerosols and trace gases were injected into the lower stratosphere. The authors note that the mutual intensification of fire and convective activity and the subsequent transport to the upper troposphere and lower stratosphere have a so far under-appreciated effect on the radiation and chemistry in this region. Similarly, Jost et al. [78] recently presented measurements of boreal forest fire emissions deep in the stratosphere and concluded that the observed particles and trace gas enhancements potentially affected the cloud properties and the chemistry in the upper troposphere and lower stratosphere.

4 Transport Modeling of Boreal Forest Fire Emissions

Model studies are an essential tool to learn more about the transport pathways of boreal forest fire emissions and to understand their chemical transformation during the transport processes. The ability of models to determine impact areas and pollution concentrations downwind of forest fires is important for the assessment of atmospheric pollutants from biomass burning and their health effects [79].

4.1 Modeling of Small-Scale Plume Processes

As wild fires have their own dynamics and convection, the regional and global impacts of their emissions on the atmospheric composition depend on small scale processes like the non-linear production of O_3 close to the fire [80], the possible formation of clouds within the convective column associated with the fire and the subsequent washout of the emissions [81]. It would therefore be desirable to apply small-scale models that compute chemistry, meteorology, and scavenging in the plume close to the fire. Such models can be submodules of regional- and global scale models, or can be the basis for developing parameterizations needed by the larger scale models [82]. A numerical model that simulated the transport of particles from a fire and their interaction with clouds was presented by Penner et al. [83], Clark et al. [84] and Grishin et al. [85] showed results from models of coupled fire dynamics and meteorology. Westphal and Toon [86] used a numerical model of meteorology, aerosols, and radiative transfer to study the impact of a smoke plume on atmospheric processes and successfully compared their results with satellite data. Recently, Trentmann et al. [87] used a high-resolution tracer model to simulate the transport of aerosol emissions from a prescribed fire, and Jost at al. [78] simulated chemical processes in a biomass burning plume by using a simple dilution box model. Such studies help to increase our knowledge on the processes within biomass burning plumes and improve the representation of forest fires within large-scale models.

4.2 Long-Range Transport Modeling

If a biomass burning plume is detected by lidar, surface or aircraft observations, the first question that arises is: where does this plume come from? A number of studies used trajectory models as a simple and useful tool to investigate the origin of observed aerosols and trace gases, e.g., [8, 9, 12, 88]. Waibel et al. [77] detected highly elevated CO concentrations in the upper troposphere and lowermost stratosphere over Europe during several research flights in August 1994. Backward trajectories starting at the flight tracks showed that the trajectory pathways passed locations in Canada where severe forest fires burned at the same time. Similarly, by using backward trajectories, Hsu et al. [56], Siebert et al. [17], and Fromm et al. [15] found that an aerosol cloud observed in the stratosphere over Greenland and Sweden in summer 1998 could be traced back to forest fire emissions originating in Canada (see below).

However, while traditional trajectory calculations provide information on the origin of observed polluted air masses, they do not give information on the dispersion or even chemical transformation during the transport. Therefore, it is desirable to do modeling studies with tracer transport models (TTMs) and CTMs. TTMs use passive tracers and help to identify the source regions, the transport pathways, and the dispersion of pollutants, e.g., [89]. CTMs are used to obtain information on the chemical loss and production of different atmospheric trace constituents during the transport processes, e.g., [90].

The uncertainties in the information on the emissions from boreal forest fires make it difficult to model the transport of boreal forest fires. Therefore, in contrast to the tropics where emission information is more accurate and where a number of CTM studies have been performed on a regional and global scale in recent years, e.g., [91–93], only a few long range transport modeling case studies and no climatology with TTMs and CTMs exist for boreal forest fire emissions to date. Photochemical modeling in combination with a Lagrangian plume model was done to quantify the influence of biomass burning emissions on the regional O_3 budget in the Arctic [94], and a CTM was used to find out to what extent biomass burning contributes to the tropospheric NO_x over high northern latitudes [95]. However, both studies considered only one- to two-day transport of the biomass burning plume. Galanter et al. [96] investigated the impacts of both tropical and extra-tropical biomass burning on tropospheric CO, NO_x, and O_3 using a global CTM. They found that boreal fires are a major source of CO and NO_x in the northern high latitudes during summer and fall.

The first study that documented and quantified long range transport of boreal wildfire emissions was presented by Wotawa and Trainer [11]. By using a TTM they successfully simulated the transport of boreal forest fire emissions from the Northwest Territories in Canada over more than 3500 km during summer 1995. It was shown that over a period of two weeks these emissions were the dominant source for highly elevated CO and O_3 concentrations measured over the south eastern United States, a region with one of the highest anthropogenic emission rates worldwide. McKeen et al. [97] extended the study by Wotawa and Trainer [11] by incorporating wild fire emission estimates for CO, NO_x, and nonmethane hydrocarbons into a CTM, which was specifically designed to simulate ozone photochemistry in the continental United States. They found that the Canadian wildfires of summer 1995 contributed 10 to 30 ppbv O_3 enhancement over the background over a large region in central and southeastern United States.

Another severe fire year was 1998. Large forest fires burned in Central America and produced elevated levels of aerosols at the Southern Great Plains site in the United States. Cheng et al. [98] recently applied the potential source contribution function (PSCF) modeling technique [99] to this case. In combination with a trajectory analysis they could correctly determine the origin and the transport pathways of smoke particles from the fires. For the Canadian forest fires in August 1998, Forster et al. [13] and Spichtinger et al. [14] documented how emissions from these fires were advected over the Atlantic to Europe. These two studies will be discussed in more detail in the next section, which presents recent cases and the related model studies of long range transport of emissions from severe boreal forest fires.

4.3 Case Studies of Long-Range Transport

The burning season 1998 was severe in both Canada and Siberia with a total of 17.9 million ha of fire, which is assumed to be the highest of the decade [6]. For instance, in the Canadian Northwest Territories more than 1000 fires burned more than one million ha of forest between August 5 and 11, the highest one-week sum reported by the Canadian Forest Service throughout the 1990s.

The aerosol and trace gas emissions from the 1998 fires were subject to large horizontal and vertical transport, which could be observed by lidar, aircraft, surface stations, and satellites over the entire Northern Hemisphere, e.g., [100]. It was estimated that the fires contributed about 25–30% of the atmospheric methane anomaly of 1998 [101]. Fromm et al. [15] studied POAMIII and SAGE data for the burning season, May to October 1998, and found an enhanced aerosol layer in the northern hemispheric lower stratosphere from August to October and brief enhancements in May and July. By using isentropic backward trajectories and AVHRR satellite data, Kato et al. [8] found that enhanced tropospheric CO and O_3 concentrations over Japan in 1998 were due to the fires in Siberia.

In more detail, the transport of the emissions from the severe Canadian fires at the beginning of August 1998 has been investigated by several studies. Spichtinger et al. [14] used Total Ozone Mapping Spectrometer (TOMS) aerosol index data, Global Ozone Monitoring Experiment (GOME) NO_2 columns, and TTM simulations of a NO_x tracer to document how the forest fire plume traveled from the Canadian Northwest Territories over Greenland and the Atlantic ocean towards Europe. On August 5, the aerosol and NO_2 maxima were located just west of Greenland (Fig. 2a,c) and agree well with the patterns shown in the transport simulation (Fig. 2b). On August 9, the plume had already reached Europe and NO_2 enhancements were even visible at the west coast of Ireland. At the same day, the

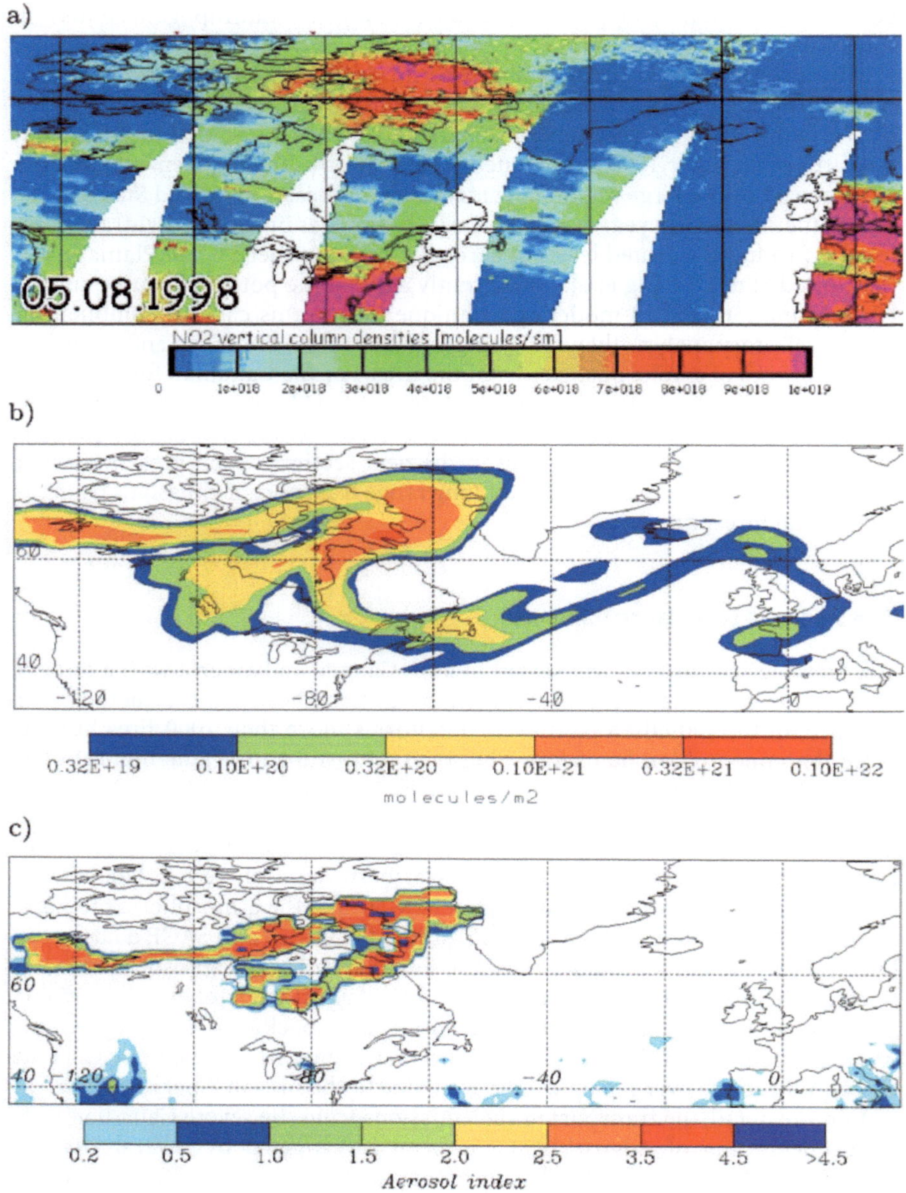

Fig. 2a–c Maps of: **a** tropospheric NO_2 slant column densities; **b** FLEXPART NO_x tracer columns at 18 UTC; **c** TOMS aerosol index on 5 August 1998. (Figure taken from [14])

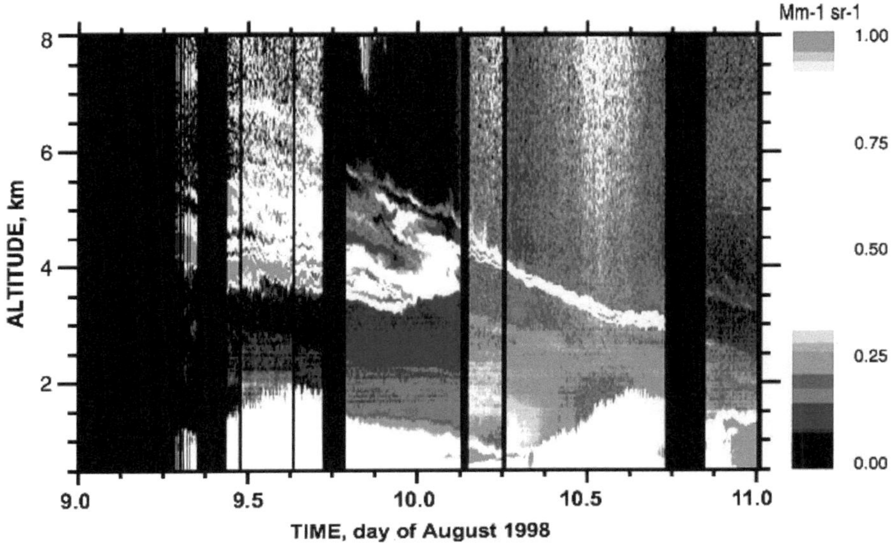

Fig. 3 Time-height contour of the 532-nm particle backscatter coefficient at Lindenberg on Aug 9 and 10, 1998. The measurements were taken with the IfT multiwavelength lidar during LACE 98 with a time resolution of 15 m. The data were smoothed with a running average of 2.5 min in time and 75 m in height. (Figure taken from [13])

aerosol plume has been observed by lidar over Lindenberg, Germany [13, 102, 103] (Fig. 3). Although wet and dry deposition occurred during transport, the aerosol layer was so dense that meteorological observers falsely reported cirrus clouds. A few days later chemical, physical and optical aircraft observations were performed in the same plume over Greece [104, 105].

During the second half of August 1998, the surface observations at Mace Head, Ireland, showed substantial CO enhancements over the background (Fig. 4). By simulating an anthropogenic and a forest fire CO tracer with a TTM, Forster et al. [13] confirmed, that substantial amounts of Canadian forest fire emissions were transported across the Atlantic within about a week and polluted large regions over Europe during the second half of August. While only 32% and 10% of the CO enhancements were caused by anthropogenic European and North American emissions, respectively, 58% of the CO enhancements were due to Canadian forest fire emissions (Fig. 4).

In 2003, the burning season in Russia started unusually early already in March/April. Strong fires were burning in the south of the Russian federation. Until the end of May more than 14 million ha were destroyed by fire, which is substantially more than the 11.7 million ha that were burned during 2002 and the 9.5–11.5 million ha of 1998 [21]. Well documented by satellite images the forest fire plume was transported around the whole northern hemisphere (Figs. 5 and 6). Ongoing work of Damoah et al. [106] will characterize and quantify long-range transport processes of these fire emissions by combining satellite data and TTM simulations.

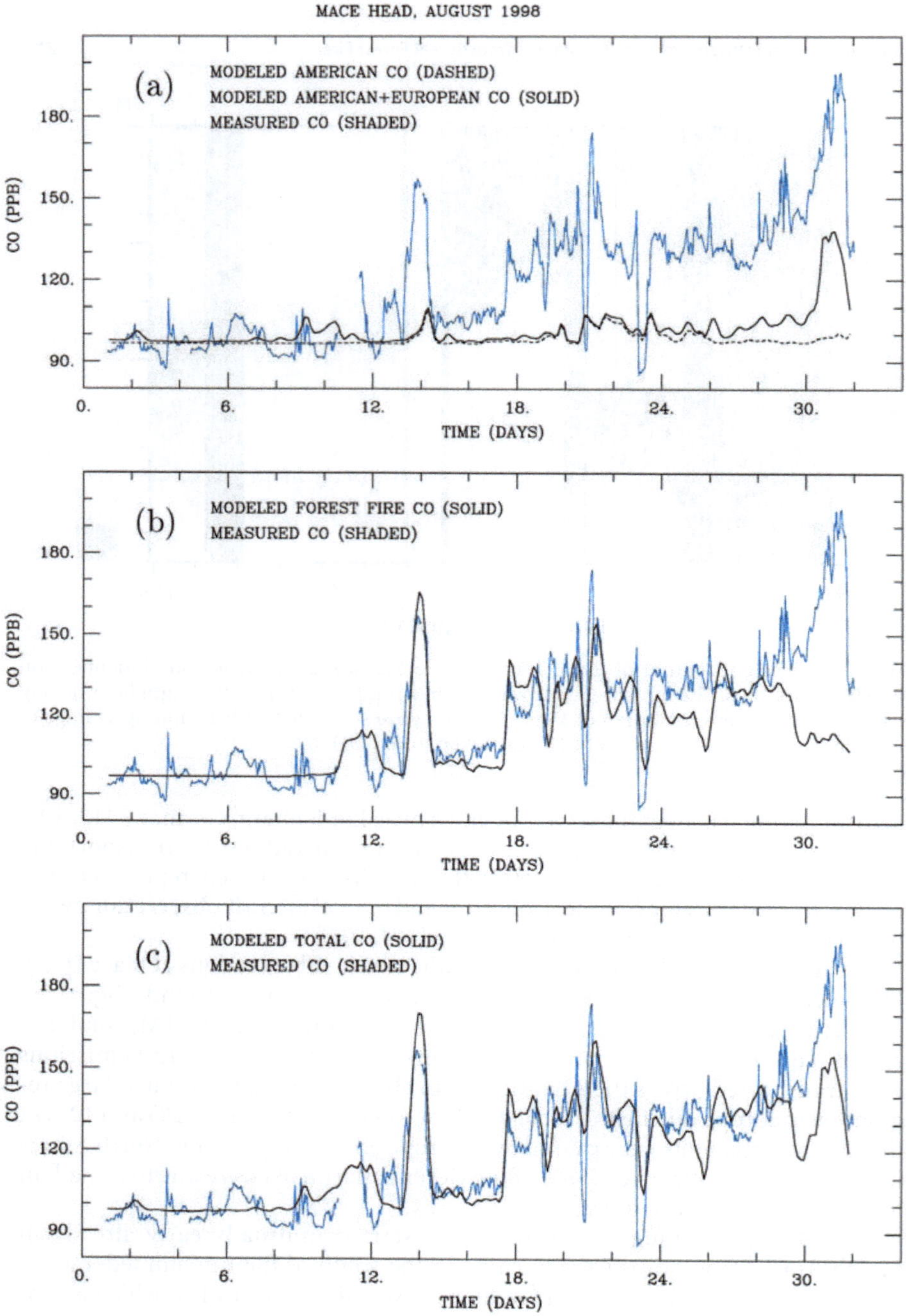

Fig. 4a–c Measured and simulated CO concentrations in pptv at Mace Head during August 1998: **a** model results for the American anthropogenic CO tracer and the American plus European anthropogenic CO tracer; **b** modeled CO concentrations for the forest fire emissions; **c** modeled CO concentrations from American and European emissions plus forest fires. A background of 97 ppbv was added to the simulated CO concentrations. (Figure taken from [13])

Fig. 5 The smoke plume over the Aleutians and Alaska detected by Seawifs (May 21th, 2003)

Fig. 6 The smoke plume stretching from the Atlantic over Southern Norway and Sweden to Finland detected by Seawifs (May 27th, 2003)

5 Summary and Conclusions

During recent years it has been noticed that boreal biomass burning emissions substantially contribute to the atmospheric trace gas and aerosol composition and chemistry. Boreal forests are a large source of carbon and thus the burning of boreal forests has an important impact on climate. If temperatures and dryness will increase in the future climate, the occurrence of boreal fires will increase and the atmospheric impact of their emissions will be even more intense.

Due to the convection associated with a fire, the fire emissions can be transported to high altitudes and subsequently advected over regional and even hemispheric scales. Observations by satellites, surface in situ, lidar, and aircraft measurements have shown evidence of pollution from boreal fires in regions far away from their source and even in the lower stratosphere. In order to assess the atmospheric impact of pollutants from fires, it is necessary to know how much trace substances are released by the fires and to understand the transport pathways and the chemistry of the released substances. For the latter, tracer transport and chemistry transport models are very useful tools. As the occurrence of boreal wildfires is highly variable in space and time though, accurate observational data on fire emissions, which are an important input for the models, are difficult to obtain. At present, the number of surface, lidar, and aircraft measurements of pollutants from boreal biomass burning increases, and the quality and availability of satellite data improves. In addition, research projects specifically dealing with the boreal forest, e.g., [67, 107], help to close the gaps in knowledge and data availability. However, still more information and consistent data bases are needed to improve the quality of model results related to the atmospheric impact of boreal forest fires.

6 References

1. Stocks BJ, Fosberg MA, Lynham TJ, Mearns L, Wotton BM, Yang Q, Jin JZ, Lawrence K, Hartley GR, Mason JA, McKenney DW (1998) Climatic Change 38:1
2. Dixon RK, Browns S, Houghton RA, Solomon AM, Trexler MC, Wisniewski J (1994) Science 263:185
3. Fan S, Gloor M, Mahlman J, Pacala S, Sarmiento J, Takahashi T, Tans P (1998) Science 282:442
4. Kim Y, Tanaka N (2003) J Geophys Res 108:8154
5. Lavoue D, Liousse C, Cachier H, Stocks BJ, Goldammer JG (2000) J Geophys Res 105: 26871
6. Kasischke ES, Bruhwiler LP (2002) J Geophys Res 108:8146
7. Fishman J, Watson CE, Larsen JC, Logan JA (1990) J Geophys Res 95:3599
8. Kato S, Pochanart P, Hlrokawa J, Kajii Y, Akimoto H, Ozaki Y, Obi K, Katsuno T, Streets DG, Minko NP (2002) Atmos Environ 36:385
9. Tanimoto H, Kajii Y, Hirokawa J, Akimoto H, Minko NP (2000) Geophys Res Lett 27: 4073
10. Cooper OR, Moody JL, Thornberry TD, Town MS, Carroll MA (2001) J Geophys Res 106:24289

11. Wotawa G, Trainer M (2000) Science 288:324
12. Chan CY, Chan LY, Harris JM, Oltmans SJ, Blake DR, Qin Y, Zheng YG, Zheng XD (2003) J Geophys Res 108:4015
13. Forster C, Wandinger U, Wotawa G, James P, Mattis I, Althausen D, Simmonds P, O'Doherty S, Jennings SG, Kleefeld C, Schneider J, Trickl T, Kreipl S, Jager H, Stohl A (2001) J Geophys Res 106:22887
14. Spichtinger N, Wenig M, James P, Wagner T, Platt U, Stohl A (2001) Geophys Res Lett 28:4579
15. Fromm M, Alfred J, Hoppel K, Hornstein J, Bevilacqua R, Shettle E, Servranckx R, Li ZQ, Stocks B (2000) Geophys Res Lett 27:1407
16. Fromm MD, Servranckx R (2003) Geophys Res Lett 30:1542
17. Siebert J, Timmis C, Vaughan G, Fricke KH (2000) Ann Geophys Atmos Hydrospheres Space Sci 18:505
18. Andreae MO, Merlet P (2001) Global Biogeochem Cycles 15:955
19. Andreae MO, Artaxo P, Fischer H, Freitas SR, Gregoire JM, Hansel A, Hoor P, Kormann R, Krejci R, Lange L, Lelieveld J, Lindinger W, Longo K, Peters W, de Reus M, Scheeren B, Dias MAFS, Strom J, van Velthoven PFJ, Williams J (2001) Geophys Res Lett 28:951
20. Crutzen PJ, Andreae MO (1990) Science 250:1669
21. Conard SG, Sukhinin AI, Stocks BJ, Cahoon DR, Davidenko EP, Ivanova GA (2002) Climatic Change 55:197
22. Conard SG, Ivanova GA (1997) Environ Pollut 98:305
23. Flannigan MD, Bergeron Y, Engelmark O, Wotton BM (1998) J Vegetation Sci 9:469
24. Goldammer JG (2002) Working Group 4 on Wildland Fire. Report of 2nd meeting UN International Strategy for Disaster Reduction (ISDR)
25. Cofer WR, Levine JS, Winstead EL, Cahoon DR, Sebacher DI, Pinto JP, Stocks BJ (1996) J Geophys Res 101:23597
26. Chapman WL, Walsh JE (1993) Bull Amer Meteorol Soc 74:33
27. Lange L, Hoor P, Helas G, Fischer H, Brunner D, Scheeren B, Williams J, Wong S, Wohlfrorn KH, Arnold F, Strom J, Krejci R, Lelieveld J, Andreae MO (2001) J Geophys Res 106:27777
28. Cahoon DR, Stocks BJ, Levine JS, Cofer WR, Pierson JM (1994) J Geophys Res 99:18627
29. Page SE, Siegert F, Rieley JO, Boehm HDV, Jaya A, Limin S (2002) Nature 420:61
30. Scholes M, Matrai PA, Andreae MO, Smith KA, Manning MR (2003) Biosphere-atmosphere interactions. In: Brasseur G, Prinn R, Pszenny A (eds) Atmospheric chemistry in a changing world. An integration and synthesis of a decade of tropospheric chemistry research. Springer, Berlin Heidelberg New York, p 19
31. Sierra Club of Canada (1996) Forest fires and climate change, available from http://www.sierraclub.ca/national/programs/atmosphere-energy/climate-change/climfire.html
32. Davidenko EP, Eritsov A (2003) The fire season 2002 in Russia. Russian Federation Fire 2002 special, part II IFFN 28
33. Stocks BJ, Mason JA, Todd JB, Bosch EM, Wotton BM, Amiro BD, Flannigan MD, Hirsch KG, Logan KA, Martell DL, Skinner WR (2003) J Geophys Res 108:8149
34. United Nations Economic Commission for Europe (UN-ECE) (1999) Timber Bull LII ECE/TIM/BULL/52/4
35. Conard SG, Hao WM, Ribau A, McRae D, Sukhinin AI, Ivanova GA (2000) Estimating and monitoring effects of area burned and fire severity on carbon cycling, emissions, and forest health and sustainability in Central Siberia. http://leluc.gsfc.nasa.gov/products/pdfs/GOFC/GOFC-Conard 2000.pdf
36. Svidenko A, Goldammer JG (2001) Int For Fire News 24:41
37. Amiro BD, Stocks BJ, Alexander ME, Flannigan MD, Wotton BM (2001) Int J Wildland Fire 10:405
38. Amiro BD, Todd JB, Wotton BM, Logan KA, Flannigan MD, Stocks BJ, Mason JA, Martell DL, Hirsch KG (2001) Can J For Res – Rev Can Rech For 31:512
39. Sukhinin AI (2003) The 2002 fire season in the Asian part of the Russian Federation. Russian Federation Fire 2002 special, part III IFFN 28

40. Canadian Forest Service (2002) Forest Fire Management in Canada, IFFN 27:2
41. Goldammer JG (2003) The wildland fire season 2002 in the Russian Federation. Russian Federation 2002 special, part IIFFN 28
42. Conny JM, Slater JF (2002) J Geophys Res 107:4116
43. Bannikov M (2003) Russian Federation Fire 2002 Special, part IV IFFN 28
44. Wotawa G, Novelli PC, Trainer M, Granier C (2001) Geophys Res Lett 28:4575
45. Ichoku C, Kaufman YJ, Giglio L, Li Z, Fraser RH, Jin JZ, Park WM (2003) Int J Remote Sensing 24:1669
46. Li Z, Nadon S, Cihlar J (2000) Int J Remote Sensing 21:3057
47. Li QB, Jacob DJ, Bey I, Palmer PI, Duncan BN, Field BD, Martin RV, Fiore AM, Yantosca RM, Parrish DD, Simmonds PG, Oltmans SJ (2002) J Geophys Res 107:4166
48. Lasaponara R, Cuomo V, Macchiato MF, Simoniello T (2003) Int J Remote Sensing 24: 1723
49. Kaufman YJ, Ichoku C, Giglio L, Korontzi S, Chu DA, Hao WM, Li RR, Justice CO (2003) Int J Remote Sensing 24:1765
50. Wooster MJ, Strub N (2002) Global Biogeochem Cycles 16:1009
51. Asakuma K, Kuze H, Takeuchi N, Yahagi T (2002) Atmos Environ 36:1531
52. Torres O, Bhartia PK, Herman JR, Ahmad Z, Gleason J (1998) J Geophys Res 103: 23321
53. Torres O, Bhartia PK, Herman JR, Sinyuk A, Ginoux P, Holben B (2002) J Atmos Sci 59: 398
54. O'Neill NT, Eck TF, Holben BN, Smirnov A, Royer A, Li Z (2002) J Geophys Res 107:4125
55. Daniel JS, Solomon S (1998) J Geophys Res 103:13249
56. Hsu NC, Herman JR, Gleason JF, Torres O, Seftor CJ (1999) Geophys Res Lett 26:1165
57. Kaufman YJ, Fraser RS (1997) Science 277:1636
58. Christopher SA, Chou J, Zhang JL, Li X, Berendes TA, Welch RM (2000) Geophys Res Lett 27:2197
59. Jaegle L, Jacob DJ, Wang Y, Weinheimer AJ, Ridley BA, Campos TL, Sachse GW, Hagen DE (1998) Geophys Res Lett 25:1705
60. Goode JG, Yokelson RJ, Ward DE, Susott RA, Babbitt RE, Davies MA, Hao WM (2000) J Geophys Res 105:22147
61. Friedli HR, Radke LF, Lu JY, Banic CM, Leaitch WR, MacPherson JI (2003) Atmos Environ 37:253
62. Cofer WR, Winstead EL, Stocks BJ, Goldammer JG, Cahoon DR (1998) Geophys Res Lett 25:3919
63. Liousse C, Penner JE, Walton JJ, Eddleman H, Chuang C, Cachier H (1996) Modeling of biomass burning aerosols. In: Levine JS (eds) Biomass burning and global change. MIT Press, Cambridge, Mass, London, p 492
64. Seiler W, Crutzen PJ (1980) Climatic Change 2:207
65. Delmas R, Lacaux JP, Brocard D (1995) Environ Monit Assessment 38:181
66. Yamasoe MA, Artaxo P, Miguel AH, Allen AG (2000) Atmos Environ 34:1641
67. FIRESCAN Science Team (1996) Fire in ecosystems of boreal Eurasia: The Bor Forest Island Experiment Fire Research Campaign Asia-North (FIRESCAN). In: Levine JS (eds) Biomass burning and global change. MIT Press, Cambridge, Mass, London, p 848
68. Duncan BN, Martin RV, Staudt AC, Yevich R, Logan JA (2003) J Geophys Res 108:4100
69. Kajii Y, Kato S, Streets DG, Tsai NY, Shvidenko A, Nilsson S, McCallum I, Minko NP, Abushenko N, Altyntsev D, Khodzer TV (2002) J Geophys Res 107:4745
70. French NHF, Kasischke ES, Williams DG (2002) J Geophys Res 108:8151
71. Isaev AS, Korovin GN, Bartalev SA, Ershov DV, Janetos A, Kasischke ES, Shugart HH, French NHF, Orlick BE, Murphy TL (2002) Climatic Change 55:235
72. Yokelson RJ, Susott R, Ward DE, Reardon J, Griffith DWT (1997) J Geophys Res 102: 18865
73. Bertschi I, Yokelson RJ, Ward DE, Babbitt RE, Susott RA, Goode JG, Hao WM (2003) J Geophys Res 108:8472
74. Pisaric MFJ (2002) J Paleolimnol 28:349

75. Banta RM, Olivier LD, Holloway ET, Kropfli RA, Bartram BW, Cupp RE, Post MJ (1992) J Appl Meteorol 31:1328
76. Rothermel RC (1991) Predicting behaviour and size of crown fires in the northern Rocky Mountains. Res Pap INT-438
77. Waibel A, Fischer H, Wienhold F, Siegmund P, Lee B, Ström J, Lelieveld J, Crutzen PJ (1999) Chemosphere-Global Change Sci 1:233
78. Jost C, Trentmann R, Sprung D, Andreae MO, McQuaid JB, Barjat H (2003) J Geophys Res 108:8482
79. Tapper N, Hess G (1999) Health guidelines for vegetation fire events. WHO 122
80. Poppe D, Koppmann R, Rudolph J (1998) Geophys Res Lett 25:3823
81. Radke LF, Hegg DA, Hobbs PV, Nance JD, Lyons KK, Laursen KK, Weiss RE, Riggan PJ, Ward DE (1991) Particulate and trace gas emissions from large biomass fires in North America. In: Levine JS (ed) Global biomass burning: atmospheric, climatic, and biospheric implications. MIT Press, Cambridge, Mass, p 209
82. Crutzen PJ, Carmichael GR (1993) Modelling the influence of fires on atmospheric chemistry. In: Crutzen PJ, Goldammer JG (eds) Fire in the environment: the ecological, atmospheric, and climatic importance of vegetation fires. Wiley, New York, p 89
83. Penner JE, Bradley MM, Chuang CC, Edwards LL, Radke LF (2003) A numerical simulation of the aerosol-cloud interaction and atmospheric dynamics of the Hardiman Township, Ontario, prescribed burn. In: Levine JS (ed) Global biomass burning: atmospheric, climatic, and biospheric implications. MIT Press, Cambridge, Mass, p 420
84. Clark TL, Jenkins MA, Coen J, Packham D (1996) J Appl Meteorol 35:875
85. Grishin AM (1996) Mathematical modelling of forest fires. In: Goldammer JG, VV Furyaev (eds) Fire in ecosystems of boreal Eurasia. Kluwer Academic Publishers, Dordrecht, Boston, London, p 285
86. Westphal DL, Toon OB (1991) J Geophys Res 96:22379
87. Trentmann J, Andreae MO, Graf HF, Hobbs PV, Ottmar RD, Trautmann T (2002) J Geophys Res 107:4013
88. Rogers CM, Bowman KP (2001) J Geophys Res 106:28357
89. Stohl A, Eckhardt S, Forster C, Huntrieser H, Heland J, Schlager H, Aufhoff F, Arnold F, Cooper OR (2002) J Geophys Res 108:4370
90. Li ZQ, Fraser R, Jin J, Abuelgasim AA, Csiszar I, Gong P, Pu R, Hao W (2003) J Geophys Res 108:4076
91. Martin RV, Jacob DJ, Logan JA, Bey I, Yantosca RM, Staudt AC, Li QB, Fiore AM, Duncan BN, Liu HY, Ginoux P, Thouret V (2002) J Geophys Res 107:4351
92. Staudt AC, Jacob DJ, Logan JA, Bachiochi D, Krishnamurti TN, Sachse GW (2001) J Geophys Res 106:32571
93. Bey I, Jacob DJ, Logan JA, Yantosca RM (2003) J Geophys Res 108:4162
94. Jacob DJ, Wofsy SC, Bakwin PS, Fan SM, Harriss RC, Talbot RW, Bradshaw JD, Sandholm ST, Singh HB, Browell EV, Gregory GL, Sachse GW, Shipham MC, Blake DR, Fitzjarrald DR (1992) J Geophys Res 97:16421
95. Fan SM, Jacob DJ, Mauzerall DL, Bradshaw JD, Sandholm ST, Blake DR, Singh HB, Talbot RW, Gregory GL, Sachse GW (1994) J Geophys Res 99:16867
96. Galanter M, Levy H, Carmichael GR (2000) J Geophys Res 105:6633
97. McKeen SA, Wotawa G, Parrish DD, Holloway JS, Buhr MP, Hubler G, Fehsenfeld FC, Meagher JF (2002) J Geophys Res 107:4192
98. Cheng MD, Lin CJ (2001) J Geophys Res 106:22871
99. Ashbaugh LL, Malm WC, Sadeh WZ (1985) Atmos Environ 19:1263
100. Browell EV, Butler CF, Kooi SA, Fenn MA, Harriss RC, Gregory GL (1992) J Geophys Res 97:16433
101. Dlugokencky EJ, Walter BP, Masarie KA, Lang PM, Kasischke ES (2001) Geophys Res Lett 28:499
102. Wandinger U, Muller D, Bockmann C, Althausen D, Matthias V, Bosenberg J, Weiss V, Fiebig M, Wendisch M, Stohl A, Ansmann A (2002) J Geophys Res 107:8125

103. Fiebig M, Petzold A, Wandinger U, Wendisch M, Kiemle C, Stifter A, Ebert M, Rother T, Leiterer U (2002) J Geophys Res 107:8130
104. Formenti P, Boucher O, Reiner T, Sprung D, Andreae MO, Wendisch M, Wex H, Kindred D, Tzortziou M, Vasaras A, Zerefos C (2002) J Geophys Res 107:4451
105. Formenti P, Reiner T, Sprung D, Andreae MO, Wendisch M, Wex H, Kindred D, Dewey K, Kent J, Tzortziou M, Vasaras A, Zerefos C (2002) J Geophys Res 107:4450
106. Damoah R, Stohl A, Spichtinger N, Forster C (2004) Atmos Chem Phys Discuss 4:1449
107. Hinzman LD, Fukuda M, Sandberg DV, Chapin FS, Dash D (2003) J Geophys Res 108: 8153

The Handbook of Environmental Chemistry Vol. 4, Part G (2004): 277–294
DOI 10.1007/978-3-540-40037-0

Intercontinental Transport of Dust: Historical and Recent Observational Evidence

Rudolf B. Husar

Center for Air Pollution Impact and Trend Analysis (CAPITA), Washington University, St. Louis, MO, USA
E-mail: rhusar@me.wustl.edu

Abstract The scientific examination of the intercontinental dust transport has a long history, a vigorous present, and a promising future. Unlike the study of man-made pollution and biomass smoke, the fundamental causes of dust production, the long-range transport as well as the factors governing the removal processes of windblown dust were well established by about 1900. Much of the past century was devoted to the quantification of these processes using surface concentration monitors, aircraft sampling, and long-range transport modeling tools. By the 1990s, several new satellite sensors allowed a global-scale observation of the dynamic dust pattern. The aerosol climatologies derived from satellite data now highlight, in a semi-quantitative manner, the global distribution of dust sources. The dynamic daily aerosol data from several satellites, along with routine surface monitors now facilitates the quantitative documentation of individual intercontinental dust transport events from Sahara to Europe and the Americas, as well as from the Gobi desert to North America. In fact, the spectacular dust transport events visualized through the new satellite sensors along with the good performance of dust transport simulation and prediction models provide the most compelling evidence of intercontinental aerosol transport. Furthermore, the detection, tracking and the full explanation of these large dust events have catalyzed the emergence of ad hoc international internet-based virtual research communities. It is hoped that such collaboration over the next decades will yield a full quantification of intercontinental dust transport.

Keywords Dust storms · Dust transport · Air pollution · Long-range transport · Air quality

1
Introduction

The intercontinental transport of anthropogenic pollutants is a recent topic of scientific inquiry. This is not the case for dust. The long-range transport of dust has been recognized and studied for at least a century from many perspectives using a variety of scientific analytical tools.

From the point of view of a geologist, wind-blown dust is one of the geological agents that shape the surface of the Earth. Agronomists view wind blown dust as the means of delivering plant nutrients to the fields or as a process that erodes the fertile soil. For oceanographers, atmospheric dust is also the source of critical oceanic nutrients. Atmospheric scientists study dust because of the influences of these aerosols on the Earth's radiative balance both directly and indirectly through cloud modification. Finally, from the point of air pollution, dust is a component of the pollutant mix that affects human health, reduces human or electronic vision or otherwise influences human welfare. In many of these fields, particularly geology and agronomy extensive quantitative literature exist on dust emissions and long-range dust transport, some dating back to the 1800s. Two admirable summaries of the early scientific literature are by E. Free, a soil scientist, in 1911 [1], and N. Kempf, an atmospheric scientist, in 1914 [2].

Skipping about 100 years, the 1990s were the beginning of the global-scale observational revolution. Remote sensing by several satellites such as the Total Ozone Mapping Spectrometer (TOMS) [3], and AVHRR (Advanced Very High Resolution Radiometer) [4] along with the growth of surface monitoring networks [5] now allows global-scale monitoring of dust emissions, transport and the analysis of their spatio-temporal patterns. Since 2000, more advanced, dedicated satellites are beginning to contribute new data on atmospheric dust. Concurrently, several global simulation models became operational that forecast the dynamic dust emissions and the resulting pattern of concentrations and depositions [6, 7].

Many of the satellite and surface monitoring network data, as well as the model forecasts, are now posted on the Internet and conveniently accessible to the interested research community. In fact, over recent years, real-time observation of the onset, evolution, transport, and long-range impact of aerosol events has become a major topic connecting the global atmospheric science communities through the Internet [8, 9].

This brief and selective review addresses the long-range dust transport by combining old and new observations on dust events. It pays tribute to the historical literature (prior to 1910), which contains many of the original scientific observations and the explanations related to dust. This review also uses the recent (post-1990) observational evidence from satellites, mainly as a graphical illustration of the phenomena reported by the historical literature. This selective treatment essentially ignores the immense high-quality research contributed during much of the twentieth century. For that literature, readers are directed to several summary monographs and review articles [10–12].

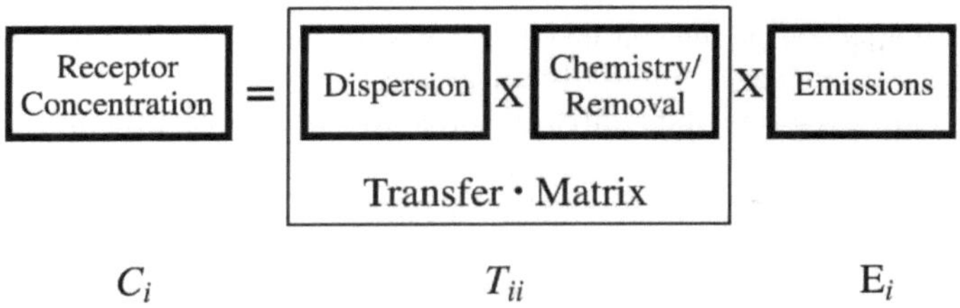

Fig. 1 Key processes that determine the source-receptor relationship

2 Intercontinental Transport Analysis Framework

Inter-continental source receptor relationships (SRR) are governed by the same natural processes as local-scale source-receptor relationships (Fig. 1). The concentration at a receptor site, C_j, is related to the emission rate at location i, E_i, through the transfer matrix T_{ij}. T_{ij} represents the fraction of the emission from source i that arrives at or is deposited at receptor j. T_{ij} incorporates the roles of atmospheric transfer processes including horizontal and vertical dispersion, chemical or physical transformations, as well as dry and wet removal. The indices i and j represent the quantized spatial distribution of these parameters. All three parameters, discussed below in more detail, also depend on time on various scales: secular (over 10 years), seasonal, synoptic (3–10 days) and diurnal. The quantification of SRR requires a full understanding of all three parameters.

For aerosol species such as wind blown dust, SRR can be established either by source oriented dispersion models or by receptor oriented chemical tracer methods. While tremendous progress has been made applying both techniques, a robust quantification of intercontinental dust transport dynamics is still a major challenge.

3 Global Dust Emissions

The origin of atmospheric dust was a hotly disputed issue in the seventeenth to eighteenth centuries (see a review of the pre-1900 scientific literature [13]). However, by the late nineteenth century it was established 'beyond doubt' that dust emissions occur when dry sandy surfaces are exposed to strong winds [1, 2]. It was also recognized that the haze due to wind-blown dust was different from smoke and other aerosols. For example, Hann in 1906 referred to the haze due to windblown dust as 'mechanical turbidity' (mechanische Truebung) since it is caused by mechanical forces [14]. Early explorers also recognized that the Sahara and Gobi deserts were two main global sources of atmospheric dust, but the quantification of global dust emission rates over the past century remained elusive.

In recent reassessments of global dust emissions, the patterns were estimated largely by the availability and broad use of the 'Absorbing Aerosol Index' derived from the TOMS ozone monitoring satellite [11, 15]. The new semi-quantitative satellite data helped refine the spatial pattern of dust emissions, but the emissions densities (gm^{-2} $year^{-1}$) are still highly uncertain.

The dust source regions that contribute most to the inter-continental transport are the Sahara desert in North-West Africa and the Gobi desert in East Asia. Further discussion on large-scale dust transport pattern is given in the Sahara and Gobi dust event sections below. While there is considerable evidence for man-induced desertification, Goudie and Middleton concluded that the largest global dust source regions (Sahara) are not perturbed by human activities since they occur over un-inhabited areas [16].

4
Horizontal and Vertical Dust Transport

The winds that lift the dust particles from the surface arise either from synoptic-scale pressure fields (1000 km, 3–5 days) or from local (orographic) effects. The thermal winds driven by the surface heating are diurnal and much smaller in scale. Dust storms are highly 'episodic', i.e., they occur intermittently as "events" but cause high aerosol concentrations over large areas. The intense dust outbreaks over West Africa exhibit 5–10 day periodicity. In other areas dust storms may be more sporadic, rare occurrences.

Satellite and photographic observations show that near the source, dust clouds may have two distinctly different patterns. *Dust plumes* resemble streaky pattern of smoke plumes where the particulate matter is emitted from a small point source and dispersed horizontally and vertically as a cone-shaped plume (Fig. 2a). The second near-source dust pattern is the *dust front* (Fig. 2b). At the leading edge of the advancing dust 'wall', the dust immediately rises to high elevations, ready for long-range transport. The synoptic, micrometeorological and dust-suspension mechanisms associated with the two dust dispersion patterns are not well understood by this writer.

a  b

Fig. 2a, b Near-source dust emission patterns: **a** dust plumes dispersing as streaks from point sources; **b** dust fronts are broad advancing 'walls of dust'

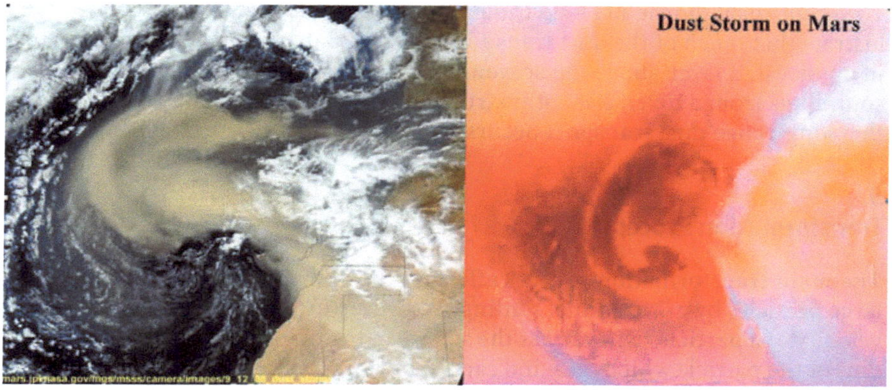

a b

Fig. 3 **a** Dust outbreaks resembling jets over West Africa (February, 2001). **b** Dust outbreak on Mars

The shapes of dust clouds in many of the outbreaks resemble fluid jets forcefully injected into a quiescent environment. Interestingly, the visualization of these peculiar atmospheric flow patterns by the dust particles (Fig. 3a, b) reveals that similar 'dust jets' also occur on Mars.

A key aspect of intercontinental dust transport is the vertical distribution of the transported dust layer. It influences both the transport speed and the removal rate of dust. Dust within the boundary layer is subject to intense dry and wet removal processes. Removal is particularly efficient in shallow marine boundary layers. It was recognized by early researchers that, if dust is raised into upper air layers, "velocities and distances of transfer are much greater, not just on account of greater vertical fall but because of the greater velocity of the air currents at higher altitudes" [1]. Hence, it is unlikely that significant intercontinental dust transport can occur at elevations below 1–2 km. Rather, continental-scale dust transport occurs mainly in the free troposphere at 2–10 km elevation where the transport is swift and cloud removal processes are weak.

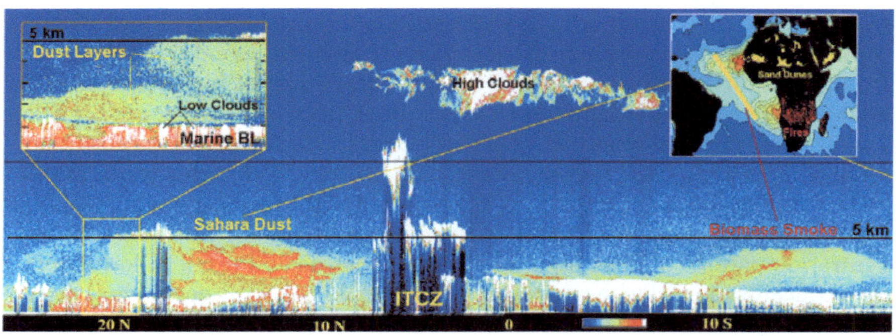

Fig. 4 Vertical distribution of Sahara dust in the Atlantic trade winds above the marine boundary layer observed through the LITE space-borne lidar

Recent space-borne LITE lidar data from the Space Shuttle (Fig. 4) clearly illustrate that much of the dust transport from the Sahara to the Americas occurs in an elevated layer above the scavenging boundary layer [17]. Vertical aerosol soundings from a diverse set of instruments (lidar, aircraft soundings) confirm that long-range dust transport occurs preferentially in elevated layers [17–19].

5 Transformation Processes

The main transformation process affecting dust during transport is the loss of large particles by gravitational settling. According to Stokes' formula, the settling velocity of a particle "will vary as the square of the diameter; or inversely, the radius of the particle which the wind can support will vary as the square root of the velocity" [1]. Based on his own experiments, Udden in 1898 concluded that the largest size of quartz particles that can be sustained in the air by strong winds is about 100 μm in diameter [20].

As Free in 1911 notes [1], "There is a small fraction of airborne dust which is fine enough to remain more or less permanently in suspension… and the distances covered by such material are consequently very great, though well established instances are rare on account of the difficulty of determining the source of the material. Examination of the dust itself fails to indicate the place of origin and one must rely on indirect evidence such as meteorological data". Occasionally evidence for the transport distance was furnished by tracing the path of a storm by observations made along the path. Udden in 1896 in his examination of western U.S. dust storms recorded observable dust transport distances of at least 400 miles [21].

The lifetime and transport distance of intercontinental dust (1–10 μm) is determined largely by the wet removal processes through clouds and rain. Hence, the transport distance of emitted dust is strongly influenced by the water cycle. Recently developed microwave sensors on polar satellites measure both the columnar water vapor concentration [gm^{-2}] as well as the precipitation rate [$gm^{-2}\,h^{-1}$] over the oceans. The ratio of these parameters yields the characteristic atmospheric lifetime of water vapor that is relevant to the dust removal. The data show an order of magnitude variation in the water removal rate ranging from 0.002 to over 0.02 h^{-1}. This corresponds to water residence times between 2 and 20 days.

In the northern hemisphere extratropics, precipitating cloud systems emanate from both North America and East Asia. A narrow belt of high removal in the equatorial Pacific and Atlantic is within the inter-tropical convergence zone. The lowest water removal rates can be observed over the subtropical oceans in both hemispheres with water lifetimes exceeding 20 days. Both water vapor and the water-soluble chemical species are transported through these regions without appreciable wet removal. The water removal rate spatially modulates the concentration pattern of dust: high water removal areas will result in low ambient dust concentrations due to wet scavenging. It would be highly desirable to use the currently operating global dust models to evaluate the specific roles of emission, transport, and removal processes in determining the dust concentration pattern during long-range transport.

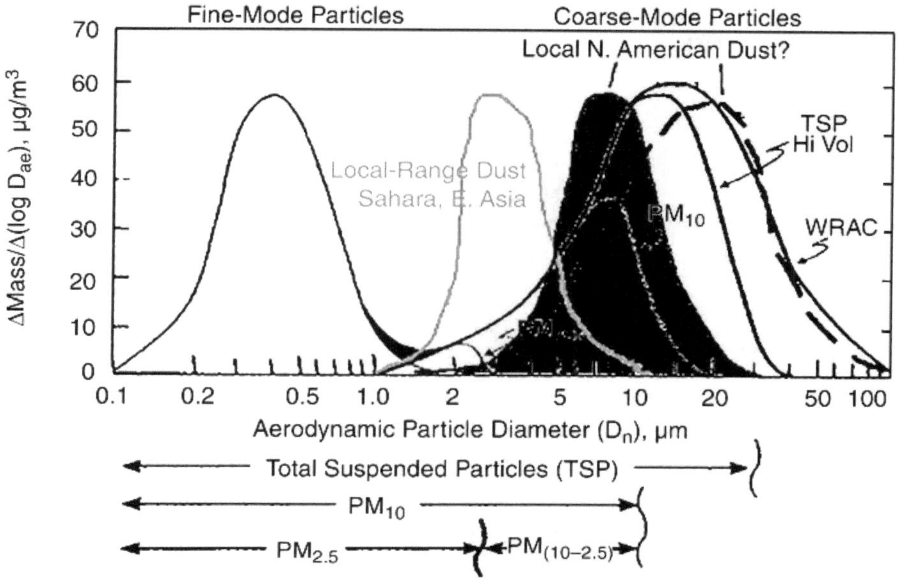

Fig. 5 Typical size distribution of local and long-range transported dust. The *shaded areas* represent local and long-range transport dust mass distribution. The *outer curves* show typical bimodal atmospheric aerosol size spectra and size cuts for TSP, PM10 and PM2.5

Recent size distribution data (Fig. 5) show that the mass of long-range transported dust mass is virtually all in the sub-10 μm diameter range. At the other extreme, there is only a small fraction below about 1 μm since strong molecular forces prevent the mechanical dispersion of sub-micron particles. As a consequence, the long-range transported dust mass distribution is in the 1–10 μm diameter range, with a mass mode in the 3–5 μm range. This size-selective sorting and the remarkably uniform grain of eolian dust deposits were pointed out by Udden in1898 [20].

The basic mechanisms governing dust suspension by frictional forces of the winds have been studied since the late 1800s [1]. Modern dust models [6, 7] use similar mechanisms as well as wind and precipitation data from meteorological models to simulate or forecast dust emissions, transport, and deposition.

Early microscopic examination of the long-range transported dust suggested a desert origin as it consists of "very fine splinters of quartz and still finer claylike dust often gathered into flocks of aggregates, which is probably the final debris from the disintegration of the feldspathic and similar minerals. The reddish color is probably due to ferruginous matter" [1]. Figure 6a shows a modern electro-micrograph of dust particles [22].

The early chemical composition data on sirocco (dry wind from Sahara) dust collected at various locations in Europe (an example is shown in Fig. 6b) were "in perfect accord with the Saharan hypothesis due to the similarity with the Sahara dust collected in Tunis". A table (Fig. 6b) of the chemical mass balance of Sahara dust deposited in Fiume (now Rijeka, Croatia) was reported by Barac in 1901 [23].

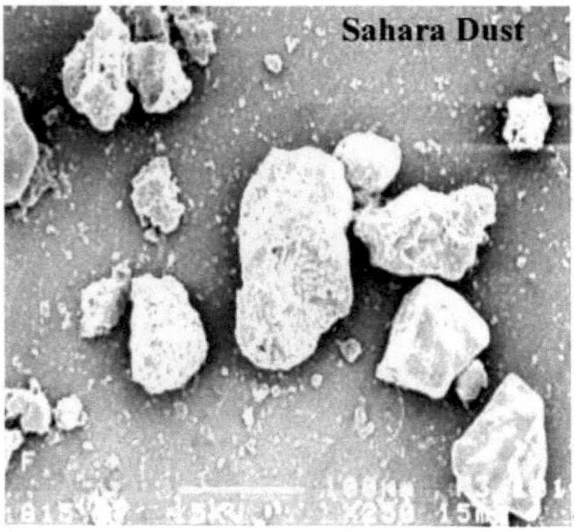

Dust Composition

Sahara Dust, Fiume 1901-03-10

Component	% Mass
SiO_2	49.5
Al_2O_3	12.1
CaO	11.5
Fe_2O_3	10
CO_2	9
Mn_3O_4	2
MgO	0.4
Organics	5.5
Traces	0.2

Fig. 6 a Scanning electron micrograph of Sahara dust. b Chemical composition of deposited Sahara dust (Barac 1901)

Some of the modern dust chemical composition data can distinguish various dust sources by the fine structure of their respective chemical composition. The new data also show that the chemical composition of the particles measured thousands of kilometers from the source changes very little during multi-day transport [8]. However, other measurements show that dust particles are often coated with sulfate, particularly in polluted regions such as the Mediterranean or eastern China [24]. Cloud processes (gas scavenging and liquid phase oxidation of SO_2) are probably responsible for the sulfate coating. However, sulfate is also found on the dust particles that have not passed through clouds [25]. This interaction between dust and the surrounding chemical environment is in dire need of quantification.

6
Global Dust Concentration Climatologies

The global distribution of atmospheric aerosol including wind-blown dust could not be determined by the early researchers. Hence, over the twentieth century the estimates of the amounts and the spatial pattern of global aerosols varied widely. In the 1970s and 1980s many of the global aerosol estimates were performed using models with some success. By the 1990s, the emergence of satellite remote sensing allowed the creation of observation-based global aerosol climatology using scattering measured by the AVHRR sensor [4], and absorbing properties obtained from the TOMS sensor [3].

Since then, several other satellite-derived global aerosol climatologies have been reported (Fig. 7). Virtually all the satellite aerosol mapping sensors are different and respond to different aspects of the eight-dimensional aerosol data space (x, y, z, t, particle size, composition, shape, and mixing). Thus, they can be considered complementary, rather than redundant characterizations of the global aerosol system. Virtually all these satellite sensors produce columnar integrals of the aerosol at a given location. The climatology derived from the Multi-angle Imaging SpectroRadiometer (MISR) sensor uses scattering at many forward and

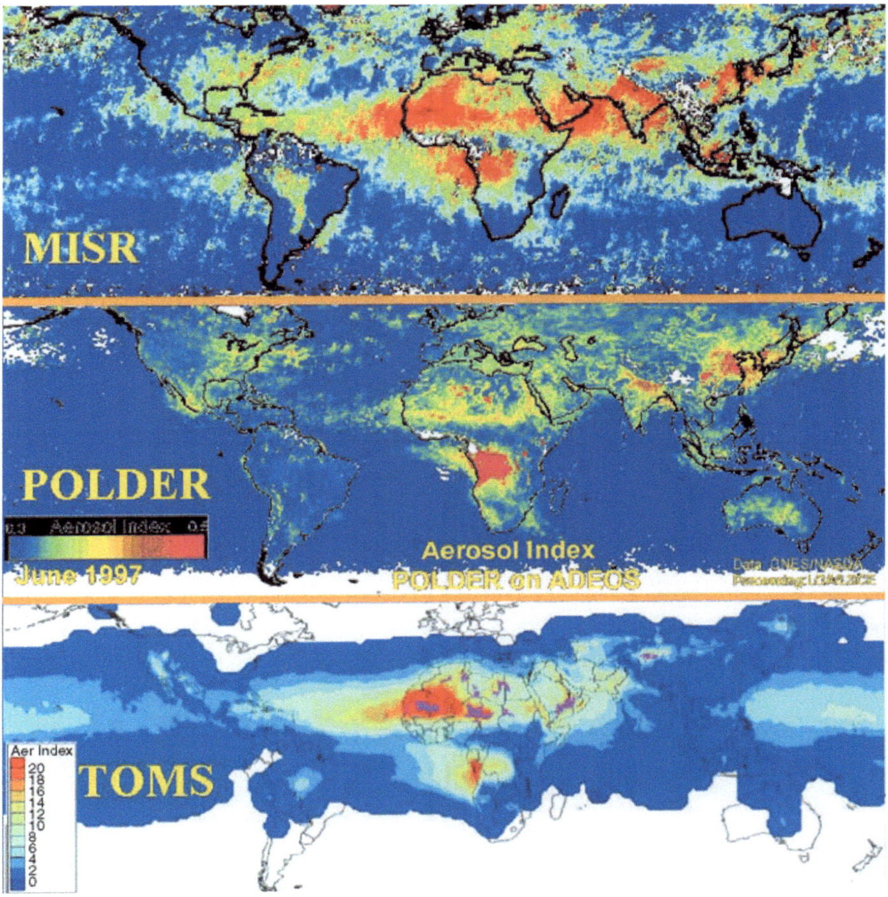

Fig. 7a–c Global aerosol pattern for July derived from three satellite sensors: **a** MISR; **b** POLDER; **c** TOMS

backward scattering angles [26]. The MISR data for July (Fig. 7a) indicates West-Saharan dust as the dominant feature of the global aerosol optical thickness, AOT. Intense dust AOT is also evident over the Arabian Peninsula and the Arabian Sea.

The aerosol maps derived from the POLarization and Directionality of the Earth Reflectances (POLDER) sensor, which uses the aerosol polarization signal, is most sensitive to small particles, such as biomass smoke [27]. The POLDER aerosol map (Fig. 7b) shows only a slight indication of dust in Africa and Asia but it accentuates the biomass burning regions of Central Africa and the haze over Eastern China and Northern India.

The qualitative index derived from the TOMS instrument is most sensitive to dust and smoke aerosols but not to non-absorbing sulfate haze. The TOMS aerosol index also depends on the height of the aerosol layer as well as on the brightness of the underlying surface in the UV. The global map of TOMS aerosol

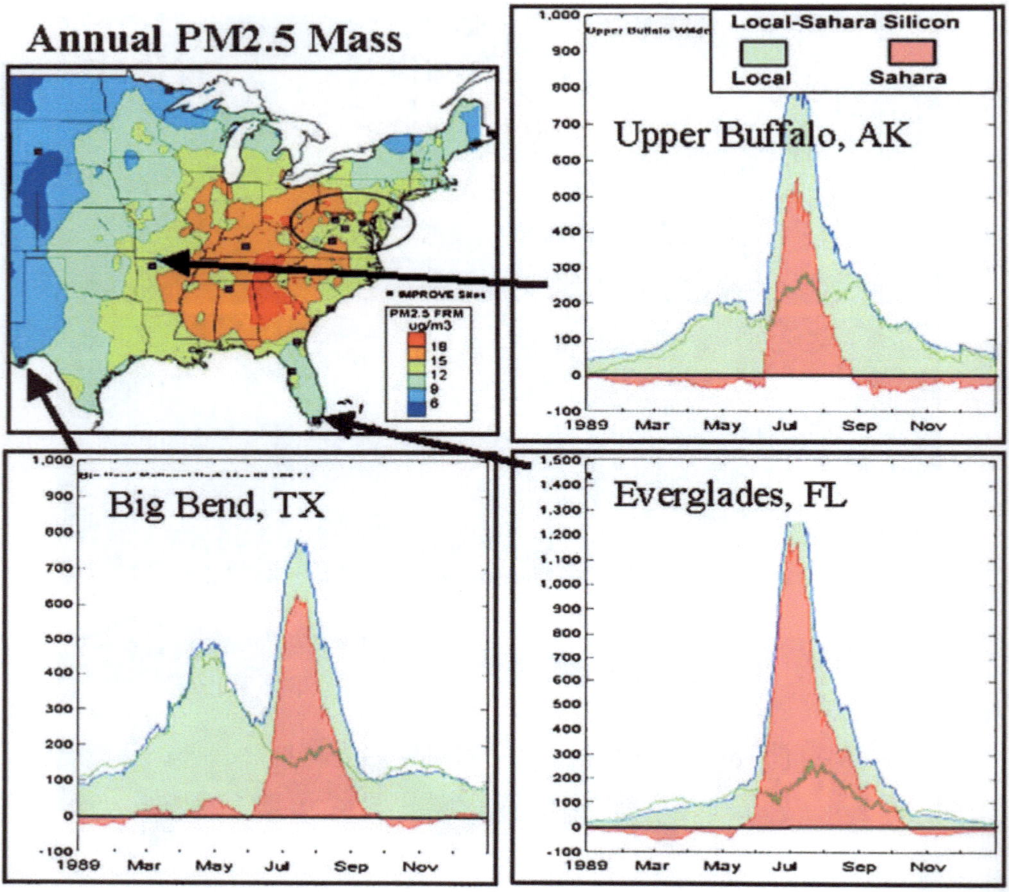

Fig. 8 **a** Quantitative estimate of the Sahara contribution to the fine particle dust concentration over the southeastern US. **b** Sahara dust pall over Great Britain and France based on SeaWiFS sensor

index for July (Fig. 7c) is dominated by the Western Sahara dust aerosol. Figure 7c also shows the surface roughness derived from the NSCAT scatterometer (3 cm wavelength) [28]. The low roughness areas (purple) indicate the smooth surfaces of sand dunes. The majority of the sandy regions are over the Sahara Desert, Arabian Peninsula, Kazakhstan as well as over the Taklamatan and Gobi deserts in Asia. It is quite compelling to observe in Fig. 7c that the sandy surfaces are located in the center of high TOMS AI regions. This essentially confirms that the detected sandy surfaces are the sources of the windblown dust.

Both the MISR and the TOMS aerosol maps in Fig. 7 show that in July, the Sahara dust plume is discernable for about 10,000 km transport distance, heading toward the Caribbean and Central America. This dust transport pathway in the summer has been observed by early observers in the 19th century as well

Fig. 8b

as by currently active researchers. When the Sahara dust plume reaches the Caribbean, part of the intercontinental dust plume curves northward into the Eastern US. Chemical 'fingerprints' of Sahara dust obtained from chemically resolved aerosol samples can be used to estimate the Sahara impact on the US air quality [29]. The seasonal charts in Fig. 8a represent the estimated Sahara dust impact on the total fine particle dust concentration over the southeastern U.S. Thus, in this case, quantitative estimation of intercontinental dust transport was possible using currently available data and techniques.

7
Documented Intercontinental Dust Transport Events

Inter-continental dust events were first studied in Europe in connection with Sahara dust-fall events. Since dust transport is highly episodic, much of the scientific understanding has been obtained by analyzing specific dust events in detail. The first known scientific description of large-scale dust transport is that of Wendelin in 1646 [30]. Ehrenberg in 1871 listed 340 dust events over Europe and N. Africa occurring prior to 1847 and updated the list with another 193 occurrences in 1870 [31]. In the bibliography by Stuntz and Free in 1911 articles by 120 different researchers were cited reporting on Saharan dust events between 1870 and 1910 [32]. It seems fair to say, that the high level of scientific attention to African dust around the 1900s, exceeded the current research effort by a wide margin.

Early trend analysis of the Sahara dust events occurring between 1782 and 1898 has shown a tendency for the dustfalls to occur most frequently in those years when the Sahara is the driest [33]. The dates for the dry years in the Sahara were obtained from Bruckners drought index. Hence, the existence of a climate-dust link has been recognized for at least a century.

Figure 8b shows a recent Sahara dust incursion to northern Europe recorded by the Sea-Viewing Wide field-of-View Sensor (SeaWiFS) satellite on March 15, 2003. A thick yellow dust cloud is seen over England and France. Operational forecast models [6, 7] show that the dust source was over North Africa. This dust transport event, like many other recent dust events, has attracted a great deal of admiration particularly due to the real-time availability of spectacular color satellite images. However, just like most other current dust events, it was not analyzed quantitatively for its key physical and chemical features.

Fortunately, the March 2003 event strongly resembled the February 1903 dust fall episode that was well documented in the contemporary literature, 100 years ago [34–37]. The 1903 February 21–22 dust event impacted northern Europe including southern England, northern France, Holland, Germany, and Denmark. For the early researchers, the most conclusive evidence regarding the Sahara origin of the dust was derived from the meteorological sources. Using the measured pressure fields they were able to map the trajectory of several dust bearing storms and traced them back to a point of origin in northern Africa. Airmass history analysis (using pressure charts) of this event indicate that the Sahara dust reached northern Europe through a loop over the Atlantic, which was consistent with the absence of excess dust in southern Europe. Also, consistent with the looping trajectory is that the dust arrived over England half a day before it reached Germany. According to Herrmann in 1903 "the European dust storm of February 1903 seems to have traveled mainly in the higher strata and with a velocity of 50 miles/h". Microscopic analysis of the February 1903 dust fall was also consistent with its desert origin of the dust [34].

The most intensively analyzed Sahara dust transport event to Europe appears to be the one that occurred in March 1901. It was chronicled and summarized in an admirable monograph by Hellmann and Meinardus in 1901 [35] and by a dozen of other researchers [e.g., 37, 38]. In a remarkable meteorological analysis, back trajectories were calculated for both sea level and at 2500 m, which clearly show the path of the dust cloud from Sahara to Central Europe. They report that a part of the dust storm of March 11, 1901 got elevated into higher strata, above the zone of rain formation at 1–2 km [35]. Three days later, it had subsided low enough to be caught and removed by the rain of the 15th [32].

The most noticeable impact of the dust event was the heavy dust-fall that blanketed much of Europe. The early researchers have performed detailed dust deposition flux analysis. The dust fall data for the March 1901 event were collected from hundreds of locations in Europe. The measured deposition fluxes ranged between 1 and 11 $g\ m^{-2}$. The dust fall values were largest in southern Europe and decreased toward the north. From the figures of Hellmann and Meinardus in 1901 [35] the European area covered with dust was 440,000 km^2, while the total estimated dust fall amount was about 1.8 Tg. The resulting deposition flux, averaged for the area of the event was 4.8 $g\ m^{-2}$. At the specific gravity of 2 $g\ cm^{-3}$ this corresponds to a dust layer thickness 0.24 mm. Based on these data free in 1911 speculates that if storms comparable to the March 1901 event would occur every five years, the dust layer thickness would be a remarkable 4.8 mm per century [1]. It would be interesting to compare these early estimates of Sahara dust deposition to the values derived from modern modeling tools. Unfortunately most current models do not explicitly include adequate size distribution and settling flux calculations for such purpose.

The impressive analysis of this 1901 Sahara dust event was accomplished by a remarkable international collaboration process. Within a day of the event, the dust fall has attracted the attention of atmospheric scientists, as well as the general public through newspaper reports. Given the evidence that the severe event covered much of central Europe, a public announcement was published in the German 'Federal Register' and reprinted in newspapers in several countries, with a request for additional reports beyond the meteorological observations. The international public responded promptly sending in hundreds of reports and actual dust samples to the central Meteorological Service for analysis. The physico-chemical analysis of the dust samples and the interpretation of the observer-reports were performed in various laboratories in Germany, Austria, Hungary, and Switzerland. The entire collaborative process was concluded by the publication of a summary report by Hellmann and Meinardus in 1901 [35], as well as a dozen of other reports by individual investigators [e.g., 37, 38] in the same year as the event occurred.

8 Transport of Asian Dust

Many of the historical records have now been analyzed to reconstruct the historical pattern of Asian dust events. The most notorious Asian dust storms originate from the Gobi Desert. Meteorological records of visibility and dust-fall indicate that over China and Korea dust events have occurred throughout the thousands of years of recorded history [39]. Typical reports of east Asian dust transport are also found in ship logs: "On April 2, 1892 there fell on a ship 95 miles west of Nagasaki a yellow dust which must have come from the interior of China and have been carried by the wind to the place where it was observed, a distance of at least 1000 miles" [40]. A similar dust transport was reported by Pumpelly in 1879 [41]. There are also several reports that the Australian dust clouds have reached New Zealand, a distance of about 1500 miles [42–43].

Compelling geological evidence of global scale transport of Asian dust arises from the chemical and radiological analysis of samples in the Greenland ice core and Hawaiian soils. The chemical and radiological fingerprints of deposited dust at both locations were most consistent with the composition of the Asian dust sources.

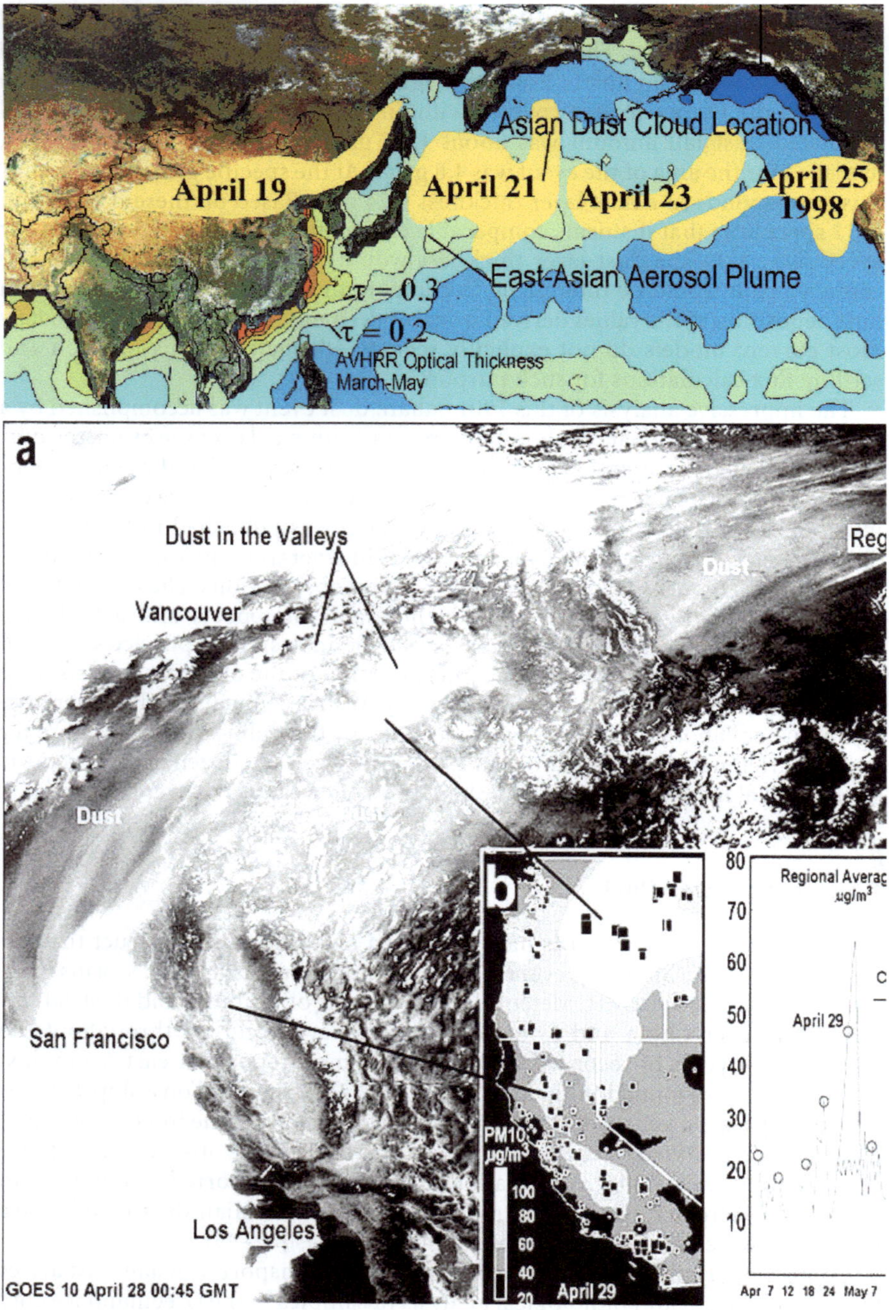

a

b

Fig. 9 **a** Trans-Pacific transport path of Asian dust to North America in April 1998. **b** GOES satellite image of Asian dust over the west coast of North America, April 1998 and the measured concentration of PM10. **c** SeaWiFS image of the "perfect dust event" on April 7, 2001

Fig. 9c

The transport events of desert dust from Asia to the North Pacific atmosphere is well documented and results in a maximum in Asian aerosol loading each spring [44–49]. The inter-continental transport of Asian dust is carried eastward by the Northern-Pacific 'conveyor belt'. The main features of the springtime transport pathways from Asia are described in Chap. 5 of this book.

A vivid illustration of trans-Pacific transport of Asian dust was reported by a large international group of investigators [8]. On April 15 and 19 1998, two intense dust storms were generated over the Gobi Desert by springtime cold low pressure systems descending from the northwest. The dust was detected and its evolution followed by its yellow color on SeaWiFS satellite images, routine surface-based monitoring, and through serendipitous observations. The April 19 dust cloud crossed the Pacific Ocean (Fig. 9a) within five days, subsided to the surface along the mountain ranges between British Columbia and California and, impacted severely the optical and the concentration environment of the region. Over Asia, the dust clouds increased the albedo over the cloudless ocean and land by up to 10–20% but it reduced the near UV cloud reflectance, causing a yellow coloration of all surfaces. Over the North American west coast, the dust layer has increased the spectrally uniform optical depth to about 0.4, reduced the direct solar radiation by 30–40%, doubled the diffuse radiation and caused a whitish discoloration of the blue sky. On April 29, the average excess surface-level dust aerosol concentration over the valleys of the West Coast was about 20–50 μg/m^3 with local peaks >100 μg/m^3. (Fig. 9b). The dust mass mean diameter was 2–3 μm and the dust chemical fingerprints were evident throughout the West Coast and extended to Minnesota. According to [8] "the April 1998 dust event has impacted the surface aerosol concentration two to three times more then any other dust event since 1988", implying that the it was a unique, rare event. Ironically, in April of 2001 a large Gobi dust event (Fig. 9c), dubbed as 'Perfect Dust Storm' resulted in Asian dust concentrations over North America that exceeded the impact of the 1998 event by a large margin. Evidently, this was Nature's reminder that generalizations based on tenuous record should always be applied with caution.

A noteworthy aspect of the report on the April 1998 Asian Dust event [8] was the ad hoc nature of the web-based international collaboration. The dust event was initially monitored by a few observers, but when it was evident that the Asian dust cloud was reaching North America, an interactive website was set up (http://capita.wustl.edu/Asia-FarEast/) to share observations and exchange ideas. Most participants exposed some of their data and preliminary reports on their web sites. A dedicated event website supported a user-maintained central catalog of web resources along with an open discussion forum. The first phase of the virtual workgroup activity was completed with a preliminary summary of the dust event that was web-published within a few days after the event. Subsequently, 12 peer-reviewed papers were prepared by various participating groups and published in a special dust issue of the Journal of Geophysical Research (Volume 106, 2001). The demonstrated collaboration through sharing of data and ideas and mutually supportive analysis was reminiscent of the swift, broad ad hoc international collaboration during the March 1901 Sahara dust event over Europe [35]. Will the next ad-hoc international collaborative dust analysis effort have to wait until 2100?

9 Summary and Conclusions

The study of intercontinental dust transport has a rich history, an interesting present and a promising future. Quantitative estimates based on solid observations begun a century ago. Presently, the combination of source-transport oriented observations using satellites and chemical fingerprinting at receptors in combination with dynamic models provide powerful set of tools for understanding, documenting, and even predicting the intercontinental transport of dust.

However, a fully quantitative source-receptor relationship for inter-continental transport of dust is not in hand. A particularly nagging problem is the definition and quantitative estimation of dust emissions. Although models simulate the spatial pattern of dust, the observational evidence to support the model estimates is very sporadic, particularly on the vertical structure of dust layers.

An outstanding future issue is the two-way interaction of atmospheric dust with the physical and chemical climate. It is evident that changes in the physical climate due to drought enhance dust emissions but the relationship is not well fortified by observations. Much more uncertain is the impact of dust on the climate system. Does dust heat or cool the atmosphere? Maybe both? If this is sorted out, can the two-way climate-dust-climate interactions be incorporated into global circulation models? The interaction with the chemical climate is also significant. There is now compelling evidence that dust particles are not inert but interact with a number of chemical processes as catalysts and as carriers of chemical species, like sulfate, that can accumulate on dust during atmospheric transport. These two-way couplings make the atmospheric dust a full and active participant in the dynamics and chemistry of the atmospheric system.

A prudent symbiotic combination of observation- and model-based approaches and global-scale scientific collaboration can accelerate the scientific progress. It is hoped that over the next decade, both the climatological and short-term characterization of intercontinental dust transport will be advanced to a satisfactory level by the collaborating global atmospheric research community.

Acknowledgement This report was prepared with the support from the National Science Foundation grant # ATM-0113868 "ITR/AP: Collaboration Through Virtual Groups".

10 References

1. Free EE (1911) The movement of soil material by the wind, U.S. Department of Agriculture, Bureau of Soils, Bulletin No. 68
2. Kempf N (1914) Doctors Dissertation vor der Kgl. Technischen Hochschule zu Muenchen, Verlag von F.C.W. Vogel, Leipzig
3. Herman JR, Bhartia PK, Torres O, Seftor C, Celarier E (1997) J Geophys Res 102:16911
4. Husar RB, Prospero JM, Stowe LL (1997) J Geophys Res 102 D14:16889
5. Holben BN, Eck TF, Slutsker I, Tanre D, Buis JP, Setzer A, Vermote E, Reagan JA, Kaufman YJ, Nakajima T, Lavenu F, Jankowiak I, Smirnov A (1998) Remote Sens Environ 66:1
6. Westphal DL (2000) Description of the NAAPS Global Aerosol Model (http://www.nrlmry.navy.mil/aerosol/Docs/globaer_model.html)
7. Nickovic S, Kallos G, Papadopoulos A, Kakaliagou O (2001) J Geophys Res 106(D16):18113
8. Husar RB, Tratt DM, Schichtel BA, Falke SR, Li F, Jaffe D, Gassó S, Gill T, Laulainen NS, Lu F, Reheis MC, Chun Y, Westphal D, Holben BN, GueymardC, McKendry I, Kuring N, Feldman GC, McClain C, Frouin RJ, Merrill J, DuBois D, Vignola F, Murayama T, Nickovic S, Wilson WE, Sassen K, Sugimoto N, Malm WC (2001) J Geophys Res 106(D16):18317 (http://capita.wustl.edu/Asia-FarEast/)
9. Prospero JM (2000) maintains an e-mail subscription-based mailing list, the 'dustlist', where he periodically summarizes interesting features of the daily global aerosol scene, such as large dust and smoke plumes, anthropogenic haze events, etc.
10. Morales C (ed) (1979) Sahara dust. SCOPE 14, Scientific Committee on Problems of the Environment, Wiley & Sons, Chichester
11. Prospero JM, Ginoux P, Torres O, Nicholson SE, Gill TE (2002) Rev Geophys 40, Article No 1002
12. Goudie AS (2001) Ann Arid Zone 40:303
13. Husar RB (2000) Atmospheric aerosol science before 1900. In: Preining O, Davis EJ (eds) History of aerosol science. Verlag der Oesterreichischen Akademie der Wissenschaften, Wien, pp 25
14. Hann J (1906) Lehrbuch der Meteorologie Vienna (cited in Free 1911 and Stuntz and Free 1911)
15. Washington R, Todd M, Middleton NJ, Goudie AS (2003) Ann Assoc Am Geograph 93:297
16. Goudie AS, Middleton NJ (2001) Earth Sci Rev 56:179
17. Winker DM, Couch RH, McCormick MP (1996) IEEE Proc 84:164
18. Tratt DM, Frouin RJ, Westphal DL (2001) J Geophys Res 106(D16):18371

19. Murayama T, Sugimoto N, Matsui I, Arao K, Iokibe K, Koga R, Sakai T, Kubota Y, Saito Y, Abo M, Hagiwara N, Kuze H, Kaneyasu N, Imasu R, Asai K, Aoki K (1998) Proc SPIE 3504:8
20. Udden JA (1898) Augustana Library Publications 1 (cited in Free 1911 and Stuntz and Free 1911)
21. Udden JA (1896) Pop Sci Mon 49:655 (cited in Free 1911 and Stuntz and Free 1911)
22. Volten H, Munoz O, Rol E, de Haan JF, Vassen W, Hovenier JW, Muinonen K, Nousiainen T (2001) J Geophys Res 106(D15):17375
23. Barac M (1901) Mon Wea Rev July 1901:31
24. Levin Z, Ganor E, Gladstein V (1996) J App Meteorol 35:1511
25. Yin Y, Wurzler S, Levin Z, Reisin TG (2002) J Geophys Res 107(D23):4724
26. Martonchik JV, Diner DJ, Kahn RA, Ackerman TP, Verstraete ME, Pinty B, Gordon HR (1998) IEEE Trans Geosci Remote Sens 36:1212
27. Deuze JL, Breon FM, Devaux C, Goloub P, Herman M, Lafrance B, Maignan F, Marchand A, Nadal F, Perry G, Tanre D (2001) J Geophys Res 106(D5):4913
28. Long DG, Drinkwater MR, Holt B, Saatchi S, Bertoia C (2001) EOS Trans Am Geophys Union 82:503 23 Included EOS Electronic Suppl (http://www.agu.org/eos_elec/010126e.html)
29. Husar RB (2004) Global and local dust over North America (in preparation)
30. Wendelin G (1646) Pluvia pupurea bruxellensis Bruessel (cited in Free 1911 and Stuntz and Free 1911)
31. Ehrenberg CG (1871) Abhk Presuss Akad Wiss Berlin 1–150 (cited in Free 1911 and Stuntz and Free 1911)
32. Stuntz SC, Free EE (1911) Bibiliography of eolian geology. Department of Agriculture, Bureau of Soils, Bulletin 68
33. Krebs W (1903) Annallen Hydrog 31:174 (cited in Free 1911 and Stuntz and Free 1911)
34. Herrmann E (1903) Annallen Hydrog 31:475 (cited in Free 1911 and Stuntz and Free 1911)
35. Hellmann JGG, Meinardus W (1901) Abh K Preuss Met Ins 2:93
36. Mill HR, Lempfert RGK (1904) Q J R Meteorol Soc 30:57 (cited in Free 1911 and Stuntz and Free 1911)
37. Vanderlinden E (1901) Ciel et terre 22:257 (cited in Free 1911 and Stuntz and Free 1911)
38. Valentin J (1902) Sitzung Kaiserl Akad Wiss Vienna 111, IIa:727 (cited in Free 1911 and Stuntz and Free 1911)
39. Chun Y (2003) (personal communication)
40. Milne J (1892) Nature 46:128 (cited in Free 1911 and Stuntz and Free 1911)
41. Pumpelly R (1879) Am J Sci 17:133 (cited in Free 1911 and Stuntz and Free 1911)
42. Marshall P (1903) Nature 68:223 (cited in Free 1911 and Stuntz and Free 1911)
43. Chapman F, Grayson HJ (1903) Vict Nat 20:17 (cited in Free 1911 and Stuntz and Free 1911)
44. Shaw GE (1980) J Appl Met 19:1254
45. Duce RA, Unni CK, Ray BJ, Prospero JM, Merrill JT (1980) Science 209:1522
46. Parrington JR, Zoller WH, Aras NK (1983) Science 220:195
47. Uematsu M, Duce RA, Prospero JM, Chen L, Merrill JT, McDonald RL (1983) J Geophys Res 88:5343
48. Merrill JT, Uematsu M, Bleck R (1989) J Geophys Res 94:8584
49. Bodhaine BA (1995) J Geophys Res 100:8967

The Handbook of Environmental Chemistry Vol. 4, Part G (2004): 295–320
DOI 10.1007/978-3-540-40037-0

Prospects for International Management of Intercontinental Air Pollution Transport

Terry J. Keating[1] · J. Jason West[1,2] · Alexander E. Farrell[3]

[1] U.S. Environmental Protection Agency, Office of Air & Radiation, 1200 Pennsylvania Ave NW, Mail Code 6103A, Washington DC, 20460 USA
E-mail: keating.terry@epa.gov; west.jason@epa.gov

[2] American Association for the Advancement of Science Fellow

[3] University of California, Energy Resources Group, 310 Barrows Hall, Berkeley, CA94720-3050 USA
E-mail: afarrell@socrates.berkeley.edu

Abstract The recognition of the intercontinental transport of air pollution continues a historic trend towards greater awareness and management of air pollution over larger spatial scales. Intercontinental transport contributes to a "tightening vise" on air quality management in industrialized nations with background concentrations rising, standards becoming more stringent, local controls becoming more difficult to achieve, and emissions in other parts of the world rapidly increasing. While several regional and global regimes currently exist for addressing issues of transboundary air quality, important gaps remain for intercontinental transport. International cooperative regimes at all levels – binational, regional, hemispheric, and global – should be encouraged to address intercontinental transport. Activities which increase the capacity in developing nations for managing domestic air pollution problems are the most important actions to be taken in the short term. In addition, international research efforts are needed to quantify source-receptor relationships between nations and to develop integrated analysis tools that connect international transport, climate change, global energy infrastructure, and economic development.

Keywords Management · Policy Regimes · Assessment · Capacity building

List of Abbreviations

AMAP	Arctic Monitoring and Assessment Program
ASEAN	Association of Southeast Asian Nations
CFCs	Chlorofluorocarbons
CTMs	Chemical transport models
EMEP	Cooperative Programme for Monitoring and Evaluation of the Long-Range Transmission of Air Pollution in Europe
EANET	East Asia Network
GHGs	Greenhouse gases
INDOEX	Indian Ocean Experiment
LRTAP	Long Range Transboundary Air Pollution
NAPAP	National Acid Precipitation Assessment Program
NRC	National Research Council (U.S.)
OECD	Organization for Economic Cooperation and Development
$PM_{2.5}$	Particles less than 2.5 µm in diameter
PM_{10}	Particles less than 10 µm in diameter
POPs	Persistent organic pollutants
RAINS	Regional Air Pollution Information and Simulation
TSP	Total suspended particulates
UNEP	United Nations Environment Program
UNFCCC	United Nations Framework Convention on Climate Change
U.S.	United States of America
VOCs	Volatile organic compounds

1 Introduction

The intercontinental transport of air pollutants is a rapidly advancing field of atmospheric science, as reflected in the other chapters of this book. As our knowledge of transport evolves, our political institutions and policies to address the problem must evolve as well, raising important questions. What are appropriate policy responses to this growing scientific understanding of air pollution? What can be learned from environmental policy formation generally, and from existing international efforts to control cross-border air pollution specifically, that will be relevant for predicting and designing a policy response? Are existing international regimes sufficient for addressing intercontinental transport, or are new regimes required?

The purposes of this chapter are to discuss how the evolution of science affects the evolution of environmental policy and to analyze alternative policy responses for the case of intercontinental air pollutant transport. In particular, we discuss the difficulties of developing international regimes for addressing transboundary air pollution problems and experiences gained historically through such regimes. Here we use the word 'regime' to mean "social institutions consisting of agreed-upon principles, norms, rules, procedures, and programs that govern the interactions of actors in specific issue areas [1]," suggesting that a regime can be more than a treaty, including the whole process of interaction by which a treaty might be formed. Thus, regimes include political as well as scientific activities, such as cooperative research programs, standards and calibration organizations, and scientific assessment bodies.

We begin by setting the historical context for understanding new scientific findings about intercontinental transport, addressing how the spatial scale of air pollution has changed historically and the forces that are now pushing towards greater international linkages. We identify some of the existing international regimes for addressing the international transport of air pollutants, consider how international environmental policy regimes evolve, and identify the characteristics that are thought to be favorable for success. In light of this discussion, we consider the prospects for effectively addressing intercontinental transport through policy actions and regimes at the national, regional, and hemispheric or global scales. Finally, we discuss how air pollution science can contribute knowledge and analysis that will be relevant and useful in informing this process.

2 Changing Views of Air Pollution: the Historical Context for Intercontinental Transport

2.1 The Evolution of the Spatial Scale of "Air Pollution"

The history of air pollution science and management has been one of gradually evolving views of what air pollution is and what components of air pollution are relevant for policy action to control emissions. This shift in emphasis among different components of air pollution is a result of 1) the improvement in our scientific understanding of air pollution through time, often driven by new measurement and observation techniques, creating an awareness of new problems, and 2) the success of emissions control efforts, effectively addressing some problems and allowing other problems to come to the forefront of policy awareness.

Figure 1 shows the evolution of the problems and pollutants emphasized in air pollution science and management in the United States. The earliest efforts at air pollution management were local efforts focused on the mitigation of smoke, soot, and odors [2]. As these readily perceptible problems came under control in the 1960s, air pollution control efforts shifted focus to the control of ozone, SO_2,

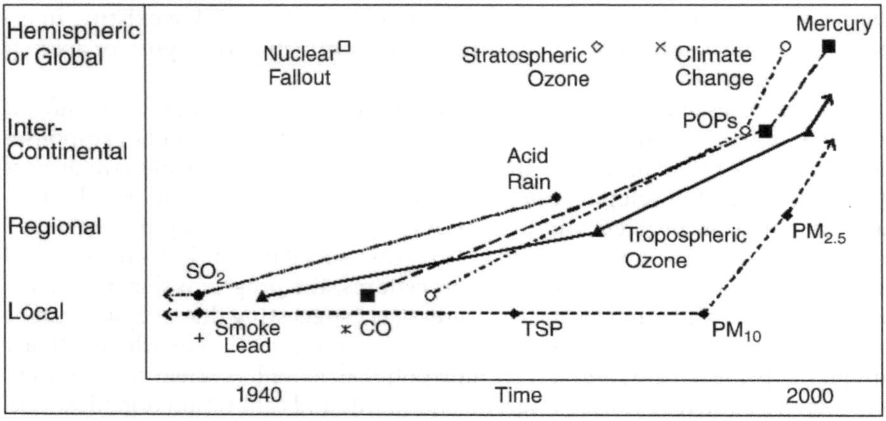

Fig. 1 Evolution of the perceived spatial scale of air pollution problems in the United States. Points indicate approximately when scientific consensus emerged that public policy action should be considered. While some pollutants have always been conceived of as local problems, some pollutants like O_3 and particles (first TSP, then PM_{10}, and then $PM_{2.5}$) have been reassessed over time as relevant on larger spatial scales. Other problems were conceived of originally at broader scales, such as acid rain at the regional scale, and there are several problems that have always been viewed globally

CO, NO_2, lead, and total suspended particulates (TSP). In the 1970s, awareness of these issues led to national ambient air quality standards, industrial controls, the removal of lead from gasoline, and the addition of catalytic converters on vehicles. Beginning in the 1970s and extending through the 1980s, attention focused on understanding acidifying deposition, eventually leading to significant SO_2 emissions controls. In the 1990s, problems such as CO came under control with changes in fuels and vehicle controls. Ozone, however, remained a stubborn problem, and the effects of changes in emissions of NO_X and volatile organic compounds (VOCs) became understood as subtle and difficult to predict [3]. Meanwhile, epidemiology developed the strongest evidence so far of the health effects of air pollution, identifying particles – first PM_{10}, then fine particles ($PM_{2.5}$) – as most responsible [4]. In the near future, there is a clear trend towards increasing concerns over emissions of mercury, airborne toxics and persistent organic pollutants (POPs), and greenhouse gases (GHGs) that contribute to climate change. In addition, the results of continuing research on the health effects of particles may focus control efforts on specific chemical components and their sources.

As this shift in focus between different pollutants has taken place, our concept of the spatial scales relevant for pollutant transport through the atmosphere, and thus air quality management, has also changed. Initially, air pollution was conceived as a localized phenomenon and air pollution control efforts were focused on the urban scale. The resolution of urban "smoke" problems, as well as the gradual demographic changes toward more regional development (itself driven in part by urban air pollution), made air pollution more regional. In the 1970s and 1980s, the perceived spatial scale of air pollution changed remarkably through studies that revealed that acidifying deposition is strongly regional, and even international, with transport between the US and Canada and between the nations of Europe [5, 6]. The regional nature of ozone, which was originally understood as an urban problem beginning in the 1950s, became a focus of management efforts in the 1990s as downwind jurisdictions discovered that they could not attain ambient standards because of emissions in upwind jurisdictions [7]. As emphasis on particle control has shifted to fine particles, the spatial scale of management efforts has also shifted from the urban to regional scale.

While our understanding of the geographical scale of conventional air pollution has expanded over time, other atmospheric problems were conceived of, from their inception, as global problems. These include studies of the transport of radioactive particles, climate change, and stratospheric ozone depletion. Measurements of chlorofluorocarbons (CFCs) in remote regions demonstrated that transport of long-lived species occurs [8]. Studies of the effects of CFCs on stratospheric ozone put atmospheric transport in a clearly global and very-long-term perspective, and led to the Vienna Convention and the Montreal Protocol, one of the earliest and most successful global agreements managing the global atmospheric commons [9].

Recently, the understanding of the transport of conventional air pollutants on international and now intercontinental scales, has taken a major step forward through the scientific research described in this volume. Satellite imagery, as well as analysis of surface observations, has provided vivid illustrations of individual events in which Saharan dust is transported to the Caribbean and Asian dust is transported to western North America. Analyses of the trends in ozone observed at remote sites have provided evidence of the hemispheric transport of ozone and the increasing hemispheric burden of ozone in the free troposphere [10].

This same evolution of the perceived spatial scale of air pollution has occurred concurrently in Europe, though with slightly different emphasis on different problems, such as the earlier recognition of acidifying deposition as a regional problem in Europe. Developing nations are observed to follow the same historical progression, but fall at different places along this path depending on the status of economic development and the severity of air pollution problems. While many developing nations are at the stage of controlling smoke and other urban pollution problems, as the United States was in the mid-1900s, they do so now with a more complete understanding of the picture painted in Fig. 1, and of the relative priorities and complex relationships between air pollution problems.

2.2
The Tightening Vise of Air Pollution Management

Four current pressures create a "tightening vise" that is increasingly faced by air pollution managers in industrialized nations. First, air pollution management has historically emphasized the control of local sources of pollution, with a relative lack of control of regional sources. Through time, this has led regional and international sources to contribute proportionally more to air pollution problems. Consequently, there has been an increasing emphasis on the control of pollutants that are transported over longer distances, as air quality managers look to force

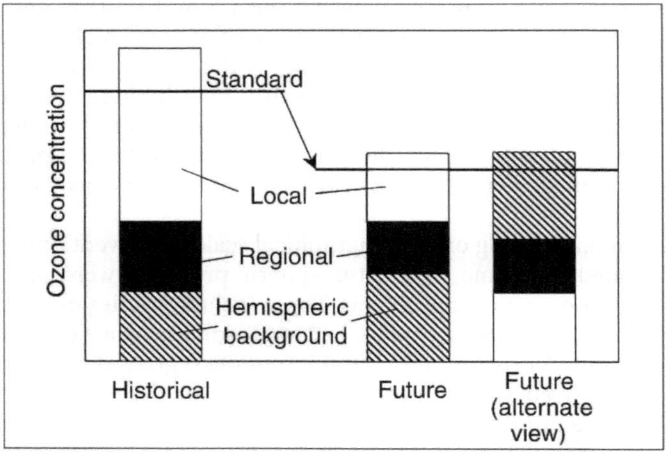

Fig. 2 The "tightening vise" of air pollution management, shown here (schematically) for ozone from the perspective of an industrialized nation. From the historical view to the future, air pollution managers succeed in reducing their local contribution to ozone, and work regionally with other jurisdictions to reduce the regional contribution. However, the hemispheric background increases, while their air quality standard becomes more stringent. In the alternate view of the future, the hemispheric background, not the local pollution, pushes ozone above the standard, and the local air quality authorities blame other nations for their exceeding of the standard

upwind jurisdictions to control their share of emissions. This same trend is now extending to the intercontinental transport of pollution.

Second, having already exhausted the cheapest and easiest controls, further local emissions reductions come at increasingly higher economic and political marginal costs relative to controls in upwind areas, although in some cases technological innovation may reduce the actual control cost.

Third, while the cost of local control is increasing, air quality standards have become more stringent, reflecting an increased environmental awareness and improved understanding of health effects. This third pressure is manifested in the new ambient standards for ozone and fine particles in North America and for ozone in Europe.

The fourth pressure arises from the rate and spatial pattern of global development, which is causing emissions of air pollutants and their precursors to increase most rapidly in the developing nations of Asia, Africa, and Latin America. In addition to industrial emissions, often from inefficient and uncontrolled processes, emissions from biomass burning and windblown dust (caused in part by human contributions to desertification) also contribute, sometimes dramatically, to international pollutant transport.

The net result is that the regional or global background contribution to pollutant concentrations is growing, while standards are becoming increasingly stringent and local pollution controls are becoming increasingly expensive and difficult to achieve (see Fig. 2). Thus, air pollution managers are increasingly in a "tightening vise" of pressures from the industrialized world and growing emissions from the developing world. These pressures create an increased motivation for industrialized nations to help decrease emissions overseas. These foreign emissions, however, not only derive from different sources, but their management involves a very different set of actors with different priorities and technical and regulatory capabilities.

2.3 Overlapping Problems, Multi-Pollutant Strategies, and Co-Benefits

Figure 1 suggests that the scope of air pollution concerns on the local, regional, and global scales is complex, and that the relative importance of different problems varies with spatial scale. These concerns are also interrelated to a significant degree. One example is that emissions of SO_2 contribute to fine particle formation on a local and regional scale, contribute to acid deposition regionally, and influence the global climate.

As discussed in other chapters of this book, the nature of intercontinental pollutant transport is different for different pollutants, although many management concepts are common. In most cases, intercontinental transport contributes only a fraction of the pollution at a given location, with the remainder coming from local or regional sources, as portrayed in Fig. 2. For ozone and mercury, the contribution of intercontinental transport occurs mainly through the build-up of concentrations in the free troposphere on a hemispheric scale. For fine particles, on the other hand, transport during discrete meteorological events may be more important. For ozone and mercury, therefore, emissions throughout the hemisphere may be important; while for fine particles there is a more clear direction of flow from source continents to receptor continents. Interestingly, the emissions that one would choose to control can differ depending on the scale of influence one cares about. For ozone, controls on emissions of VOCs will reduce local concentrations, while NO_X controls are likely to have more regional benefits [3]. For the global background of ozone, however, recent modeling results suggest that controlling emissions of methane in addition to NO_X will be most effective, while changes in emissions of VOCs are less relevant [11].

There are also close linkages between air pollution – especially the long-range transport of air pollution – and climate change. Ozone is a GHG, and fine particles also influence climate by altering the Earth's radiation budget regionally to cause a net cooling or warming [12]. The linkage to the long-range transport of pollutants is particularly important because urban-scale pollution occurs over too small a scale to have a meaningful influence on the global climate. In the case of particles, there has been recent interest in controlling emissions of black carbon, both because of its benefits for human health and as a way of more quickly reducing human contributions to climate change [13].

In addition to the many scientific linkages between air pollution and climate, important policy linkages result from the fact that many air pollutants and GHGs share common sources. Many actions to address emissions of some pollutants may also affect emissions of other pollutants, such as the effect of GHG mitigation on aerosol concentrations and climate [14, 15]. Likewise, there has been increased recognition of the "co-benefits" of GHG mitigation in terms of reduced air pollution [16, 17], and studies to plan the control of GHGs and air pollutants simultaneously [18].

Together, understanding of these scientific and policy linkages has led to an acknowledgment of a complex pollution control landscape, where there are multiple sources of pollution, causing emissions of multiple pollutants, in turn causing multiple impacts that become manifest in inter-related ways on local, regional, and global scales. Likewise, policy linkages extend to other goals, such as providing transportation and energy, which can overlap with environmental goals. Interest in multi-pollutant strategies comes not only from governments, which want more efficient policies, but also from industry, which wants to have more long-term certainty in regulations to aid business planning.

Given this complex landscape, the main question for air quality managers in industrialized nations becomes increasingly: how to plan local air quality management strategies, accounting for international pollutant transport, together with climate change and other economic and social priorities at the same time? Meanwhile, air quality managers will increasingly be inclined to pursue emissions controls overseas, where those emissions controls will be tied together with overlapping environmental and development priorities.

3 Status of Current International Air Pollution Control Regimes

To manage effectively the international or intercontinental aspects of air pollution, some form of an international regime is necessary. International regimes that address transboundary air pollution can be found at the binational, regional, and global scales. Some existing regimes are listed chronologically in Table 1.

At the binational level, existing regimes include cooperative agreements between neighboring nations, which identify joint goals and obligations, such as the Canada-U.S. Air Quality Agreement [29] or the La Paz Agreement between Mexico and the U.S. [30]. Binational regimes may also take the form of technical cooperation between more distant nations, such as existing cooperative agreements between the U.S. and China [31] or between Norway and Poland [32].

At the regional level, examples of existing multinational regimes range from initial agreements acknowledging shared interests, such as the Malé Declaration; to regional scientific cooperation, such as the East Asia Network (EANET) and the Arctic Council's Arctic Monitoring and Assessment Program (AMAP); and well-developed policy regimes, such as the Convention on Long-Range Transboundary Air Pollution (LRTAP Convention). Signed in 1979 and encompassing the United States, Canada, and all the nations of Europe, the LRTAP Convention is one of the most successful international environmental regimes. Over time, it has developed a robust analytical support structure that includes a number of working groups, task forces, and international cooperative programs, including the Cooperative Programme for Monitoring and Evaluation of the Long-Range Transmission of Air Pollution in Europe (EMEP). Eight protocols have been negotiated under the LRTAP Convention addressing financing of scientific cooperation and obligations to reduce emissions related to acidification, ozone, POPs, heavy metals, and eutrophication.

Table 1 Some international agreements addressing transboundary air pollution

Agreement	Geographic region	Pollutants addressed	Reference
1979: Convention on Long-Range Transboundary Air Pollution (LRTAP)	United States, Canada, and 47 European Nations		[19]
1984: Geneva Protocol on Long-Term Financing of the Cooperative Programme for Monitoring and Evaluation of the Long-Range Transmission of Air Pollution in Europe (EMEP)			
1985: Helsinki Protocol on the Reduction of Sulphur Emissions or their Transboundary Fluxes By At Least 30% (1st Sulphur)		SO_2	
1988: Sophia Protocol Concerning the Control of Emissions of Nitrogen Oxides or their Transboundary Fluxes (NO_X)		NO_X	
1991: GenevaProtocol Concerning the Control of Emissions of Volatile Organic Compounds or their Transboundary Fluxes (VOC)		VOCs, O_3	
1994: Oslo Protocol on Further Reduction of Sulphur Emissions (2nd Sulphur)		SO_2	
1998: Aarhus Protocol on Persistent Organic Pollutants (POPs)		15 POPs	
1998: Aarhus Protocol on Heavy Metals (Metals)		Hg, Pb, Cd	
1999: Gothenburg Protocol to Abate Acidification, Eutrophication, and Ground-level Ozone (Multi-Effects)		SO_2, NO_X, VOCs, NH_3, O_3	
1985: Vienna Convention on the Protection of the Ozone Layer	Global (185 ratifications)		[20]
1987: Montreal Protocol on Substances that Deplete the Ozone Layer		CFCs, and other O_3 depleting substances	

Table 1 (continued)

Agreement	Geographic region	Pollutants addressed	Reference
1992: United Nations Framework Convention on Climate Change (UNFCCC)	Global (188 ratifications)	GHGs	[21]
1997: Kyoto Protocol			
1995: ASEAN Cooperation Plan on Transbounary Pollution	10 Southeast Asian nations		
2002: ASEAN Agreement on Transboundary Haze Pollution		Visibility, fine particles	[22]
1996: Arctic Council	8 Arctic nations	POPs, metals	[23]
1998: East Asia Network	12 East Asian nations	SO_2, acidification	[24]
1998: Malé Declaration on Control and Prevention of Air Pollution and Its Likely Transboundary Effects for South Asia	8 nations on the Indian subcontinent		[25]
1998: Harare Resolution on the Prevention and Control of Regional Air Pollution in Southern Africa and its likely Transboundary Effects	Southern Africa nations (3 primary participants)		[26]
2001: Stockholm Convention on Persistent Organic Pollutants (POPs)	Global (151 signatories)	12 POPs	[27]
2003: UNEP Global Mercury Assessment	Global	Hg	[28]

At the global level, existing regimes range from technical cooperation under the auspices of multinational organizations, such as the United Nations Environment Program (UNEP) and the Organization for Economic Cooperation and Development (OECD), to multilateral treaties, such as Vienna Convention and Montreal Protocol for the protection of stratospheric ozone, the United Nations Framework Convention on Climate Change (UNFCCC) and the Kyoto Protocol, and the Stockholm Convention on Persistent Organic Pollutants.

While these existing regimes provide possible foundations for building future regimes to address intercontinental transport, additional institutional infrastructure will be required. Not all regions of the world are engaged in an international environmental policy regime that has the potential to address transboundary flows of air pollution, and some existing regimes are only in the early stages of development. Intercontinental pollutant transport can occur on spatial scales that exceed those of existing regional air pollution regimes. However, given differences in the transport characteristics of pollutants, not all intercontinental pollutants may be appropriate to address through global regimes. Before discussing what future regimes may be most appropriate, we first consider how international regimes are constructed and what characteristics lead to their success.

4 Constructing and Implementing International Regimes for Air Pollution

The development of an international environmental policy regime is a social process, and like the process of developing scientific knowledge, this process takes time. Each of the regimes identified in the previous section or in Table 1 are at different stages of development. These and other international environmental policy regimes have developed through what has been called the co-production of science and policy [33]. Co-production is the idea that knowledge and political order are constructed, through social processes, in ways that enable each to support the legitimacy of the other. For example, government decision-makers support scientific research that they believe will help them structure policy options. Scientists focus their research to provide information that they believe will be of most use to the decision-makers. The research results, along with other considerations, constrain the policy options to be considered by the decision-makers. Once a policy framework is constructed, future research is focused to provide further information that is useful within this framework.

Studies of the development of international regimes suggest co-production does not usually follow a linear path of problem recognition leading to a search for solutions and then a policy choice among those solutions. From their comparative history of social responses to climate change, stratospheric ozone depletion, and acid rain, Clark et al. [34] suggest that the development of international regimes can be described in terms of separate streams of activity related to six management functions: risk assessment, monitoring, option assessment, goal and strategy formation, strategy implementation, and evaluation. The six streams of activity progress simultaneously, feeding each other. Each of the streams of activity may involve significant international cooperation. Periodically, the streams intersect where activities are brought together by conferences, assessments, or decision processes. The evolution of a social or policy response to a given issue through these six streams of activity may take long periods of time, even decades.

"Our research has shown that the process of building capacity to address global environmental risks needs time. This is irreducibly true, since it is not primarily the amount of resources (human resources and money) that is of importance but rather the generation of the coordination, cooperation, and trust needed to create an effective management process" ([34], p 191).

During this slow evolution, attention to the issue by the public and decision-makers will rise and fall. It is helpful to know where one is in the cycle of waxing and waning public attention to constructively move the process towards a policy response [35].

While the development of international regimes may not follow a linear pattern, based on the literature on international environmental policy regimes, we have divided the development of regimes into two phases [36–38]. The first phase, which we will call the framing phase, focuses on organizing knowledge and politics around an issue and building new institutions. Policies developed during this phase tend to be "least common denominator" agreements that codify pre-existing positions, often those that already exist in national laws. The second phase, the implementation phase, focuses on the implementation of environmental policies and making further improvements in environmental quality. Policies developed during the second phase may be more complex, building upon continued technical cooperation and advancing common environmental goals. Some of the issues that arise in each of these two phases are discussed below.

4.1 Framing Phase

The initial phase of regime development involves building new institutions for collective action and focuses on bounding the issues under consideration and constructing conceptual problem frames. International environmental policy problems are often very complex, involving interrelated environmental phenomena and multiple human actors with multiple interests. Problem frames serve to simplify and make tractable the complexity of a problem that is the subject of scientific study, assessment, or policy debate. Like a camera lens, a problem frame not only defines the field of view, but also defines how the components of the system within the field of view and the relationships between the components are perceived [39]. Problem frames define what sorts of knowledge are relevant, and the appropriate data and analytical methods to use in gaining such knowledge [34]. In this way, a problem frame "provides us with a whole structure by integrating interests, values, actions, theory, and facts [40]."

Problem frames are socially constructed and reflect the beliefs of the participants in the process. Some beliefs, such as those related to national sovereignty, regional identity, and notions of rights and responsibilities, may be deeply held. These core beliefs evolve gradually [41], thus constructing a common problem frame between multiple participants may take a very long time.

An important step in developing a common problem frame is coming to a shared understanding of the nature of the physical systems involved. Many international environmental policy regimes have begun by creating institutions for cooperative scientific efforts. The LRTAP Convention grew out of cooperative studies of acidifying deposition in Europe initially organized by the Nordic Council and later by the OECD. This cooperation created the foundation for the formation of EMEP and the LRTAP Convention. The development of EANET has followed a similar path, beginning with cooperative scientific studies initially organized and funded by Japan, and now organized under the UNEP Regional Cooperative Centre in Thailand. In both of these cases, scientific cooperation was organized first by the downwind nations who were the recipients of transboundary flows of pollution. Once established, the coordination function was shifted to a more neutral international organization. Scientific studies conducted by individual nations that have clear environmental, economic, or political interests at stake may be perceived to be biased. The participation of representatives of different nations with different interests lends credibility to the results. Through international scientific cooperation, participants may not only develop a shared understanding of a problem, but they may also come to a new understanding of their own nations' self interests as they relate to that problem. Furthermore, cooperation on scientific studies helps to build relationships, goodwill, and trust, which can be carried into discussions of policy options.

A shared understanding of the physical problem to be addressed, however, does not have to come through participation in cooperative scientific research. In some cases, international regimes have formed due to events or scientific discoveries that have focused public attention on a particular environmental phenomenon. For example, the discovery of the stratospheric ozone hole in the 1980s spurred political action that resulted in the Vienna Convention and Montreal Protocol and the Indonesian fires of 1998 led to the ASEAN Agreement on Transboundary Haze Pollution. The signatories of the Malé Declaration in 2002 were informed by the findings of the Indian Ocean Experiment (INDOEX), an international scientific field study in which few of the Malé Declaration signatories were active participants [42].

In addition to a shared understanding of the physical systems to be addressed, it is important for the participants in an international regime to have a common understanding of the social, economic, and political implications of the environmental problem and its possible solutions. Managing transboundary air pollution can be complex because the polluter and the receptor are in different nations, which may have different legal and social norms concerning rights and responsibilities related to pollution. Some nations value a right to a clean environment; others value a right to use the environment's waste assimilative capacity. The relative priority of these rights may differ depending on whether the nation is the source or the receptor of the pollution. Some nations may see the international or global fraction of their pollution as "background," uncontrollable by domestic management; while others may blame the international contribution for pushing their air quality above the policy objective (see Fig. 2).

Issues of rights and responsibilities are further complicated by differences in the wealth of the source and receptor nations. Developing nations may assert a right to exploit an equal share of the global commons or a right to follow a path of development and pollution similar to that of industrialized nations. Global policy agreements, such as the Stockholm Convention or Montreal Protocol, have included provisions for allowances, compensation, or assistance for less developed nations.

The more similar nations' interests or situations are to begin with, the easier it is to construct a common understanding. Thus, regional agreements between nations that share common borders, are interconnected politically and economically, or have additional reasons to cooperate other than air pollution, are generally easier to reach than global agreements, which involve a larger diversity of perspectives and positions. Given their relative tractability, regional agreements can serve as stepping stones to more comprehensive global agreements. For example, international control of POPs began with technical work under OECD that led to a regional agreement under the LRTAP Convention and then to a global agreement under the Stockholm Convention [43].

In many cases, upwind nations may have little reason to participate in an international regime, at least in terms of protecting their own environment. However, many other types of issues may bring them into the regime. Developing nations may participate as a way to secure development assistance. Nations may participate to protect or pursue interests related to economic trade or to establish relationships that are useful in addressing other concerns, such as national security. The early years of the LRTAP Convention were shaped in part by Cold War politics and were a mechanism of détente between the U.S. and its Western European allies and the Soviet Bloc [44]. More recently, participation in the LRTAP Convention has been shaped by the expansion and evolution of the European Union [45].

Coming to a common problem frame may take a long time. In some cases, agreement on some contentious elements of a frame may never be reached, and the regime either breaks down or continues in a fashion that accommodates different perspectives. However, once a viable problem frame and sustainable institutions have been constructed, the international regime can move beyond an acknowledgement of a problem and the need to work together and make progress towards environmental protection.

4.2 Implementation Phase

As an international environmental regime matures, it moves from the framing phase into the implementation phase, which builds on the institutions created in the framing phase and focuses on making increasingly substantive agreements and actions possible [38]. In this phase, participants begin to focus less on the international nature of the problem and more on identifying the most efficient solutions to the environmental problem, irrespective of national borders. Several of the regimes identified below have reached this stage of development, such as the Montreal Protocol and the LRTAP Convention. The evolution of the LRTAP Convention is manifest in its protocols. The first generation of protocols, which includes the protocols of 1984, 1985, 1988, 1991, and 1998, primarily identified problems, established cooperation, and set readily attainable goals. The transition from the framing phase to the implementation phase coincided with the beginning of a second generation of protocols, including the 1994 and 1999 protocols, which established differentiated obligations based in part on analysis of environmental impacts.

While the implementation phases of mature regimes have included obligations or actions for environmental protection, sometimes with increasing stringency over time, it is legitimate to question whether the environmental protection associated with international agreements would have occurred without the regimes. In their review of international environmental policy regimes, Victor et al. find that almost all nations comply with almost all of their binding international commitments [46]. Given that governments often cannot ensure national compliance with environmental quality goals, it is not very surprising to find that most governments are very conservative in choosing what commitments to make. Therefore, they may not go beyond environmental protection efforts that have already been incorporated into their national laws. This suggests that compliance with international obligations and effectiveness in advancing environmental protection may not be the same; "compliance often simply reflects that countries can negotiate and join agreements with which they know they can comply [46]."

Although the direct effect of international regimes on environmental protection is difficult to establish in many cases, there are several ways in which participation in an international regime can affect national or local environmental policies and environmental quality. First, Victor et al. [46] note that, while national environmental protection efforts are often developed in response to key events or disasters or changing economic conditions, the existence of international regimes increases the impact of these events and enables policy advocates to seize the opportunities presented. The creation and shaping of environmental regimes is one of the few things that advocates can influence directly, unlike chance (and often unfortunate) events that can initiate major changes in policy. Second, participation in an international regime, especially those involving scientific cooperation, can change a government's perception of its own national interests, leading to a change in national policies [37]. As Young observes, "regimes are often effective in solving international problems when they can redirect the interplay of political forces within the domestic policymaking arenas of key members [1]." Third, as noted below, nations may agree to environmental protection efforts in return for action on another issue affecting their perceived self interests [47]. Finally, international regimes create a sense of peer pressure among nations. In his study of the LRTAP Convention, Marc Levy labeled this aspect of international regimes "tote-board diplomacy:" leading nations declare their intent to take strong environmental protection efforts, creating a challenge or pressure for other nations to follow [44].

4.3 Characteristics of Successful International Regimes

A number of scholars have identified characteristics of international regimes that contribute to their effectiveness in maintaining international dialogue and cooperation and, ultimately, in influencing policies at the national and international level. These characteristics include:

- Sponsorship: Effective regimes, at least in the beginning, need to have strong sponsorship from one or more nation whose interests are served by investing the resources needed to establish and maintain new institutions. Over time, other nations may share the burden as they see the value of participating in the regime, but the major sponsors can have a lasting influence on the framing of issues, the nature of institutions, and the eventual policy outcomes [48].
- Participation: In the short term, regimes with few and homogeneous participants tend to be more effective than regimes with many and heterogeneous participants. However, in the long term, it is unlikely that regimes that do not include all major contributors to the problem can remain effective [46]. Furthermore, national governments must be sufficiently concerned about the problem to actively participate and cooperate. To help raise government concern, effective regimes can create, collect and disseminate technical knowledge; link environmental issues to other issues that governments are more directly concerned about; and help magnify public pressure, both domestic and international [37].
- Coordination: Effective regimes have the capacity to coordinate efforts of the participants, often through central secretariats or administrators. Financially strong and technically capable secretariats tend to make regimes more effective and can be "critically important in a situation where several or all the cooperative parties have weak domestic administrative capacities [49]."
- Technical capacity: Effective regimes have the ability to generate and use technical analyses that are credible and relevant to all of the participants [34, 49]. Establishing credibility and relevancy in technical analyses requires transparency, participation, and a level of technical capability within each of the participating nations [50]. Effective regimes can help build national capacity by creating professional networks and providing financial assistance to aid in the transfer of technical and management expertise [37]. Furthermore, effective regimes create institutions to actively inform policy makers of the results of scientific research and technical analyses, through assessments, reports, and meetings.
- Adaptability: Effective regimes have sufficiently comprehensive and flexible agendas and structures that allow them to adapt to changing knowledge, conditions, and needs [7]. Flexibility can allow the parties within the regime to take advantage of windows of opportunity to make progress on specific issues when they open [34, 49].
- Trust: Effective regimes require multiple layers of trust between the parties [34, 50]. At the individual level, trust is developed through repeated interactions of individuals in both formal and informal settings, provided by hierarchies of workgroups, taskforces, and committees [50]. At the institutional level, trust is associated with decision-making processes that are fair, provide adequate time for participation, and include opportunities to revisit the decision, in another venue or subsequent negotiation [43, 50]. Furthermore, the regime must provide a positive contractual environment, in which the costs of ac-

quiring information and iterative bargaining are minimized and parties are held accountable for fulfilling their commitments [37]. This accountability may be achieved through reporting systems and implementation review bodies [49].

It is important for an international environmental policy regime to develop these characteristics if the regime is to evolve beyond the framing phase and into the implementation phase.

5 Prospects for Future Regimes

Given our understanding of how international environmental regimes evolve and what characteristics contribute to their success, what can we say about the likely success of regimes for addressing intercontinental transport? What actions in the short term should be recommended to develop or strengthen regimes?

There are several possible levels at which governments can address intercontinental transport, including actions taken unilaterally, bilaterally, or multilaterally on regional, hemispheric, or global scales. Actions at these different scales, however, are not equally likely to be successful in the short and long term.

5.1 National and Bilateral

At the national level, nations have incentives to reduce their own emissions, to the extent they will benefit from decreased air pollution within their own borders. The incentive to reduce emissions to address intercontinental transport is currently weakest among the developing nations, where emissions are growing rapidly. In many developing nations, awareness of environmental quality is growing, but these nations may lack the technical expertise to understand the sources and effects of air pollution, and may lack the regulatory capabilities and experience to effectively control the sources. Lacking this expertise, developing nations will not only have reduced capabilities to address their own air quality problems, but will be less likely to participate meaningfully in international regimes addressing transboundary air pollution.

Even where scientific, technical, and managerial expertise exists in industrialized nations, transferring this expertise to developing nations can be expensive, but ultimately worthwhile. Given the importance of developing nations' emissions for existing and projected intercontinental transport problems, it is implausible that international regimes to address intercontinental transport will be effective without first taking steps to build capacity in developing nations to understand and manage air pollution. Capacity-building actions should, therefore, be a short-term priority for any series of actions to address intercontinental transport. Such capacity-building, whether undertaken bilaterally with individual developing nations or multilaterally through regional or global regimes, need only focus on providing support for developing nations to address their own air pollution problems. Industrialized nations will benefit from such pollution control efforts through reduced international or intercontinental transport, and through the other benefits that motivate development support currently.

Unilateral actions by industrialized nations to improve scientific understanding of pollutant transport are important, but sponsorship of cooperative actions to develop a shared understanding can produce greater international legitimacy and can better support international agreements.

5.2 Regional

Regional regimes have proven effective at reducing international pollution among neighboring countries, and regional regimes are growing among developing nations. Although current regional regimes do not cover the geographic scope of intercontinental transport, they offer existing functional structures that can serve as a basis for further cooperation with other nations and regimes. Regional regimes have the advantage of dealing with nations that are often similar in development characteristics and have shared regional interests, which as discussed previously, is a common characteristic of successful regimes.

Because of the scale of transport, regional regimes are most likely to be effective for fine particle pollution and acid deposition, but will not be able to cover all relevant source nations for ozone, mercury, and POPs. In North America and Europe, regional agreements focusing on the control of NO_X as a regional precursor to ozone are already in place. With respect to fine particles, regional regimes should work within their boundaries to confront the causes of widespread particulate emissions from windblown dust, exacerbated by desertification, and forest fires, as well as the important industrial and urban sources of particles. Given the health effects associated with fine particles, investment in capacity building activities related to fine particle control is likely to be very beneficial.

The LRTAP Convention, which has a geographical scope stretching from North America, across Europe, and into Central Asia, provides an existing forum where meaningful progress can be made towards addressing the intercontinental transport of multiple pollutants. It is also a framework through which capacity-building activities engaging developing nations can be further encouraged. Likewise, the inception of regional pollution regimes in Asia and Africa, with a foundation in developing a shared scientific understanding of air pollution problems, is encouraging and should be supported by industrialized nations.

As pressure mounts to address intercontinental transport over larger spatial scales, there will be pressures for regional regimes to expand their boundaries, or to work together with regimes representing other regions. Both should be encouraged, as regional agreements provide useful existing structures.

In addition, environmental managers increasingly use market-based mechanisms, such as emissions trading, to reduce overall control costs. The potential for cost savings has motivated discussion of emissions trading under the Kyoto Protocol, but agreeing upon the rules for international trading has proved contentious. Trading in GHG emissions, which are long-lived and relatively homogeneous, is relatively simple compared to trading pollutants that are more heterogeneous and have strong local or regional components, such as fine particles, ozone, mercury, and POPs. Any trading system for these pollutants would have to be designed to take source-receptor relationships into account and avoid the creation of emissions hot spots.

5.3
Hemispheric and Global

Regimes on a hemispheric or global scale are the only regimes that can fully address some problems of intercontinental transport. This observation has led some scholars to speculate about the potential for a new hemispheric treaty on air pollution [51]. Hemispheric and global regimes, however, lose some of the characteristics that make regimes successful. At this scale, the commonalities of interests, shared borders, and regional objectives that support the development of regimes are weaker, while inequities between industrialized and developing nations are highlighted.

While developing nations have shown a great willingness to participate in global environmental regimes, they are less frequently willing and able to participate at the level of reducing their own emissions. Although the Montreal Protocol is a very successful global regime, for example, some developing nations still produce CFCs. Likewise, no binding emissions reduction commitments for developing nations are currently contemplated under the UNFCCC. The prospect of using a global regime to leverage meaningful emissions reductions from developing nations, therefore, is not likely to be successful in the short-term unless significant incentives are made for developing nations to participate. Incentives in the form of development aid for clean energy infrastructure can advance the developmental and environmental priorities of both developing and developed nations.

A global regime currently exists for POPs and has momentum to address this problem globally for several important pollutants. For mercury, a global regime has recently been created to focus on scientific assessment and capacity building. Given that scientific understanding of mercury transport is weak in relation to other pollutants, scientific cooperation is important at this stage. Beyond an assessment, actions in industrialized nations to control their own emissions, and capacity-building and financial support for projects that reduce mercury emissions in developing nations will be important.

No global regime exists for ozone or other traditional air pollutants. The creation of such regimes may be part of a long-term solution for these problems, but it is not clear how these regimes might evolve. Under the UNFCCC, significant actions may be taken to address emissions of the precursors of ozone and fine particles, since both affect climate change. Methane emissions are already addressed by the UNFCCC, and their control will benefit climate change efforts as well as air quality management efforts.

As mentioned earlier, however, achieving agreement among many parties in a global regime can be time-consuming. We are therefore not optimistic that global regimes can bring meaningful emissions reductions in the short term, and should therefore not take priority over the types of bilateral and regional activities discussed earlier.

6
Needs for Research to Support an Effective Policy Response

The air pollution and atmospheric science research community can contribute to the development of effective regimes to address intercontinental transport by focusing their research on meeting the information needs of policy makers and by helping to communicate effectively their research results to inform policy decisions. Before discussing some specific information needs of policy makers, we briefly discuss scientific assessment processes. In recent decades, assessments have served an increasingly important role in summarizing scientific knowledge and in communicating that understanding to decision-makers – so much so that assessments are now routine parts of domestic and international environmental management [34, 52, 53].

6.1
Assessment Processes

Although assessments are often organized around the production of a report or book, it is more useful to view scientific assessments of environmental problems as communication processes [54]. This view of assessments stresses the social processes by which expert knowledge is organized, evaluated, integrated, and presented to inform decision-making, with communication occurring both among scientists and between scientists and policy-makers. Indeed, a major function of assessments is to allow an extended technical dialogue to occur among experts united by discipline or topic area, but differentiated by nationality or other interest.

Participants in assessments, as well as most scientists conducting the research that may be used in assessments, generally want their efforts to increase knowledge and improve environmental policy [55, 56]. Decision-makers in business and government want their decisions to be firmly grounded in scientifically-supported data and analysis so that they improve environmental quality [57, 58]. In part, this desire reflects a belief that scientific and engineering research produces knowledge more likely to be effective than personal opinion, political ideology, or other sources of information. Decision-makers therefore use assessments to show that policy positions are not merely a pursuit of self-interest, but are informed by objective analysis to achieve agreed upon public ends [59].

Experience with past assessments has made clear that assessments that effectively inform decisions have several common characteristics:

- Salience: the assessments address questions that are relevant for decision-making and are made available in a timely fashion, ahead of critical decision junctures.
- Legitimacy: the assessments are developed through a fair and transparent process in which relevant perspectives are represented.
- Scientific and technical credibility: the individuals, methods, and institutions that are engaged in the assessment are judged to be competent and appropriate [54].

To increase salience and legitimacy, participants in assessment processes have increasingly engaged decision-makers early in the process to provide input on relevant questions [60].

6.2 Information Needs

6.2.1 *Source-Receptor Relationships*

Since intercontinental transport involves contributions of many source nations influencing many receptor nations, developing fair and effective policies will require scientific tools capable of determining contributions at the level of individual nations. Current global chemical transport models (CTMs), however, involve large uncertainties in quantifying source-receptor relationships, especially at the national or sub-national scales. An ambitious scientific research program, addressing both physical and chemical processes, is needed over the coming decades to better characterize and quantify transport in CTMs without excessive computational burdens.

Atmospheric science on the global scale is an "observation-limited" science, where a lack of ambient observations, particularly in remote regions and above the surface, limits progress in quantification. In the coming decades, it will be necessary to design measurement campaigns so that they will be useful in evaluating global models and to increasingly use automation to collect and analyze air quality samples for multiple pollutants. Satellite-based instruments are especially promising because of their ability to provide global, long-term monitoring.

Emission inventories used as input to CTMs are themselves highly uncertain, especially in developing nations, and efforts need to be made around the globe to create accurate, gridded emissions inventories. Likewise, the development of ground-based air pollution monitoring networks in developing nations will help improve CTM capabilities. More importantly, developing better emissions inventory and ambient monitoring capabilities is critical for improving the air quality management in developing nations. Such capacity-building steps are among the most important actions that industrialized nations can take in the short term.

CTMs need not only to reproduce current air quality and transport, but also to simulate well the changes in air quality due to future changes in emissions and meteorology. Since climate change will be one of the key atmospheric changes over the next century and since ozone and fine particles are also key uncertainties for predicting global climate change, it is important to increasingly coordinate efforts with the climate research community and to represent climate change and relevant climate-air pollution feedback processes in CTMs.

6.2.2 *Control Technologies and Costs*

An important question that will drive the policy debate in industrialized nations is: To meet domestic air quality goals, is it more cost-effective to reduce emissions abroad or domestically? Even though the impact of one ton of emissions from abroad may be less than the impact of a domestic ton, a one-ton reduction abroad may be substantially less costly. In some cases, the emissions to control abroad may be different than the relevant domestic emissions – for ozone, it is most important to reduce domestic emissions of VOC or NO_X, while controlling foreign emissions of methane may be more important. If it is shown that the cost-effectiveness, expressed as cost per unit reduction in domestic concentration, is comparable for domestic and international controls, then the motivation for investment in international controls will increase substantially.

For this purpose, research into control costs in developing nations is clearly important. Since many industrial and vehicular sources are similar, emissions reduction strategies for developing nations can learn directly from past actions in industrialized nations. Other sources in developing nations, however, will require new research to develop effective control technologies and determine control costs for sources such as fires and household energy uses. Working together with developing nations to develop control technologies and management strategies, industrialized nations can help improve developing nations air quality management capabilities and bring multiple economic and environmental benefits.

6.2.3
Integrated Tools and Future Scenarios

To inform policy decisions, analytical tools are needed to combine information from modeling studies of source-receptor relationships with information on the costs of pollutant control in different nations. Such tools may use the output of global CTMs in a reduced form, for easy combination with other global, regional, and local models of energy, economics, emissions, and pollution transport. As an example, the RAINS models for Europe and Asia use reduced-form transfer matrices derived from atmospheric models to represent source-receptor relationships for a number of regional pollutants [61, 62]. Such integrated tools should be used to explore the effects of future scenarios of growth and emissions, and to identify and highlight cost-effective strategies to achieve environmental objectives across multiple spatial scales.

Given that some emission control actions may reduce several pollutants simultaneously, integrated analysis tools should be developed to integrate information across different air pollution problems, such as ozone, fine particles, mercury, and POPs, and to explore the various linkages between climate change and air pollution policies. Some integrated assessment programs have recently taken such steps, such as the inclusion of greenhouse gas emissions in the RAINS project [63], and efforts to add regional pollutants to climate integrated assessments [64].

Finally, it is important to use these tools for long-term planning of air quality, climate, and energy policies that will achieve environmental, economic, and developmental objectives. This can be done by considering a variety of future development and emissions scenarios over the next 20 to 100 years, to identify policies that are robust over a variety of plausible futures.

7 Next Steps

While further research and better information is needed, a better understanding of intercontinental air pollution transport will not be enough to address the problem. New or expanded international regimes are needed, but the development of effective regimes takes a great deal of effort and time. Given that not all efforts are equally likely to be successful, efforts to further develop regimes at all scales should be encouraged, with different results expected at each scale. While global regimes may ultimately be necessary to address global scale problems, these regimes will be the most difficult to establish. Global regimes have the inherent difficulty of dealing with many parties and highlight differences between industrialized and developing nations, particularly where industrialized nations are downwind of developing nations. Existing regional regimes provide useful foundations upon which additional cooperation could be constructed, and creative uses of these regimes should be encouraged to address transport over larger scales.

Any of these efforts will require national leaders to recognize the need to cooperate and to commit the resources, particularly from industrialized nations, necessary to improve our understanding of the problems and build institutions. In the short-term, the most important investments are those that build capacity in developing nations for quantifying and managing their own air pollution problems – such investments will have the desired effects of decreasing emissions and transport, and will prepare developing nations to more meaningfully participate in international regimes.

Acknowledgements The authors thank Stacy VanDeveer, Dale Evarts, Tracey Holloway, and Andreas Stohl for their helpful suggestions. Any opinions, findings, conclusions, or recommendations expressed in this article are the authors' and do not necessarily reflect the views of the U.S. EPA or the AAAS.

8 References

1. Young OR (1999) The effectiveness of international environmental regimes: causal connections and behavioral mechanisms In: Choucri N (ed) Global environmental accord: strategies for sustainability and institutional innovation. MIT Press, Cambridge
2. Krier J, Ursin E (1977) Pollution and policy: a case essay on California and federal experience with motor vehicle air pollution 1940–1975. University of California Press, Berkeley, CA
3. National Research Council – Committee on Tropospheric Ozone Formation and Measurement (1991) Rethinking the ozone problem in urban and regional air pollution. National Academy Press, Washington, DC
4. Schwartz J, Dockery DW, Neas LM (1996) J Air Waste Manage Assoc 46:927
5. National Acid Precipitation Assessment Program (1991) Summary Report of the U.S. National Acid Precipitation Assessment Program. National Acid Precipitation Assessment Program – Office of the Director, Washington, DC
6. Organization for Economic Cooperation and Development (1979) The OECD Programme on Long Range Transport of Air Pollutants, Measurements and Findings, 2nd edn. Organization for Economic Cooperation and Development, Paris

7. Farrell AE, Keating TJ (2002) Environ Sci Technol 36:2537
8. Lovelock JE (1971) Nature 230:379
9. Parson EA (2003) Protecting the ozone layer: science and strategy. Oxford University Press, Oxford
10. United Nations Economic Commission for Europe (UNECE) (2003) Hemispheric air pollution: trends and intercontinental transport of photo-oxidants, particles and their precursors across the Northern hemisphere (observations, models, policy implications). EB.AIR/GE.1/2003/7. Steering Body to the Cooperative Programme for Monitoring and Evaluation of the Long-range Transmission of Air Pollutants in Europe (EMEP), Geneva
11. Fiore AM, Jacob DJ, Field BD, Streets DG, Fernandes SD, Jang C (2002) Geophys Res Lett 29:1919
12. Prather M, Ehhalt D, Dentener F, Derwent R, Dlugokencky E, Holland E, Isaksen I, Katima J, Kirchhoff V, Matson P, Midgley P, Wang M (2001) Atmopsheric chemistry and greenhouse gases. In: Houghton JT, Ding Y, Griggs DJ, Noguer M, van der Linden PJ, Dai X, Maskell K, Johnson CA (eds) Climate change 2001: the scientific basis. Contribution of Working Group I to the Third Assessment Report of the Intergovernmental Panel on Climate Change. Cambridge University Press, Cambridge
13. Hansen J, Sato M, Ruedy R, Lacis A, Oinas V (2000) Proc Nat Acad Sci USA 97:9875
14. West JJ, Hope C, Lane SN (1997) Energy Policy 25:923
15. Wigley TML (1991) Nature 349:503
16. Working Group on Public Health and Fossil-Fuel Combustion (WGPHFC) (1997) Lancet 350:1341
17. Cifuentes L, Borja-Aburto VH, Gouveia N, Thurston G, Davis DL (2001) Science 293: 1257
18. West JJ, Osnaya P, Laguna I, Martinez J, Fernandez A (2003) Co-control of urban air pollutants and greenhouse gases in Mexico City. Final report, Subcontract ADC-2-32409-01, U.S. National Renewable Energy Laboratory, Washington DC
19. http://www.unece.org/env/lrtap/
20. http://www.unep.org/ozone/
21. http://unfccc.int/
22. http://www.haze-online.or.id/
23. http://www.arctic-council.org/
24. http://www.adorc.gr.jp/
25. http://www.rrcap.unep.org/issues/air/maledec/
26. http://www.york.ac.uk/inst/sei/rapidc2/apina/
27. http://www.pops.int/
28. http://www.chem.unep.ch/mercury/
29. http://www.epa.gov/airmarkt/usca/
30. http://www.epa.gov/usmexicoborder/
31. http://www.epa.gov/oia/airandclimate/byregion/chinaair.html
32. http://www.nilu.no/informasjon/nilu-polska.html
33. Jasanoff S, Wynne B (1998) Science and decisionmaking. In: Rayner S, Malone E (eds) Human choice & climate change: the societal framework, vol 1. Battelle Press, Columbus, OH, p 1
34. The Social Learning Group (2001) Learning to manage global environmental risks: a functional analysis of social responses to climate change, ozone depletion, and acid rain, vol 2. MIT Press, Cambridge, MA
35. Kingdon JW (1995) Agendas, alternatives, and public policies. HarperCollins, New York
36. Björkbom L, Lindau L (2003) The role of science for the Convention on Long-Range Transboundary Air Pollution (LRTAP) and the EU CAFE Programme (Clean Air For Europe) ASTA Annual Report 2002. IVL Swedish Environmental Research Institute, Göteborg
37. Keohane RO, Haas PM, Levy MA (1994) The effectiveness of international environmental institutions. In: Haas PM, Keohane RO, Levy MA (eds) Institutions for the Earth: sources

of effective international environmental protection (global environmental accords). MIT Press, Cambridge, MA, p 3
38. VanDeveer SD (2000) Environment 42:10
39. Litfin K (1994) Ozone discourses. Columbia University Press, New York
40. Rein M (1983) Value-critical policy analysis. In: Callahan D, Jennings B (eds) Ethics, the social sciences, and policy analyses. Plenum Press, New York
41. Sabatier PA, Jenkins-Smith HC (1993) Policy change and learning: an advocacy coalition approach. Westview Press, Boulder
42. Lelieveld J, Crutzen PJ, Ramanathan V, Andreae MO, Brenninkmeijer CAM, Campos T, Cass GR, Dickerson RR, Fischer H, de Gouw A, Hansel A, Jefferson A, Kley D, de Laat ATJ, Lal S, Lawrence MG, Lobert JM, Mayol-Bracero OL, Mitra AP, Novakov T, Oltmans SJ, Prather KA, Reiner T, Rodhe H, Scheeren HA, Sikka D, Williams J (2001) Science 291:1031
43. Eckley N (2002) Global Environl Change Human Policy Dimensions 12:15
44. Levy MA (1994) European acid rain: the power of tote-board diplomacy. In: Haas PM, Keohane RO, Levy MA (eds) Institutions for the Earth: sources of effective international environmental protection. MIT Press, Cambridge, MA
45. VanDeveer SD (forthcoming) European politics with a scientific face: framing, national participation, and capacity in LRTAP. In: Farrell A, Jaeger J (eds) The design of environmental assessments. RFF Press, Washington, DC
46. Victor DG, Raustiala K, Skolnikoff EB (1998) The implementation and effectiveness of international environmental commitments. In: Choucri N (ed) Global environmental accord. MIT Press, Cambridge, MA
47. Selin H, VanDeveer SD (2003) Mapping institutional linkages in European air pollution politics, global environmental politics
48. Murphy C (1997) Leadership and international environmental policy making. In: Brooks LA, VanDeveer SD (eds) Saving the seas: values, scientists, and international governance. Maryland Sea Grant College, College Park, MD, p 255
49. Wettestad F (2001) Glob Gov 7:317
50. Farrell AE, Keating TJ (forthcoming) Dissent and trust in assessment processes: a comparison of OTAG and LRTAP. In: Farrell AE, Jaeger J (eds) The design of environmental assessments: choices for effective processes. RFF Press, Washington, DC
51. Holloway T, Fiore AM, Hastings MG (2003) Environ Sci Technol 37:4535
52. Farrell AE, Jaeger J (forthcoming) The design of environmental assessments: choices for effective processes. RFF Press, Washington, DC
53. Mitchell R, Clark WC, Cash D, Alcock F (forthcoming) Global environmental assessments: information, institutions, and influence. MIT Press, Cambridge
54. Farrell AE, Jaeger J, VanDeveer SD (forthcoming) Introduction. In: Farrell AE, Jaeger J (eds) The design of environmental assessments: choices for effective processes. RFF Press, Washington, DC
55. Bolin B (1994) Ambio 23:25
56. Davies JC (1993) Environmental regulation and technical change: overview and observations. In: Uman M (ed) Keeping pace with science and engineering: case studies in environmental regulation. National Academy Press, Washington, DC, p 251
57. Bronk DW (1974) Science 186:116
58. Carnegie Commission on Science T, and Government (1991) Science, Technology, and Congress: Expert Advice and the Decision-Making Process. The Carnegie Corporation, New York
59. Ezrahi Y (1990) The descent of Icarus: science and the transformation of contemporary democracy. Harvard University Press, Cambridge, MA
60. NARSTO (2003) Appendix E. Preparation of this assessment. In: Particulate matter science for policy-makers: A NARSTO assessment, part 2. Electric Power Research Institute, Palo Alto, CA
61. Amann M, Bertok I, Cofala J, Gyarfas F, Heyes C, Klimont Z, Makowski M, Schöpp W, Syri S (1998) Cost-effective control of acidification and ground-level ozone – 5th Interim Re-

port to the European Commission, DG-XI, International Institute for Applied Systems Analysis, Laxenburg, Austria
62. Alcamo J, Shaw R, Hordijk L (1990) The Rains model of acidification: science and strategies in Europe. Kluwer Academic, Dordrecht
63. United Nations Economic Commission for Europe (UNECE) (2003) Linkages and synergies of regional and global emission controls, EB.AIR/GE.1/2003/4/Add.1, Steering Body to the Cooperative Programme for Monitoring and Evaluation of the Long-range Transmission of Air Pollutants in Europe (EMEP), Geneva
64. Mayer M, Wang C, Webster M, Prinn RG (2000) J Geophys Res 105:22869

Subject Index

MIX
Papier aus verantwortungsvollen Quellen
Paper from responsible sources
FSC® C105338

If you have any concerns about our products,
you can contact us on
ProductSafety@springernature.com

In case Publisher is established outside the EU,
the EU authorized representative is:
Springer Nature Customer Service Center GmbH
Europaplatz 3, 69115 Heidelberg, Germany

Printed by Libri Plureos GmbH
in Hamburg, Germany